VOLUME FORTY SEVEN

ADVANCES IN CHEMICAL ENGINEERING

Mesoscale Modeling in Chemical Engineering Part II

ADVANCES IN CHEMICAL ENGINEERING

VOLUME FORTY SEVEN

ADVANCES IN CHEMICAL ENGINEERING

Mesoscale Modeling in Chemical Engineering Part II

Edited by

GUY B. MARIN
Department of Chemical Engineering,
Ghent University,
Ghent, Belgium

JINGHAI LI
Institute of Process Engineering,
Chinese Academy of Sciences,
Beijing, P.R. China

ELSEVIER

AMSTERDAM • BOSTON • HEIDELBERG • LONDON
NEW YORK • OXFORD • PARIS • SAN DIEGO
SAN FRANCISCO • SINGAPORE • SYDNEY • TOKYO
Academic Press is an imprint of Elsevier

Academic Press is an imprint of Elsevier
225 Wyman Street, Waltham, MA 02451, USA
525 B Street, Suite 1800, San Diego, CA 92101-4495, USA
The Boulevard, Langford Lane, Kidlington, Oxford OX5 1GB, UK
125 London Wall, London, EC2Y 5AS, UK

First edition 2015

Notices

Knowledge and best practice in this field are constantly changing. As new research and experience broaden our understanding, changes in research methods, professional practices, or medical treatment may become necessary.

Practitioners and researchers must always rely on their own experience and knowledge in evaluating and using any information, methods, compounds, or experiments described herein. In using such information or methods they should be mindful of their own safety and the safety of others, including parties for whom they have a professional responsibility.

To the fullest extent of the law, neither the Publisher nor the authors, contributors, or editors, assume any liability for any injury and/or damage to persons or property as a matter of products liability, negligence or otherwise, or from any use or operation of any methods, products, instructions, or ideas contained in the material herein.

ISBN: 978-0-12-803845-1
ISSN: 0065-2377

CONTENTS

CONTRIBUTORS

Jianfeng Chen
State Key Laboratory of Organic-Inorganic Composites, and Research Centre of the Ministry of Education for High Gravity Engineering and Technology, Beijing University of Chemical Technology, Beijing, China

Xueqian Chen
State Key laboratory of Chemical Engineering, East China University of Science and Technology, Shanghai, China

Yanpei Chen
State Key Laboratory of Multiphase Complex Systems, Institute of Process Engineering, Chinese Academy of Sciences, Beijing, China

Yihui Dong
State Key Laboratory of Materials-Oriented Chemical Engineering, College of Chemistry and Chemical Engineering, Nanjing Tech University, Nanjing, PR China

Ying Hu
Department of Chemistry, East China University of Science and Technology, Shanghai, China

Hua Li
Dalian National Laboratory for Clean Energy and National Engineering Laboratory for MTO, Dalian Institute of Chemical Physics, Chinese Academy of Sciences, Dalian, PR China

Licheng Li
College of Chemical Engineering, Nanjing Forestry University, Nanjing, PR China

Youwei Li
State Key Laboratory of Organic-Inorganic Composites, and Research Centre of the Ministry of Education for High Gravity Engineering and Technology, Beijing University of Chemical Technology, Beijing, China

Guotao Liu
Department of Chemical Engineering, The State Key Laboratory of Chemical Engineering, Tsinghua University, Beijing, PR China

Honglai Liu
State Key laboratory of Chemical Engineering, East China University of Science and Technology, Shanghai, China

Yu Liu
State Key laboratory of Chemical Engineering, East China University of Science and Technology, Shanghai, China

Zhongmin Liu
Dalian National Laboratory for Clean Energy and National Engineering Laboratory for MTO, Dalian Institute of Chemical Physics, Chinese Academy of Sciences, Dalian, PR China

Linghong Lu
State Key Laboratory of Materials-Oriented Chemical Engineering, College of Chemistry and Chemical Engineering, Nanjing Tech University, Nanjing, PR China

Xiaohua Lu
State Key Laboratory of Materials-Oriented Chemical Engineering, College of Chemistry and Chemical Engineering, Nanjing Tech University, Nanjing, PR China

Yunxiang Lu
Department of Chemistry, East China University of Science and Technology, Shanghai, China

Guangsheng Luo
Department of Chemical Engineering, The State Key Laboratory of Chemical Engineering, Tsinghua University, Beijing, PR China

Xuebo Quan
School of Chemistry and Chemical Engineering, Guangdong Provincial Key Lab for Green Chemical Product Technology, South China University of Technology, Guangzhou, PR China

Mingquan Shao
State Key Laboratory of Organic-Inorganic Composites, and Research Centre of the Ministry of Education for High Gravity Engineering and Technology, Beijing University of Chemical Technology, Beijing, China

Kai Wang
Department of Chemical Engineering, The State Key Laboratory of Chemical Engineering, Tsinghua University, Beijing, PR China

Wei Wang
State Key Laboratory of Multiphase Complex Systems, Institute of Process Engineering, Chinese Academy of Sciences, Beijing, China

Wenlong Xie
State Key Laboratory of Materials-Oriented Chemical Engineering, College of Chemistry and Chemical Engineering, Nanjing Tech University, Nanjing, PR China

Jianhong Xu
Department of Chemical Engineering, The State Key Laboratory of Chemical Engineering, Tsinghua University, Beijing, PR China

Mao Ye
Dalian National Laboratory for Clean Energy and National Engineering Laboratory for MTO, Dalian Institute of Chemical Physics, Chinese Academy of Sciences, Dalian, PR China

Gaobo Yu
School of Chemistry and Chemical Engineering, Guangdong Provincial Key Lab for Green Chemical Product Technology, South China University of Technology, Guangzhou, PR China

Tao Zhang
Dalian National Laboratory for Clean Energy and National Engineering Laboratory for MTO, Dalian Institute of Chemical Physics, Chinese Academy of Sciences, Dalian, PR China

Yi Zhang
State Key Laboratory of Organic-Inorganic Composites, and Research Centre of the Ministry of Education for High Gravity Engineering and Technology, Beijing University of Chemical Technology, Beijing, China

Shuangliang Zhao
State Key laboratory of Chemical Engineering, East China University of Science and Technology, Shanghai, China

Yinfeng Zhao
Dalian National Laboratory for Clean Energy and National Engineering Laboratory for MTO, Dalian Institute of Chemical Physics, Chinese Academy of Sciences, Dalian, PR China

Jian Zhou
School of Chemistry and Chemical Engineering, Guangdong Provincial Key Lab for Green Chemical Product Technology, South China University of Technology, Guangzhou, PR China

Yudan Zhu
State Key Laboratory of Materials-Oriented Chemical Engineering, College of Chemistry and Chemical Engineering, Nanjing Tech University, Nanjing, PR China

PREFACE

Opportunities and Challenges: Both at Mesoscales

Evolution of chemical engineering science features increasing generality of knowledge from "unit operations," common for different industries, to "transport phenomena," common in these operations, and is now at a transitional period with challenges in sustainable development and to its knowledge base. Designing materials, scaling-up reactors, and system optimization continue to be the main tasks for chemical engineers. However, upgraded knowledge and enabling technologies are needed to revolutionize R&D capability, which present challenges and opportunities to chemical engineering community.

Opportunities are always obscured by challenges. Among many challenges currently in chemical engineering, mesoscale phenomena at different levels of chemical engineering are recognized to be the most critical. At the material level, chemists and material scientists know molecular structures and properties of bulk materials very well, but they are still unable to manipulate the mesoscale structures in between. At the reactor level, chemical engineers have accumulated much knowledge on hydrodynamics, transport, and reaction behaviors of single particle, and on global performance of reactors; however, what happens at the mesoscale between single particles and the whole reactor is still difficult to be formulated. These two mesoscales at different levels are believed to be essential not only for resolving the current global challenges but also for upgrading the knowledge base of chemical engineering science.

Currently in chemical engineering, chemists and material scientists synthesize a variety of material structures every day, most of which, however, cannot be produced massively in industry due to the lack of understanding of the formation mechanisms of these structures and the difficulty in controlling the microenvironment for their formation in reactors. In addition, scaling-up of reactors is still a big challenge for chemical engineers. Understanding of mesoscale phenomena is critical to increase our capability in resolving these issues.

In fundamental research, average approaches are widely used, such as computational fluid dynamics (CFD), which are, however, not sufficient enough in predictability; on the other hand, discrete approaches are blooming up, based on microscale mechanisms, which are unfortunately limited to scalability in engineering applications. As a compromising strategy between these two, coarse-graining approaches received more and more attention, aiming at bridging the scales of continuum cells and the discrete moving elements, which still suffer from the lack of understanding of intrinsic mechanisms at mesoscales. This is now a common challenge not only in chemical engineering.

Gradually, people recognized that "meso" will be a big thing. Most importantly, there are many clues that a common principle for all mesoscale problems at different levels between elemental particles and the universe may exist, leading to exploration of a transdisciplinary science, mesoscience, though great efforts are needed to further clarify its rationale and importance.

In responding to these new possibilities and challenges, *Advances in Chemical Engineering* published a thematic volume in 2011 with the title: "Characterization of Flow, Particles and Interfaces," aiming at showing the dynamic behaviors at mesoscales. These two volumes further this effort to analyze these mesoscale phenomena and processes with the title: *Mesoscale Modeling in Chemical Engineering* so as to stimulate wide discussion and to promote transdisciplinarity between chemistry, material science, and chemical engineering, and even beyond. Volume I invited a few participants from the 4th International Conference on Multiscale Structures and Systems with the theme of "***Mesoscales: the key to multiscale problems***," and Volume II invited contributions from the major program on mesoscience of National Natural Science Foundation of China entitled "***Mechanisms and manipulation at mesoscales in multi-phase reaction systems***."

Although these contributions may be preliminary in addressing mesoscale problems, we hope these two volumes can play a role of "throwing out a minnow to catch a whale" and to stimulate thinking in the following questions: what are mesoscale phenomena? Why are they so important? How to relate mesoscale concepts to our research? Is it necessary or possible to have a common science for all mesoscales?

We would like to thank all authors for their contributions, and reviewers for their critical comments and suggestions. Help from Prof. Ning Yang and Dr. Xiaowei Wang are sincerely appreciated.

JINGHAI LI AND GUY MARIN

CHAPTER ONE

Unified Framework of Multiscale Density Functional Theories and Its Recent Applications

Shuangliang Zhao*, Yu Liu*, Xueqian Chen*, Yunxiang Lu†, Honglai Liu*,[1], Ying Hu†
*State Key laboratory of Chemical Engineering, East China University of Science and Technology, Shanghai, China
†Department of Chemistry, East China University of Science and Technology, Shanghai, China
[1]Corresponding author: e-mail address: hlliu@ecust.edu.cn

Contents

Abstract

Most chemical engineering processes involve complex multiphase fluid systems, and their evolution depends on the mechanism by which the inhomogeneous subsystems exchange information at different length scales. Whereas numerous theoretical methods with specific description accuracies have been developed for investigating physicochemical properties of various fluid systems, a unified theory that enables the investigation of mesoscale problems is still needed. Here, we introduce a unified

Advances in Chemical Engineering, Volume 47
ISSN 0065-2377
http://dx.doi.org/10.1016/bs.ache.2015.10.001

framework of multiscale density functional theories (DFTs). With the same physical concept and mathematical framework four different versions of DFTs, covering quantum, atomic, molecular, and polymeric DFTs, are presented complemented with their illustrative applications at individual scales. In addition, the combinations of those DFTs with each other and with other conventional theories and simulation approaches are also discussed. Finally, general discussions on the up-to-date progress of DFTs and the expectations on their further extensions are given. The introduction of this unified framework of DFTs is expected to advance the theoretical study of mesoscale problems.

ABBREVIATIONS

1D one-dimensional
2D two-dimensional
3D three-dimensional
B3LYP 3-parameter Becke–Lee–Yang–Parr (approximation)
BH Barker–Henderson
BMCSL Boublik–Mansoori–Carnahan–Starling–Leland
CS Carnahan–Starling
DCF direct correlation function
DFTs density functional theories
DPD dissipative particle dynamics
EOS equation of state
FFT fast Fourier transform
FMSA first-order mean spherical approximation
FMT fundamental measure theory
GGA generalized gradient approximation
HRFA homogeneous reference fluid approximation
HNC hypernetted chain (approximation)
HNCB hypernetted chain approximation plus bridge
HS hard-sphere
LDA local-density approximation
LJ Lennard–Jones
MBWR-EOS Modified Benedict–Webb–Rubin equation of state for Lennard–Jones system
MC Monte Carlo
MD molecular dynamics
MDFT molecular density functional theory
MFA mean-field approximation
MFMT modified fundamental measure theory
MM molecular mechanics
MOF metal-organic framework
MOZ molecular Ornstein–Zernike
OZ Ornstein–Zernike
PB Poisson–Boltzmann
PDT potential distribution theorem
PMF potential of mean force

PY Percus–Yevick
QM quantum mechanics
QDFT quantum density functional theory
RDF radial distribution function
RHNC reference hypernetted chain (approximation)
RISM reference interacting site model
SAFT statistical associating fluid theory
SDFT site density functional theory
SPC simple point charged model (for water)
SPC/E simple point charged/extended model (for water)
TFW Thomas–Fermi–Weizsacker
TIP3P three-point transferable intermolecular potential model (for water)
TPT thermodynamic perturbation theory
TPT1 first-order thermodynamic perturbation theory
WCA Weeks–Chandler–Andersen
WDA weighted density approximation
ZIF Zeolitic imidazolate framework

1. INTRODUCTION

Nowadays, energy, resources, and environment are essential as a basis of human survival and development (Hoffert et al., 2002). As the growth of human population and development of social society, the annual consumption amount of energy and resources sharply increases (Turner et al., 2007). The low efficiency of energy utilization and tremendous dissipation of natural resources not only aggravate the fast depletion of resources and fossil fuels but also cause severe eco-environmental problems (Allen et al., 2009; Macdonald et al., 2005), which, including the greenhouse effect (Ramanathan and Feng, 2009), have become the focus of global politics (IPCC, 2014). In response to these common challenges in energy, resource, and environment, a major task of chemical process intensification is to save energy, lower energy consumption, and reduce pollutants discharge. Toward this end, a persistent pursuit of modern chemical engineering is to design elaborative routes for improving the efficiency of chemical process (Charpentier, 2007).

The efficiency of chemical process is mainly associated with the match degree between the transport properties of the chemical reactors and the engineering characteristics of the chemical reactions. The transport properties involve fluid mixing and mass/heat transfer affiliated with the flow-field distribution, density distribution, and temperature profile, while the engineering characteristics of a chemical reaction typically refer to the rate of the reaction, the by-product characteristics, and the sensitivity of the temperature and concentration to reaction rate. A mismatch between those factors causes scaling effects in chemical engineering processes (Rozen and Kostanyan, 2002). To resolve these scaling effects, mechanistic understandings on the dynamic and thermodynamic properties of the chemical systems are necessitated (Charpentier, 2010).

Most chemical engineering processes involve complex multiphase systems, and their evolution depends on the mechanism by which the inhomogeneous subsystems exchange information at different length scales. For instance, the properties of interfacial regions greatly affect the efficiency of entire chemical process, especially when the volume of the continuous phase is on the nano- or microscale and the interfacial area becomes relatively large (Hu et al., 2015). The interfacial/surface properties are determined by the spatial density distribution of each component, which is highly inhomogeneous in the interfacial/surface regions (Shi et al., 2015). Generally, the local density distributions of all components provide the knowledge of microscopic structural information, which is crucial to access the macroscopic properties of an inhomogeneous fluid (gas or liquid) system. While in a further step, the microscopic structure of a fluid system is determined by both the intermolecular interaction and external potential, which, further on, is determined by the elemental interactions among electrons and nuclei. Consequently, exchanges of structural and energy information among those elements at different length scales provide key links between the elemental interactions and the properties of chemical process (Charpentier, 2007, 2010).

The investigation on information exchange between systems at different scales essentially belongs to the research scope of mesoscale problems (Li et al., 2015). In past decades, owing to the development of efficient numerical approaches and powerful computational tools, a large number of theoretical methods have been proposed, either in qualitative or in quantitative ways, to address the properties of fluid systems (Gray and Gubbins, 1985; Gray et al., 2011; Hansen and McDonald, 2013). However, most of those methods enable the investigation only in a specific length scale, and limited research progress has been made regarding the coupling of the information at different scales (Estrada-Tejedor et al., 2014; Hodak, 2014; Karplus, 2014), among which the development of hybrid quantum mechanics (QM)/molecular mechanics (MM) approach (Car and Parrinello, 1985) is very noticeable. The QM/MM approach is a molecular simulation method that combines the merits of the QM (accuracy) and MM (speed) approaches, thus allowing for the study of complex chemical systems.

As illustrated in Fig. 1, different simulation methods have virtually been developed (Chu et al., 2007; Loehnert and Belytschko, 2007; Sherwood et al., 2008) for the investigations of the model systems at different length scales and the associated phenomena at different time scales. The combination of two or more adjacent simulation methods is in principle capable of treating mesoscale problems. For instance, dissipative particle dynamics (DPD) simulation is a coarse-grained dynamics method basing on simplified Hamiltonians with solvent being treated as continuum, and thus allows for the description of mesoscopic phenomena (Groot and Warren, 1997); while the involved Flory–Huggins binary interaction parameters should be determined by the more fine-grained simulation methods (Chremos et al., 2014; Groot and Warren, 1997) or other theoretical/experimental approaches. Indeed, these "bottom-up" and "top-down" criteria often provide hybrid approaches for mesoscale investigations.

In this chapter, a unified framework of multiscale density functional theories (DFTs) is introduced, and we demonstrate that this framework is capable of supplying a versatile tool to address mesoscale phenomena. DFTs are modern statistical mechanics methods, and historically, the foundations of DFTs were first laid in 1964 by Hohenberg and Kohn (HK) (Hohenberg and Kohn, 1964) who proved that the ground state energy of any QM system could be expressed as a functional of the one-body density only. Basing

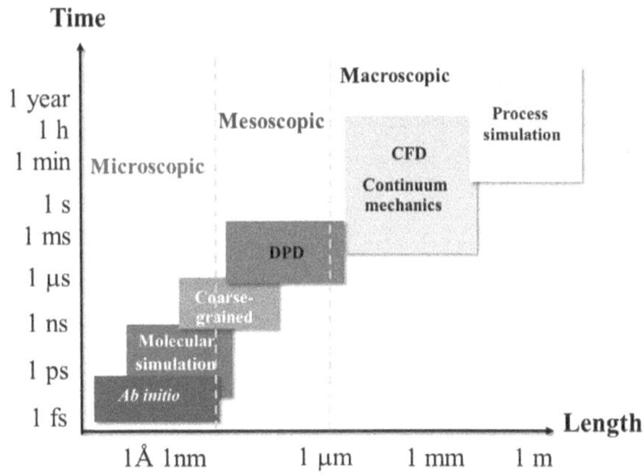

Figure 1 Multiscale computer simulation methods for investigating the properties of model systems at different length scales and/or the associated phenomena at different time scales.

on this interpretation, Kohn and Sham subsequently developed the density functional which could be applied to electronic systems (Kohn and Sham, 1965). The HK proof was extended to finite temperature by Mermin (1965), opening up the possibility of using DFT to calculate the free energy of a statistical–mechanical system. Later on, the connection of the variational method in statistical mechanics with DFT in quantum chemistry was first recognized by Ebner, Saam, and Stroud, and then the classical analog of quantum DFT (QDFT) was developed for nonuniform simple fluids (Ebner et al., 1976; Saam and Ebner, 1977). This formulism was spread to a broad audience by Evans' excellent review paper on the gas–liquid interface in 1979 (Evans, 1979). In past 40 years, the DFTs have been remarkably extended and developed for various complex systems with countless applications in condensed matter physics, chemistry, biology, and chemical engineering (Chai and Weeks, 2007; Cohen et al., 2012; Geerlings et al., 2003; Koch and Holthausen, 2001; Wu, 2006).

While DFTs have been gradually growing to a workhorse for solving theoretical problems of a very wide range, the key concept of DFTs is to formulate a thermodynamic quantity (global information) of the system under investigation as a functional of local density distribution (local information). Following thermodynamic laws, this quantity under equilibrium condition reaches its minimum (or maximum) and hence the structure information can be determined through functional minimization. Because DFTs naturally connect the information at two different length scales, they are inherently mesoscale approaches. In past decades, various versions of DFTs have been developed and extended for the investigation of complex model systems at individual description accuracy. Conventionally, two major research communities of DFTs exist, and one is dedicated to QDFT (Chai and Weeks, 2007; Cohen et al., 2012; Geerlings et al., 2003; Zhao and Truhlar, 2008), and the other is the classical analog historically called classical DFT (Evans, 1992; Wu, 2006, 2009; Wu and Li, 2007). However, because QDFT is associated with QM, and its classical analog is currently developed basing on statistical mechanics, we refer this analog as statistical DFT. Within the scope of statistical DFT, recently three distinct versions have also been developed for treating simple fluids, molecular fluids, and polymeric fluids. These versions refer to atomic, molecular, and polymeric DFTs as introduced below. Regarding the individual developments and applications of different DFTs, many excellent reviews and monographs have been contributed into the literature either illustratively (Becke, 2014; Evans, 1992; Lowen, 2002, 2003) or exhaustively (Evans, 1979,

1992). While some of them focused on the formulation and gave rigorous and clear proof for the existence and uniqueness of density functional (Becke, 2014; Evans, 1979), some of them face to physicists and illustrate the major concept of DFT (Parr and Yang, 1994; Zhou and Solana, 2009) and demonstrate the applications into various physical phenomena, and some of them focused on diverse practical applications of DFTs in chemical engineering (Lowen, 2002).

The purpose of this chapter is threefold. First, we introduce the unified framework of multiscale DFTs, and within this framework four existing versions of DFTs at different description accuracies are presented. We demonstrate that those DFTs, covering quantum, atomic, molecular, and polymeric DFTs, share the same physical concept and mathematical framework. Second, we summarize the recent progress of this unified framework including our own contributions, and demonstrate the diverse applications of DFTs on the investigations of ionic liquid, gas adsorption and storage, liquid–liquid interface properties, solvation properties, and solvent-sensitive structural behaviors of polymer brushes etc. Lastly and most importantly, weillustrate that these DFTs can be combined with each other due to the same principle, and with other theoretical methods and simulations. Those "bottom-up" or "top-down" combinations provide a versatile methodology for the investigation of different mesoscale problems.

Although we try to present this material in a rather self-contained way, the involved rigorous mathematical derivations are impossible to supply due to the page limitation, and fortunately they can be found in the relevant reviews or books chapters. Also, due to the limits of our knowledge and time, we make no attempt to cover all aspects of DFTs in the recent literature, and therefore significant contributions may have been omitted.

2. UNIFIED FRAMEWORK OF MULTISCALE DFTS

2.1 Modeling and Theory

Theoretical investigation of the properties of chemical systems relies on modeling. Modeling provides scientific and reasonable descriptions for practical systems, and these descriptions should generally be accurate to capture the major characteristics of the system under investigation, yet simple enough allowing for deep theoretical study and/or computer simulations toward the acquisition of mechanistic understanding on the fundamental processes governing these systems (Leach, 2001). Logically, the concept of modeling is different to that of theory: theory is expected to provide a general framework for predicting the behaviors of different (model) systems in a wide range, while modeling supplies theory the indispensable key access to the dynamic and thermodynamic properties of practical systems. The models for different chemical systems are generally different, and even could be distinct for the same system if the research purposes are different. Because the goal of modeling is to appropriately characterize the systems and the success of this characterization, in turn, requires the implementation of theoretical study, modeling is thus closely associated together with theory. Clearly, the agreement of theoretical prediction with the relevant experimental measurement is determined from the accuracy of both theory and modeling.

During the last 40 years, computer-based computational methods, such as molecular dynamics and Monte Carlo simulations, have evolved from advanced techniques accessible only to a few specialists, to standard calculation approaches available to both scientists and engineers. Similarly, numerical methods of solving the Schrodinger equation, implemented in computational softwares such as Gaussian or Materials studio, are now commonly used to study complex systems. Proper modeling of those systems plays more and more important role in modern scientific research.

The modeling for quantum systems is key important, and unfortunately many predicted properties from quantum calculations deviate from or are even contradictory with experimental observations due to the improper modeling of real chemical systems. For instances, it has been observed that the solvent can greatly affect the direction of reactions and the reaction efficiency (Hirata, 2003; Hirata et al., 2001); however, in many investigations associated with chemical reaction systems, the solvent surroundings are completely ignored due to the difficulty in numerical calculation when taking into account hundreds of solvent molecules (Hirata et al., 2001). It is generally believed that the catalytic molecular reaction most likely occurs at the interface between the catalyst particle and the substrate, and the geometry of this interface together with the location and distribution of defects have significant effects on the reaction efficiency. For this reason, the measured reaction efficiency is expected to be the mean value of the particular reaction efficiencies at different active sites. Without properly taking into account this configurational average, the quantum calculations upon crystal-like catalysts naturally give no quantitative predictions in comparison to apparent experimental observations although the involved theoretical approach is sufficiently accurate.

For classical fluid systems proper modeling is also important. Water probably is the most important solvent in nature while on the other hand it is also regarded as the most difficult fluid system to model because of its extraordinary properties in different aspects (Errington and Debenedetti, 2001). For examples, the density of liquid water not always increases when increasing system temperature but reaches its maximum at 277 K at 1 atm and then decreases as temperature continues to increase; its surface tension is surprisingly large comparing to other similar fluids (Wallqvist and Mountain, 1999). More than 50 explicit classical models (Guillot, 2002; Jorgensen et al., 1983) have been developed for water molecule including the extended simple point charged (SPC/E) model (Berendsen et al., 1987) and the three-point transferable intermolecular potential (TIP3P) model. Although these models have achieved great success in describing certain properties of bulk water and the water-involved systems, none of them provides accurate descriptions in all aspects of the physicochemical properties of bulk water (Vega and Abascal, 2011). Note that water molecule contains three chemical atoms, namely, one oxygen atom and two hydrogen atoms, while in many classical models more than three "sites" are introduced in a water molecule in order to better reproduce the structure and thermodynamic properties of entire bulk water. Apparently, a site in the model of a molecule does not necessarily represent a real chemical atom. Due to this subtle difference, we conventionally use the concept "site" instead of "atom" in molecular modeling. Besides those explicit models for water, water is also treated implicitly (Tomasi and Persico, 1994), e.g., as a continuum with desired dielectric constant (Tomasi et al., 2005), or the hydration effect is specifically included although water molecules are not explicitly considered (Mennucci and Cammi, 2007).

The modeling of classical liquid system not only gives the composition and geometry of the explicit component molecule but also identifies the interactions among them. In most cases, these interactions are pairwise additive, and the potential energy is simply the summation of interaction energy between pairs of molecules. Typically, these interactions include Coulombic potentials and empirical Lennard–Jones (LJ) potentials. The charge valence and the parameters involved in LJ potentials (for a classical system) are determined during the course of modeling, which are irreverent to the ensuing applications of theoretical approaches. It is reasonable to compare the accuracies of two theoretical approaches applied for the same model systems, and because simulation is a kind of exact approach, simulation results are usually adopted to rationalize the other theoretical approaches including DFTs.

The last example on classical modeling we wish to mention is on charged particles like spherical ions. For these particles, to our knowledge two typical descriptions are proposed regarding the charge distribution. The first one is the so-called primitive model, in which the charge, similar to LJ interaction, is placed on the center of particle. In contrast, in the other model, the charge is uniformly distributed on the surface of the particle, and then the image charge effect (Xu et al., 2009a, 2011) should be additionally included. Whereas there is no simple way to validate which model for ion is more realistic, we limited ourselves in the present work to primitive model for the descriptions of ions unless otherwise specified.

2.2 Microscopic Structures of Fluid Systems

For a homogeneous fluid system at equilibrium, the components are uniformly distributed in space (see Fig. 2), and the average densities are equal to $\rho_i^b = N_i/V$. Here N_i refers to the particle number of component i in the system, and V is the system volume. In this circumstance, the thermodynamic properties of the system can be described by using equation of state (EOS). As discussed below, numerous equations of state have been developed to investigate different homogeneous fluid systems. For instances, Carnahan and Starling have proposed the EOS for hard-sphere (HS) bulk systems (Carnahan and Starling, 1969), and Gubbins et al. has developed a modified Benedict–Webb–Rubin EOS for LJ systems (MBWR-EOS) (Johnson et al., 1993), and Chapman et al. developed the SAFT EOS for associating fluid (Chapman et al., 1989, 1990). In 1996, two of us developed an EOS for fluids containing chainlike molecules (Hu et al., 1996; Liu and Hu, 1998). These equations of state provide accurate estimations for the

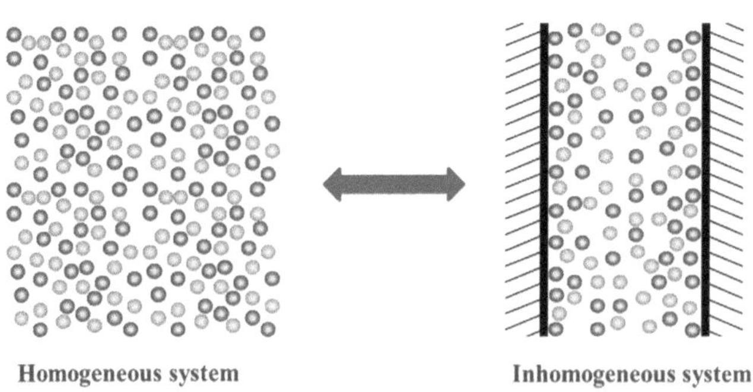

Figure 2 Illustrations for homogeneous fluid system (left) and inhomogeneous fluid system (right). The spheres in different color denote different components.

thermodynamic properties of homogeneous systems including pressures, free energies, and chemical potentials.

While for inhomogeneous fluid systems as illustrated in Fig. 2, the thermodynamic quantities can no longer directly predicted by using EOS, and instead, they can be determined by the one-body density distribution $\rho_i(\mathbf{R})$. Here $\mathbf{R}$ is the abbreviation of the set of complete variables which describes the spatial state of the concerned molecule. This generic notation stands for different variables in different circumstances. In specific, $\mathbf{R}$ refers to position $\mathbf{r}$ for a quantum particle or spherical classical particle, to $(\mathbf{r}, \theta, \phi)$ for a dipolar molecule with (θ, ϕ) being the Euler angle, to $(\mathbf{r}, \theta, \phi, \psi)$ for a rigid nonlinear molecule, and to $(\mathbf{r}_1, \mathbf{r}_2, \ldots, \mathbf{r}_M)$ for a flexible polymeric molecule with $\mathbf{r}_i$ $(i = 1, 2, \ldots, M)$ being the position of the ith out of the M sites in total in the molecule.

The one-body density distribution describes the local density of molecule at given spatial configuration $\mathbf{R}$, and it is defined as (Hansen and McDonald, 2013)

$$\rho_i(\mathbf{R}) = \left\langle \sum_{j=1}^{N_i} \delta(\mathbf{R}_j - \mathbf{R}) \right\rangle. \tag{1}$$

Here the subscript i represents molecule species. The bracket stands for the ensemble average and $\delta(x)$ is the Dirac delta function. The distribution of one-body density reflects the microscopic structure of the corresponding fluid system. To illustrate, Fig. 3A depicts a typical reduced one-body density distribution of simple solvent (blue spheres (dark gray in the print

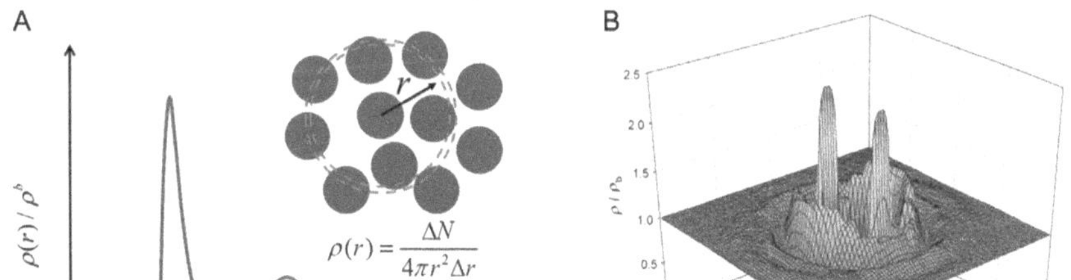

Figure 3 Illustrations of 1D and 2D density profiles: (A) typical one-dimensional local density distribution $\rho(r)$ for spherical solvent molecules around a spherical solute; (B) three-dimensional local density distribution $\rho(\mathbf{r})$ on X–Y plane.

version) in figure inset) around a spherical solute (red sphere (dark gray in the print version)). Due to the spherical symmetry of the solute, the local density of solvent depends solely on the distance to the center of the solute, and thus is a one-dimensional (1D) function. The reduced local density is defined as $g(r) = \rho(r)/\rho^b$, and conventionally called radial distribution function (RDF). Because at the remote place far from the solute, the local density should recover to the bulk density ρ^b, namely, the RDF recovers to 1 as the distance r increases. Another important quantity called total correlating function $h(r) = g(r) - 1$ is introduced as the net reduced density. The appropriate integration of $\rho(r)$ over the distance r can yield the numbers of solvent molecule within the first coordination shell (calls hydration shell when solvent is water), which determines the degree of solventphobicity and solventpholicity. Figure 3B illustrates the reduced solvent distribution around a nonspherical solute on x–y plane, and such distribution usually describes the microscopic contour of fluid density around a target object. Generally, the total correlation function depends on the positions of two particles instead of on their distance, thereafter is a three-dimensional (3D) function once one particle is fixed at the origin. Similarly, a higher-dimensional total correlation function can be defined when the local density $\rho(\mathbf{R})$ depends not only on position $\mathbf{r}$ but also on the orientation of molecule.

2.3 Multiscale DFTs

As mentioned, conventionally QDFT and statistical DFT are developed separately. While QDFT mainly addresses the electron structures of many-body system, statistical DFT focuses on the microscopic structure of thermodynamic systems. Depending on the complexity, those classical fluid systems can be divided into simple systems composed of spherical particles (or atoms) such as inert gas or other similar model systems, molecular systems composed of rigid molecules which contain more than one sites such as SPC/E bulk water, and polymeric system like polymers which is composed of flexible molecules containing a few sites. For all three types of classical fluid systems, grand potential functional can be formulated as a functional of density distribution function. The formulations of grand potentials are quite different for different types of fluid systems, and three versions of statistical DFTs are therefore developed which are atomic DFT, molecular DFT (MDFT), and polymeric DFT. As illustrated in Fig. 4 and as demonstrated below, QDFT together with those three major versions of

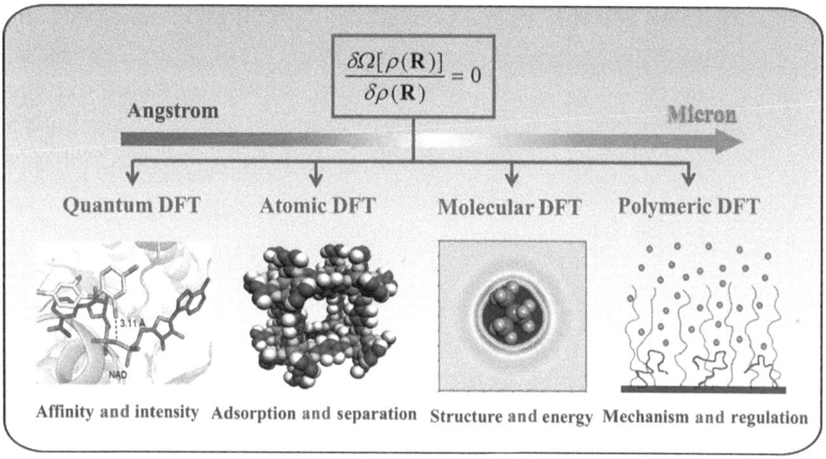

Figure 4 Schematic graph for the unified framework of density functional theories.

statistical DFT share the same framework and are applicable for the model systems at different length scales.

The existence and uniqueness of density functional for a specific quantum or classical system have been proven by variational approach (Evans, 1979), which shows that there exists a unique density distribution $\rho(\mathbf{R})$ corresponding to the applied external potential $V^{\text{ext}}(\mathbf{R})$. The grand potential (or equivalently the Helmholtz free energy) of the system can be formulated as a functional of density distribution function $\rho(\mathbf{R})$ subject to an arbitrary external potential. By minimizing this functional, one can obtain the one-body distribution $\rho(\mathbf{R})$ and thereafter the corresponding thermodynamic properties.

Although the essence of DFT is quite simple, the formulation of the functional is challenging. For different systems in which the natures of the molecular interaction are distinct, the formulations of the grand potentials can be different. In grand canonical ensemble, the grand potential functional Ω and the intrinsic free energy functional for a fluid system under the external potential $V_i^{\text{ext}}(\mathbf{R})$ are related by

$$\Omega[\rho_i(\mathbf{R})] = F[\rho_i(\mathbf{R})] + \sum_i \int d\mathbf{R}\left[\rho_i(\mathbf{R})\left(V_i^{\text{ext}}(\mathbf{R}) - \mu_i\right],\right. \tag{2}$$

where $F[\rho_i(\mathbf{R})]$ is the Helmholtz free energy functional, and μ_i is the chemical potential of component i in the uniform system which is in contact with the system under investigation and thus shares the same chemical potential of each component and the system temperature T. As noted below, this bulk chemical potential μ_i can be calculated through EOS. In general, the Helmholtz free energy can be decomposed into two contributions, and one corresponds to the noninteraction contribution (namely, the ideal contribution) and the other calls excess contribution stemming from the intermolecular interactions, i.e.,

$$F[\rho(\mathbf{R})] = F^{\mathrm{id}}[\rho(\mathbf{R})] + F^{\mathrm{ex}}[\rho(\mathbf{R})]. \tag{3}$$

Generally the exact expression of excess free energy functional is unattainable (Lowen, 2002), to which approximation should be made. In recent years, the pursuit of accurate formulation of $F^{\mathrm{ex}}[\rho(\mathbf{R})]$ has become an important research direction especially for theoreticians. Many approximations have been proposed, among which the local-density approximation (LDA), weighted density approximation (WDA), generalized gradient approximation (GGA), etc., are widely applied.

At equilibrium, the grand potential reaches its minimum, and the equilibrium density profile $\rho(\mathbf{R})$ can be determined by minimizing the grand potential functional,

$$\frac{\delta\Omega[\rho_i(\mathbf{R})]}{\delta\rho_i(\mathbf{R})} = 0, \tag{4}$$

which yields the so-called Euler–Lagrange equation for local structure $\rho_i(\mathbf{R})$, and simultaneously the thermodynamic properties can be obtained. Equations (2) and (4) constitute the unified framework for different versions of DFTs.

3. APPLICATIONS OF MULTISCALE DFTs

The four versions of DFTs, namely, quantum, atomic, molecular, and polymeric DFTs, are introduced individually within this unified framework, and the selected applications of those DFTs are also presented for illustration purpose.

3.1 Quantum DFT and Its Applications

Computational chemistry is an important branch of chemistry, and it adopts methods of theoretical chemistry, incorporated into efficient computer programs, to investigate the structures, and thermochemical properties of molecules and solids. Numerous methods have been developed, among which

QDFT initiated by HK (Evans, 1979; Hohenberg and Kohn, 1964) and then extended by Mermin (1965) appears to be a remarkably successful tool. In QDFT, the electron density $\rho(\mathbf{r})$ is the central quantity rather than the wave function. Use of electron density over the wave function greatly reduces the dimensionality, and this enables DFT to be readily applied to much larger system. We present here the main formulism of QDFT for electron systems, and the extensions to other quantum systems are expected to be straightforward.

For an electron system at temperature T and volume V with fixed electron number N, following Hohenberg–Kohn–Mermin theorem the grand potential of the electron system under an external potential $V(\mathbf{r})$ follows (Geerlings et al., 2003)

$$\Omega[\rho(\mathbf{r})] = F[\rho(\mathbf{r})] + \int \rho(\mathbf{r})(V(\mathbf{r}) - \mu)\mathrm{d}\mathbf{r}. \tag{5}$$

Here $\rho(\mathbf{r})$ denotes the electron local density. $F[\rho(\mathbf{r})]$ is the free energy functional, and as stated above it can be split into the noninteraction part (F^{id}) and the excess part (F^{ex}) accounting for the electron–electron interaction. Note that here the bulk chemical potential μ is not an independent thermodynamic parameter and it is determined by the electron total number. Namely, the bulk chemical potential involved in the Euler–Lagrange equation for $\rho(\mathbf{r})$ should satisfy the electron number conversation equation $\int \rho(\mathbf{r})\mathrm{d}\mathbf{r} = N$.

For temperature $T \ll T_F$ with $T_F = E_F/k_{\mathrm{B}}$ being the Fermi temperature defined via the Fermi energy E_F, the electronic entropy makes relatively insignificant contribution, and then we have $F^{\mathrm{id}}[\rho(\mathbf{r})] = E^{\mathrm{id}}[\rho(\mathbf{r})]$ (kinetic energy functional) (Hong, 2002; Xu and Hansen, 1998). It should be pointed out that conventionally the energy functional $E[\rho(\mathbf{r})]$ instead of grand potential functional of a quantum system is minimized (Becke, 2014; Burke, 2012; Koch and Holthausen, 2001; Parr and Yang, 1994). The energy functional is associated with the HK functional F_{HK} through $E[\rho(\mathbf{r})] = F_{\mathrm{HK}}[\rho(\mathbf{r})] + \int \rho(\mathbf{r}) V(\mathbf{r})\mathrm{d}\mathbf{r}$. By noting the ensemble difference, the grand potential functional is consistent with the energy functional at low temperature limit (Geerlings et al., 2003).

Remarkably, accurate forms have been developed for the kinetic energy functional (Chai and Weeks, 2007; Foley and Madden, 1996; Goodpaster et al., 2010; Ho et al., 2008; Zhou et al., 2005). A simple analytical form given by the Thomas–Fermi plus square-gradient (Weizsacker) correction approximation follows (Xu and Hansen, 1998),

$$E^{\text{id}}[\rho(\mathbf{r})] = C_k \int [\rho(\mathbf{r})]^{5/3} d\mathbf{r} + \frac{\lambda}{8} \int \frac{|\nabla\rho(\mathbf{r})|^2}{\rho(\mathbf{r})} d\mathbf{r}, \tag{6}$$

with $C_k = \frac{3(3\pi^2)^{2/3}}{10}$ and $\frac{1}{9} < \lambda < 1$. The excess free energy functional reads

$$F^{\text{ex}}[\rho(\mathbf{r})] = F_H^{\text{ex}}[\rho(\mathbf{r})] + F_X^{\text{ex}}[\rho(\mathbf{r})] + F_C^{\text{ex}}[\rho(\mathbf{r})], \tag{7}$$

where F_H^{ex} and F_X^{ex} are the so-called Hartree and exchange functionals, respectively. The Hartree functional accounting for the direct interaction contribution reads

$$F_H^{\text{ex}}[\rho(\mathbf{r})] = \frac{1}{2} \int d\mathbf{r} \int d\mathbf{r}' \frac{\Delta\rho(\mathbf{r})\Delta\rho(\mathbf{r}')}{|\mathbf{r} - \mathbf{r}'|} \tag{8}$$

with $\Delta\rho(\mathbf{r}) = \rho(\mathbf{r}) - \rho^b$ being the density deviation with respect to the bulk electron density ρ^b. The exchange correlation accounting for the indistinguishability of two electrons, and it can be expressed as

$$F_X^{\text{ex}}[\rho(\mathbf{r})] = C_X \int [\rho(\mathbf{r})]^{4/3} d\mathbf{r}. \tag{9}$$

In above equation, the constant $C_X = \frac{-3(6/\pi)^{1/3}}{4}$ a.u. The only term in Eq. (7) for which no explicit exact expression can be given is of course the correlation contribution F_C^{ex}, and a set of approximations toward its formulation have been developed. Typically, three kinds of approximations have been proposed: (i) LDA which leads to a functional depending only on the local density distribution. LDA approximation works relatively well for systems with smooth density distributions, and (ii) GGA and its extensions (Janesko, 2015), which are formulated with the local density of electron and its gradient, and they generally works well for hydrogen bonded and other weakly bonded systems (Janesko, 2015), and (iii) hybrid approximation which empirically combines two different kinds of approximations to treat F_X^{ex} and F_C^{ex}. For example, the most popular B3LYP is the three-parameter combination of Becke'88 for treating exchange contribution and Lee–Yang–Parr approximation for treating correlation contribution, and this approximation provides satisfactory descriptions for thermochemical properties, reaction barrier heights, etc. (Cohen et al., 2012).

In the past decades, with the advent of modern computer and the computation packages such as Gaussian (Frisch et al., 2009), VASP (Hafner, 2008), and material studio (Accelrys, n.d) etc., QDFT has been extensively applied to address the properties of chemical systems including molecular structure (i.e., the positions of the constituent atoms in a chemical molecule), electronic charge distribution on each atom, absolute and relative interaction and bond energies, dipoles and higher multipole moments, reaction barriers, and binding affinities, etc. Excellent reviews are available for those practical applications (Burke, 2012; Cohen et al., 2012; Geerlings et al., 2003; Zhao and Truhlar, 2008), and here we illustrate one of our recent applications in ionic liquids.

Ionic liquids, the salts in the liquid state under ambient condition, have experienced a comet-like boost in the last few years due to their promising applications in many aspects as powerful solvent and electrically conducting fluids. The QDFT has been successfully applied to study the adsorption of ionic liquids on various surfaces, such as on graphene, boron-nitride sheets, and on noble metal surfaces (Ghatee and Moosavi, 2011; Shakourian-Fard et al., 2014, 2015). It was also employed for studying the mechanism of many organic reactions, e.g., Diels–Alder reaction, Markovnikov addition, and S_N2 rearrangement, in ionic liquids (Chiappe and Pomelli, 2013; Kirchner, 2010). By using QDFT, we have studied noncovalent interactions in four halogenated ionic liquids (Li et al., 2013), i.e., 2-bromo/iodo and 4,5-dibromo-/diiodo-1,3-dimethylimidazolium trifluoromethanesulfonates. Several intermolecular interactions, such as C–H···O hydrogen bonds, C–X···O halogen bonds, and electrostatic interactions with anions located over the imidazolium cation, were found in the ion pairs, in good agreement with the X-ray crystal structures of corresponding organic salts. In addition, we also investigated the interactions of these halogenated imidazolium cations with CO_2 via the DFT method and revealed that CO_2 can interact with halogen atoms in these cations through C–X···O bonds (Li et al., 2012; Zhu et al., 2011). Therefore, these halogenated ionic liquids may have a high CO_2 solubility.

More recently, we have utilized QDFT (B3LYP) to explore the structures, energetics, and acidity/basicity of metal-containing ionic liquids (Wu et al., 2014, 2015). It was found that chlorometallate-based ionic liquids show stronger acidity than conventional ionic liquids, due to the increased values of the most-positive-surface electrostatic potential ($V_{s,\mathrm{max}}$). However, these ionic liquids exhibit weaker basicity, because of the decreased values of the most-negative-surface electrostatic potential ($V_{s,\mathrm{min}}$) and the higher magnitude of the lowest-surface average local ionization energy ($\bar{I}_{s,\min}$), as shown in Fig. 5.

Catalytic reaction is an important application direction of QDFT in recent decades, among which the controlling of catalytic selectivity is one of main challenge of modern chemical engineering. In order to achieve the optimum selectivity, it is very beneficial to explore the intrinsic relation between catalyst structure (for example, size and shape) and properties (activity, selectivity). At present, it is still a computational challenge to directly simulate the catalytic properties over different sizes of metal nanoparticles. In order to solve this problem, Wang et al. (Cai et al., 2015; Gao et al., 2015; Lyu et al., 2014) developed a simple method based on QDFT calculations, microkinetic modeling, and the relationship between the reaction sites over an ideal polyhedral metal catalyst for building the relation between size and selectivity. The predicted optimum catalyst sizes in a series of reactions (for example, furfural conversion (Cai et al., 2015), CO_2 reduction (Gao et al., 2015), and selective hydrogenation of halogenated nitrobenzene (Lyu et al., 2014)) are in nice agreement with experimental ones.

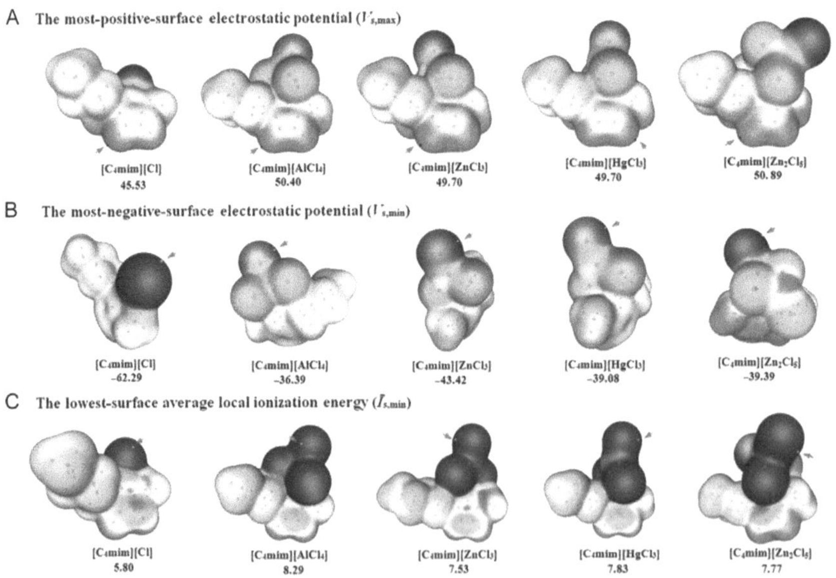

Figure 5 Molecular surface properties of chlorometallate ionic liquids and [C_4mim][Cl]. *After Wu et al. (2015).*

3.2 Atomic DFT and Its Applications

The classical fluid systems are characterized with simplified Hamiltonian in which the semiempirical pairwise-additive interaction between two particles are included. Those semiempirical interactions substantially arise from the Pauli exclusion of two electrons at the same quantum state and from the electrostatic interactions among electrons and nuclei. As a result, both repulsion and attraction appear in the two-body interaction. Specifically, when all involved particles are spherical ones like atoms, ions, or coarse-grained beads, the systems are called simple fluids. Obviously, the pair interactions in simple fluid systems are simply distance-dependent. Toward the investigation of simple fluid systems, atomic DFT is developed. A notable merit of atomic DFT is that the contributions to the free energy functional from different interaction parts can be treated separately. To demonstrate, below we present the DFT investigations for the simple systems of HS fluids, LJ fluids, and charged systems.

3.2.1 Hard-Sphere Fluids

For representing the repulsive nature of two-body interaction, HS model is widely adopted in liquid state theories. As depicted in Fig. 6, the HSs are defined simply as impenetrable spheres which cannot overlap with each other in space.

Mathematically, the interaction between two HSs can be expressed as

$$u(r) = \begin{cases} \infty & r < d_{12} \\ 0 & r > d_{12} \end{cases}. \tag{10}$$

Here d_{12} is the closest distance that two HSs can approach. If all HSs are of same size of d, we have $d_{12} = \frac{(d+d)}{2} = d$ for additive systems. HS system has

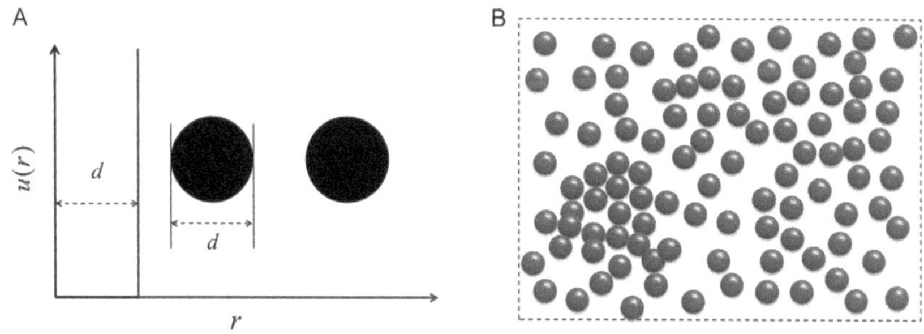

Figure 6 (A) Hard-sphere potential and (B) schematic graph for colloidal system.

proven to be an excellent primitive model to capture the properties of colloidal systems (see Fig. 6B) in which the entropy dominates (Yethiraj and van Blaaderen, 2003).

The properties of HS systems can be accurately investigated by using fundamental measure theory (FMT) (Rosenfeld, 1989) or its variants (Roth et al., 2002; Yu and Wu, 2002). The FMT is a special version of DFT developed uniquely for HS systems, and it is originally proposed by Rosenfeld (1989) by combining the scaled particle theory (Ashbaugh and Pratt, 2006; Holovko and Dong, 2009) and the diagrammatic expansion of the free energy, and since its foundation, several modifications (Roth et al., 2002; Yu and Wu, 2002) have been proposed which improve the accuracy of this geometrical theory especially under the condition of high packing fraction. The packing fraction η is defined as $\eta = \frac{\rho^b \pi d^3}{6}$ with ρ^b being the corresponding bulk number density of the HS system and d being the diameter of the HS. The larger is the value of η, and more dense the HS system is. The maximum value of η can be estimated by referring to a crystal solid with face-centered-cubic structure, which represents the most packed solid with packing fraction $\eta = \frac{\pi\sqrt{2}}{6}$.

Among these variants, the white bear version developed by Roth et al. (2002) or equivalently the modified FMT (MFMT) by Yu and Wu (2002) is most reputed. Although both modifications are developed by two individual research groups at almost the same time, they are completely equivalent (Roland, 2010). The basic idea in these modifications is that more accurate EOS for bulk HS system can be incorporated into the free energy functional.

Similar to Eqs. (2) and (3), the grand potential of a HS system in grand canonical ensemble with temperature T, volume V, and chemical potential $\mu_{\rm hs}$ reads

$$\Omega[\rho(\mathbf{r})] = F^{\rm id}[\rho(\mathbf{r})] + F^{\rm ex}_{\rm hs}[\rho(\mathbf{r})] + \int \rho(\mathbf{r})(V^{\rm ext}(\mathbf{r}) - \mu_{\rm hs})d\mathbf{r}. \qquad (11)$$

In above equation, $V^{\rm ext}(\mathbf{r})$ is the external potential at position $\mathbf{r}$. $F^{\rm id}$ denotes the Helmholtz free energy for noninteraction contribution. Because the position of a classical particle is completely decoupled with its momentum, the potential energy of a classical model system is independent of the kinetic energy in Hamiltonian. Hereby, the integration over momenta in grand partition function can be carried out analytically for noninteracting system, which gives the ideal part of chemical potential, i.e.,

$\mu^{id} = k_B T \ln(\rho^b \Lambda^3)$ (Hansen and McDonald, 2013). Here Λ is the de Broglie thermal wavelength coming from the integration over momenta, and it reads $\Lambda = h/\sqrt{2\pi m k_B T}$ with h being the Plank constant and m the mass of particle. Typically, the thermal wavelength is 3.3 Å for hydrogen gas (H_2), 0.46 Å for methane (CH_4), and 0.42 Å for nitrogen gas (N_2).

The ideal part of Helmholtz free energy functional can be obtained by integrating the ideal chemical potential in its local form, namely $\mu^{id}(\mathbf{r}) = k_B T \ln(\rho(\mathbf{r})\Lambda^3)$, over the one-body density distribution $\rho(\mathbf{r})$, and it follows

$$F^{id}[\rho(\mathbf{r})] = k_B T \int d\mathbf{r}\rho(\mathbf{r})\left[\ln(\rho(\mathbf{r})\Lambda^3) - 1\right]. \tag{12}$$

The excessive part of Helmholtz free energy over the ideal contribution, i.e., F_{hs}^{ex}, accounts for the HS repulsive interaction, and following MFMT, it can be determined through the excess free energy density Φ^{hs} by (Yu and Wu, 2002)

$$\beta F_{hs}^{ex} = \int \Phi^{hs}[n_\alpha(\mathbf{r})]d\mathbf{r}. \tag{13}$$

Here $\beta = \frac{1}{k_B T}$, and Φ^{hs} is a function of six parameters n_α with $\alpha = 0, 1, 2, 3, V_1, V_2$, out of which the first four parameters are scalars and the last two are vectors,

$$\begin{aligned}\Phi^{hs} &= -n_0 \ln(1-n_3) + \frac{n_1 n_2 - \mathbf{n}_{V_1}\cdot\mathbf{n}_{V_2}}{1-n_3} \\ &+ \frac{n_3 + (1-n_3)^2 \ln(1-n_3)}{36\pi n_3^2 (1-n_3)^2}\left(n_2^3 - 3n_2\mathbf{n}_{V_2}\cdot\mathbf{n}_{V_2}\right).\end{aligned} \tag{14}$$

The local value of parameter n_α is determined by the local density

$$n_\alpha(\mathbf{r}) = \int \rho(\mathbf{r}')\omega^{(\alpha)}(|\mathbf{r}-\mathbf{r}'|)d\mathbf{r}' \tag{15}$$

with $\omega^{(\alpha)}$ is the weighting function reading

$$\begin{cases}\omega^{(2)}(r) = \pi d^2 \omega^{(0)}(r) = 2\pi d\omega^{(1)}(r) = \delta(d/2 - r) \\ \omega^{(3)}(r) = \theta(d/2 - r) \\ \boldsymbol{\omega}^{(V_2)}(\mathbf{r}) = 2\pi d\boldsymbol{\omega}^{(V_1)}(\mathbf{r}) = \omega^{(2)}(r)\dfrac{\mathbf{r}}{r}\end{cases}, \tag{16}$$

where $\theta(r)$ is the Heaviside step function. At homogeneous state, the local density is uniformly distributed and recovers to the bulk number density ρ^b. In that circumstance, the six parameters recover to the trivial constants (Roland, 2010), viz., $n_0 = \rho_0$, $n_1 = \frac{\rho_0 d}{2}$, $n_2 = \rho_0 \pi d^2$, $n_3 = \eta$, and $\mathbf{n}_{V_1} = \mathbf{n}_{V_2} = 0$.

At equilibrium, by noting $\mu_{\text{hs}} = \mu_{\text{hs}-b}^{\text{ex}} + k_B T \ln\left(\rho^b \Lambda^3\right)$ the Euler–Lagrange equation reads

$$\rho(\mathbf{r}) = \rho^b \exp\left[-\beta V^{\text{ext}}(\mathbf{r}) - \beta\left(\mu_{\text{hs}}^{\text{ex}}(\mathbf{r}) - \mu_{\text{hs}-b}^{\text{ex}}\right)\right]. \tag{17}$$

The excess bulk chemical potential $\mu_{\text{hs}-b}^{\text{ex}}$ for HS system can be accurately evaluated by using Carnahan–Starling (CS) EOS (Carnahan and Starling, 1969). The local excess chemical potential $\mu_{\text{hs}}^{\text{ex}}(\mathbf{r})$ can be derived from the functional derivative of excess Helmholtz free energy with respect to the local density distribution $\rho(\mathbf{r})$, which follows

$$\begin{aligned}\beta\mu_{\text{hs}}^{\text{ex}}(\mathbf{r}) &= \frac{\beta\delta F_{\text{hs}}^{\text{ex}}}{\delta\rho(\mathbf{r})} = \int d\mathbf{r}' \frac{\delta\Phi^{\text{hs}}}{\delta\rho(\mathbf{r})} \\ &= \sum_\alpha \int d\mathbf{r}' \frac{\delta\Phi^{\text{hs}}}{\delta n_\alpha(\mathbf{r}')} \frac{\delta n_\alpha(\mathbf{r}')}{\delta\rho(\mathbf{r})} \\ &= \sum_\alpha \int d\mathbf{r}' \frac{\delta\Phi^{\text{hs}}}{\delta n_\alpha(\mathbf{r}')} \omega^{(\alpha)}(\mathbf{r}' - \mathbf{r})\end{aligned} \tag{18}$$

The combination of Eqs. (14) and (15) gives the expressions for $\frac{\delta\Phi^{\text{hs}}}{\delta n_\alpha(\mathbf{r})}$ as follows

$$\frac{\delta\Phi^{\text{hs}}}{\delta n_0(\mathbf{r})} = -\ln(1 - n_3); \tag{19}$$

$$\frac{\delta\Phi^{\text{hs}}}{\delta n_1(\mathbf{r})} = \frac{n_2}{1 - n_3}; \tag{20}$$

$$\frac{\delta\Phi^{\text{hs}}}{\delta n_2(\mathbf{r})} = \frac{n_1}{1 - n_3} + \left(\frac{\ln(1 - n_3)}{n_3} + \frac{1}{(1 - n_3)^2}\right)\frac{n_2^2 - \mathbf{n}_{V_2}{}^2}{12\pi n_3}; \tag{21}$$

$$\begin{aligned}\frac{\delta\Phi^{\text{hs}}}{\delta n_3(\mathbf{r})} = &-\left(\frac{\ln(1 - n_3)}{18\pi n_3^3} + \frac{1 - 3n_3 + (1 - n_3)^2}{36\pi n_3^2(1 - n_3)^3}\right)\left(n_2^3 - 3n_2\mathbf{n}_{V_2}{}^2\right) \\ &+ \frac{n_0}{1 - n_2} + \frac{n_1 n_2 - \mathbf{n}_{V_1} \cdot \mathbf{n}_{V_2}}{(1 - n_3)^2};\end{aligned} \tag{22}$$

$$\frac{\delta\Phi^{\mathrm{hs}}}{\delta\mathbf{n}_{V_1}(\mathbf{r})} = -\frac{\mathbf{n}_{V_2}}{1-n_3}; \tag{23}$$

$$\frac{\delta\Phi^{\mathrm{hs}}}{\delta\mathbf{n}_{V_2}(\mathbf{r})} = -\frac{\mathbf{n}_{V_1}}{1-n_3} - \left(\frac{\ln(1-n_3)}{n_3} + \frac{1}{(1-n_3)^2}\right)\frac{n_2\mathbf{n}_{V_2}}{6\pi n_3}. \tag{24}$$

The combination of Eqs. (17) and (18) gives the final equations to calculate the one-body density distribution. Typically, with a given external potential the one-body density distribution $\rho(\mathbf{r})$ can be determined by solving Eq. (17) iteratively or with a minimization procedure, and therefore the thermodynamic properties can be achieved straightforwardly.

The extension of FMT (or its variants) to additive HS mixture is simple, and the local value of parameter n_α should summarize all contributions from different components, i.e.,

$$n_\alpha(\mathbf{r}) = \sum_i n_{\alpha,i}(\mathbf{r}) = \sum_i \int \rho_i(\mathbf{r}')\omega_i^{(\alpha)}(\mathbf{r}-\mathbf{r}')\mathrm{d}\mathbf{r}'. \tag{25}$$

Here $\rho_i(\mathbf{r})$ is the local density distribution of HS component i. The weighting function $\omega_i^{(\alpha)}$ takes the same mathematical expression as in Eq. (16) with the involved HS diameter being replaced by d_i. The extension to nonadditive HS mixture is a little bit tricky, and a few methods (Ayadim and Amokrane, 2010; Matthias, 2004, 2011) have been developed, and the accuracies of these extensions are rather striking. Recently, the FMT has been extended to sticky HS fluids (Hansen-Goos and Wettlaufer, 2011) and to nonspherical hard-convex systems (Hansen-Goos and Mecke, 2009; Hendrik and Klaus, 2010).

Figure 7 plots the surface pressure as a function of the separation H between two paralleled slit wall filled with one-component HS fluid. The external potential for the confined HS fluid is zero if $0<z<H$ and otherwise infinitely large, where z is the distance to one of the slit walls. Generally, the surface pressure, i.e., P_W, can be calculated from the integration of one-body density distribution multiplying the external force over the distance z. In the circumstance of hard-wall, the external force recovers to a Dirac delta function and thus the surface pressure is directly related to the fluid contact density $\rho(z=0)$. The predicted results from MFMT and original FMT are compared with simulation results. This comparison shows that FMT, especially the modified version, can yield very accurate results.

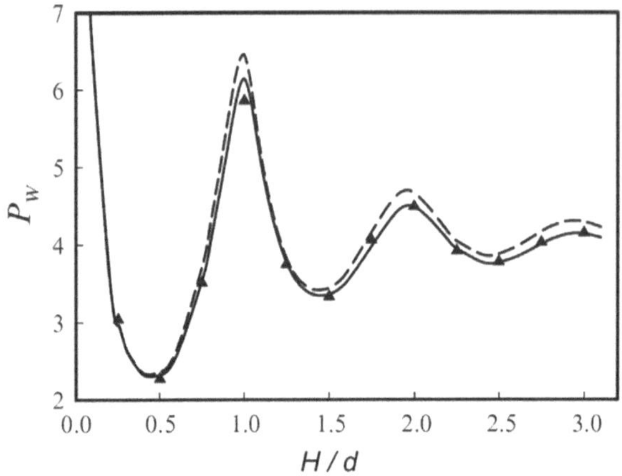

Figure 7 Predicted surface pressure by MFMT (solid line) and original FMT (dashed line) as a function of the separation H between two parallel hard walled filled with HS fluid with reduced bulk density $\rho^b d^3 = 0.7$. The symbols are from simulation. *After Yu and Wu (2002).*

The numerical calculation in MFMT involves 3D integrations as included in Eq. (18), toward which various numerical methods (Douglas Frink and Salinger, 2000; Levesque et al., 2012) have been developed for improving the calculation efficiency. Knepley et al. demonstrated that convolution of Eq. (18) could be performed by using fast Fourier transform (FFT) (Knepley et al., 2010). Note that the Fourier transforms of $\omega^{(\alpha)}(\mathbf{r})$ can be derived analytically.

The development of FMT for inhomogenous HS system is striking. Traditionally, ideal gas systems have been widely employed as reference systems, especially by physicists, to rationalize the theories which are proposed for interacting complex systems, and this is because, on one hand, a general theory applicable for an interacting system should be of course applicable for its reference ideal gas system after taking appropriate limits, and on the other hand, the ideal gas reference systems are usually trivial enough so that the exact properties of those systems can be achieved. Following the same logic, the HS systems supply more practical reference systems than ideal gas, at least from numerical perspective owing to the development of FMT.

3.2.2 Lennard–Jones Fluids

Nearly 150 years ago, Van der Waals argued that an attraction should be introduced in addition to the repulsion between two atoms (Van der Waals, 1873). Later on, LJ interaction model is introduced to characterize the empirical interaction between two neutral atoms at distance r, which follows,

$$u_{\mathrm{LJ}}(r) = 4\varepsilon\left[\left(\frac{\sigma}{r}\right)^{12} - \left(\frac{\sigma}{r}\right)^{6}\right]. \tag{26}$$

Here ε and σ are the energy and size parameters, respectively. Typically, the LJ parameters are $\sigma = 3.4\,\text{Å}$ and $\varepsilon = 1.0$ kJ/mol for argon (Hansen and McDonald, 2013). The reduced LJ interaction in terms of reduced separation is plotted in Fig. 8A.

Following thermodynamic perturbation theory (TPT) (Shukla, 1987; Zhou and Solana, 2009), the LJ interaction can be generally decomposed into the summation of a repulsion part and a perturbative attraction part

$$u_{\mathrm{LJ}}(r) = u^{\mathrm{rep}}(r) + u^{\mathrm{att}}(r). \tag{27}$$

Whereas in principle, there are numerous ways to decompose the LJ interaction, conventionally two popular methods exist: Barker–Henderson (BH) method (Barker and Henderson, 1967) and Weeks–Chandler–Anderson (WCA) method (Weeks et al., 1971) as schematically depicted in Fig. 8A and B, respectively. For instance, in BH method the two complementary parts follow (Barker and Henderson, 1967)

$$u_{\mathrm{BH}}^{\mathrm{rep}}(r) = \begin{cases} u_{\mathrm{LJ}}(r) & r \le \sigma \\ 0 & r > \sigma \end{cases}, \tag{28}$$

and

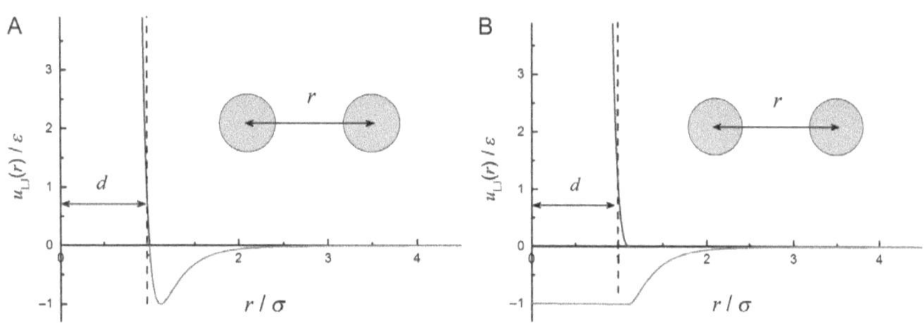

Figure 8 Lennard–Jones interaction (the combination of blue (dark gray in the print version) and red (gray in the print version) curves) and two typical decompositions with blue lines (dark gray in the print version) represent repulsion and red lines (gray in the print version) for attraction: (A) BH method; (B) WCA method.

$$u_{\rm BH}^{\rm attr}(r)=\begin{cases}0 & r<\sigma\\ 4\varepsilon\left[\left(\frac{\sigma}{r}\right)^{12}-\left(\frac{\sigma}{r}\right)^{6}\right] & r>\sigma\end{cases}. \tag{29}$$

The excess Helmholtz free energy functional for the LJ system can be formulated by using several approaches, among which the first-order mean spherical approximation (FMSA) is formally simple. The FMSA is developed by Tang (2003, 2004) basing on functional expansion approach (Hansen and McDonald, 2013) (also named density-expansion method in some work (Wu, 2009)), by which the excess Helmholtz free energy functional of an inhomogeneous system can be expanded around that of the corresponding bulk system with respect to the local density, i.e.,

$$\begin{aligned}F^{\rm ex}[\rho(\mathbf{r})]=&F_b^{\rm ex}(\rho^b)+\mu_b^{\rm ex}\int\Delta\rho(\mathbf{r})\mathrm{d}\mathbf{r}\\&-\frac{k_{\rm B}T}{2!}\int\Delta\rho(\mathbf{r})c(|\mathbf{r}-\mathbf{r}'|)\Delta\rho(\mathbf{r}')\mathrm{d}\mathbf{r}\mathrm{d}\mathbf{r}'+F^{\rm B}[\rho(\mathbf{r})]\end{aligned} \tag{30}$$

with $F_b^{\rm ex}$ being the free energy of the reference bulk system. The second and third terms in above equation account for the first- and second-order functional expansions because we have (Evans, 1979)

$$\left.\frac{\delta F^{\rm ex}}{\delta\rho(\mathbf{r})}\right|_{\rho(\mathbf{r})=\rho^b}=-k_{\rm B}Tc^{(0)}=\mu_b^{\rm ex}, \tag{31}$$

$$\left.\frac{\delta F^{\rm ex}}{\delta\rho(\mathbf{r})\delta\rho(\mathbf{r}')}\right|_{\rho(\mathbf{r})=\rho^{\rm b}}=-k_{\rm B}Tc(|\mathbf{r}-\mathbf{r}'|). \tag{32}$$

The summation of those terms beyond the second-order expansion contributes to the bridge functional, abbreviated as $F^{\rm B}[\rho(\mathbf{r})]$.

For a homogeneous LJ system, the excess Helmholtz free energy $F_b^{\rm ex}$ and the excess chemical potential $\mu_b^{\rm ex}$ can be evaluated from MBWR equations of state (Johnson et al., 1993). $c(|\mathbf{r}-\mathbf{r}'|)$ is the direct correlation function (DCF) (Hansen and McDonald, 2013) of the corresponding LJ bulk system, and it has been proven that the DCF and the total correlation function satisfy the Ornstein–Zernike (OZ) equation (Hansen and McDonald, 2013)

$$h(|\mathbf{r}-\mathbf{r}'|)=c(|\mathbf{r}-\mathbf{r}'|)+\rho^b\int c(|\mathbf{r}-\mathbf{r}''|)h(|\mathbf{r}''-\mathbf{r}'|)\mathrm{d}\mathbf{r}''. \tag{33}$$

The OZ equation constitutes the cornerstone of integral equation theory, which is a sister theory of DFT in modern liquid state statistical mechanics (Hansen and McDonald, 2013; Ratkova et al., 2015).

The knowledge of DCF is needed for the free energy functional in Eq. (30). There are several routes to obtain the DCF of a bulk LJ system. First, the OZ equation can be solved by combining an additional closure relation between the DCF and the total correlation function. The exact closure formally reads (Hansen and McDonald, 2013)

$$h(r) + 1 = \exp\left[-\beta u(r) + \rho^b \int c(|\mathbf{r} - \mathbf{r}'|)h(r')\mathrm{d}\mathbf{r}' + B(r)\right]. \qquad (34)$$

Here $u(r)$ is the intermolecular interaction and it refers to the LJ potential as far as LJ bulk system is concerned. $B(r)$ is the bridge term accounting for multibody correlation functions in the reference bulk system (Llanorestrepo and Chapman, 1992), and this term is generally unknown, to which several approximations have been developed (Henderson and Sokolowski, 1995). A simple approximation is to ignore this bridge term, which leads to the Hypernetted-Chain (HNC) closure. It has been demonstrated that the HNC approximation gives very accurate description for the systems dominated by long-range attractions. Another treatment is to approximate the bridge term with its counterpart in a reference HS system, i.e., $B_{\mathrm{hs}}^{\mathrm{ref}}(r)$

$$h(r) + 1 = \exp\left[-\beta u(r) + \rho^b \int c(|\mathbf{r} - \mathbf{r}'|)h(r')\mathrm{d}\mathbf{r}' + B_{\mathrm{hs}}^{\mathrm{ref}}(r)\right]. \qquad (35)$$

Conventionally, this is called reference HNC approximation (RHNC), which yields surprisingly accurate prediction for the structures of LJ systems. We also want to mention the mean spherical approximation (MSA)

$$\begin{cases} g(r) = 0, & r < d \\ c(r) = -\beta u(r), & r > d \end{cases}, \qquad (36)$$

which provides a much better description of the properties of square-well fluid than the HNC approximation.

Second, the DCF can be solved from Eq. (33) by injecting the total correlation function, which, in prior, is prepared from molecule simulation as demonstrated by Ramirez et al. (2002). In this route, the obtained DCF is exact supposing that the long-range behavior of $h(r)$ can be properly accounted during the involved Fourier transform. Third, Tang and his coworkers derived an analytical expression for the DCF of LJ system in the spirit that one LJ potential can be decomposed into the summation of two Yukawa potential (Tang, 2003, 2004).

Besides the derivation of the analytical result for DCF, Tang (2004) incorporated the universality hypothesis of the bridge functional that was originally proposed by Rosenfeld (Kahl et al., 1996) to account for the higher-order terms into the free energy functional., i.e., bridge functional in Eq. (30) is approximated by that for the reference HS system

$$F^B[\rho(\mathbf{r})] \approx F^B_{\text{hs}}[\rho(\mathbf{r})] = F^{\text{ex}}_{\text{hs}}[\rho(\mathbf{r})] - F^{\text{ex}}_{\text{hs}-b}(\rho^b) - \mu^{\text{ex}}_{\text{hs}-b}\int \Delta\rho(\mathbf{r})d\mathbf{r} + \frac{k_B T}{2!}\int \Delta\rho(\mathbf{r})c_{\text{hs}}(|r-r'|)\Delta\rho(\mathbf{r}')d\mathbf{r}d\mathbf{r}' \tag{37}$$

The second line in Eq. (37) is derived straightforwardly by referring to Eq. (30). The reference HS system has identical local density distribution $\rho(\mathbf{r})$ as the LJ system and the diameter of the HS can be estimated by using BH method (Barker and Henderson, 1967)

$$d_{\text{BH}} = \frac{1 + 0.2977(\beta\varepsilon)^{-1}}{1 + 0.33163(\beta\varepsilon)^{-1} + 0.0010477(\beta\varepsilon)^{-2}}\sigma. \tag{38}$$

The bridge functional of the reference HS system can be addressed by using MFMT.

The minimization of the grand potential functional gives rise to the Euler–Lagrange equation for the LJ system,

$$\rho(\mathbf{r}) = \rho^b \exp\left[-\beta V^{\text{ext}}(\mathbf{r}) + \int c(|\mathbf{r}-\mathbf{r}'|)\Delta\rho(\mathbf{r}')d\mathbf{r}' + B_{\text{hs}}(\mathbf{r})\right], \tag{39}$$

with

$$B_{\text{hs}}(\mathbf{r}) = -\frac{\delta\beta F^B_{\text{hs}}[\rho(\mathbf{r})]}{\delta\rho(\mathbf{r})} = \beta\mu^{\text{ex}}_{\text{hs}-b} - \beta\mu^{\text{ex}}_{\text{hs}}(\mathbf{r}) - \int c_{\text{hs}}(|\mathbf{r}-\mathbf{r}'|)\Delta\rho(\mathbf{r}')d\mathbf{r}'. \tag{40}$$

It should be noted that Eq. (39) is derived from DFT, and it presents the local expression similar to RHNC in Eq. (35). By noting that at bulk $g(\mathbf{r}) = \rho(\mathbf{r})/\rho^b$ recovers to $g(r)$ and the external potential recovers to the intermolecular interaction (in that case the solute is virtually a solvent particle, and the system is locally inhomogeneous but represents a homogeneous system), both equations are essentially equivalent. This suggests that DFT and integral equation theory are closely related and moreover the closure relation for the integral equation can be derived from DFT.

From both fundamental and practical perspectives, FMSA developed by Tang is significant in particular for engineering applications, and it has been employed to investigate interfacial properties, gas adsorption, and phase diagram. Figure 9 plots the reduced local density $\rho^*(z)=\rho(z)\sigma^3$ of the LJ fluid system confined in a slit pore with reduced pore width $H/\sigma=7.5$. Here, the pore width refers to the distance from one flat surface to the other one. The reduced temperature of the confined system is defined as $T^*=k_B T/\varepsilon$, and the fluid–wall interaction is described by Steele's 10-4-3 potential (Steele, 1974). The comparison of FMSA prediction (solid line) with simulation result (symbols) demonstrates that the performance of FMSA is satisfactory. The figure inset depicts the DCFs of a LJ bulk system from FMSA and its extension (Zeng et al., 2009) in comparison with the corresponding simulation result, showing that the analytical results for DCF are accurate.

Besides FMSA that is developed basing on the functional expansion, the excess Helmholtz free energy functional can also be constructed by using WDA method. Specifically, following the TPT, the excess Helmholtz free

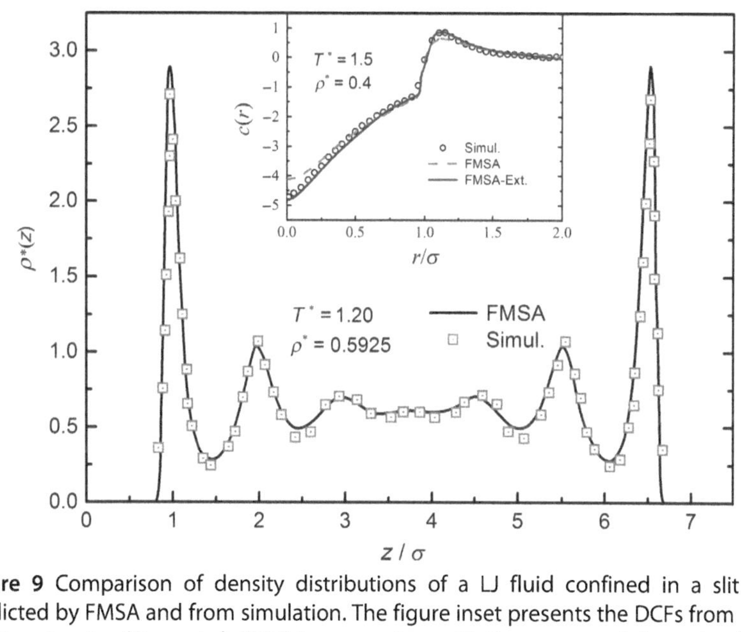

Figure 9 Comparison of density distributions of a LJ fluid confined in a slit pore predicted by FMSA and from simulation. The figure inset presents the DCFs from FMSA and its extension (Zeng et al., 2009) in comparison with the exact result from simulation. *After Tang and Wu (2004).*

energy functional can be decomposed into the summation of three contributions

$$F^{\mathrm{ex}}[\rho(\mathbf{r})] = F_{\mathrm{hs}}^{\mathrm{ex}}[\rho(\mathbf{r})] + F_{\mathrm{att}}^{\mathrm{ex}}[\rho(\mathbf{r})] + F_{\mathrm{cor}}^{\mathrm{ex}}[\rho(\mathbf{r})]. \tag{41}$$

Here $F_{\mathrm{hs}}^{\mathrm{ex}}[\rho(\mathbf{r})]$ corresponds to the contribution from the repulsive interaction, and similarly it can be addressed by using MFMT. $F_{\mathrm{att}}^{\mathrm{ex}}[\rho(\mathbf{r})]$ stands for the contribution from the attractive interaction, and it can be formulated by using the mean-field approximation (MFA) implying that $c^{\mathrm{att}}(|\mathbf{r}-\mathbf{r}'|) = -\beta u^{\mathrm{att}}(|\mathbf{r}-\mathbf{r}'|)$, i.e.,

$$F_{\mathrm{att}}^{\mathrm{ex}}[\rho(\mathbf{r})] = \frac{1}{2}\iint \rho(\mathbf{r})\rho(\mathbf{r}')u^{\mathrm{att}}(|\mathbf{r}-\mathbf{r}'|)\mathrm{d}\mathbf{r}\mathrm{d}\mathbf{r}'. \tag{42}$$

Here $u^{\mathrm{att}}(r)$ is the attractive part in LJ potential, and it is determined by Eq. (29) when the HS diameter is estimated with BH method. The last term in Eq. (41), i.e., $F^{\mathrm{cor}}[\rho(\mathbf{r})]$ is the mixing contribution stemming from the correlation between repulsion and attraction, and it can be calculated by using WDA (Tang and Lu, 1997; Yu, 2009)

$$F_{\mathrm{cor}}^{\mathrm{ex}}[\rho(\mathbf{r})] = \int \rho(\mathbf{r})f^{\mathrm{cor}}[\bar{\rho}(\mathbf{r})]\mathrm{d}\mathbf{r}, \tag{43}$$

where $\bar{\rho}(\mathbf{r})$ is the weighted density evaluated by

$$\bar{\rho}(\mathbf{r}) = \frac{3}{4\pi d^3}\int \rho(\mathbf{r}')\theta(d-|\mathbf{r}-\mathbf{r}'|)\mathrm{d}\mathbf{r}', \tag{44}$$

and $f^{\mathrm{cor}}(\rho)$ is the bulk free energy density (per number of molecules) due to the correlation effect, and it can be calculated by

$$f^{\mathrm{cor}}(\rho) = \frac{F_{\mathrm{bulk}}^{\mathrm{LJ}}(\rho)}{N} - \frac{F_{\mathrm{bulk}}^{\mathrm{hs}}(\rho)}{N} - \frac{F_{\mathrm{bulk}}^{\mathrm{MFA}}(\rho)}{N}. \tag{45}$$

In above equation, N is the number of molecules and $F_{\mathrm{bulk}}^{\mathrm{LJ}}(\rho)$ is the excess free energy for LJ bulk fluid calculated by using MBWR-EOS (Johnson et al., 1993). $F_{\mathrm{bulk}}^{\mathrm{hs}}(\rho)$ is the excess free energy for bulk HS fluid, which can be determined from CS EOS (Carnahan and Starling, 1969). $F_{\mathrm{bulk}}^{\mathrm{MFA}}(\rho)$ is the attractive term, and it follows

$$\frac{F_{\mathrm{bulk}}^{\mathrm{MFA}}(\rho)}{N} = -\frac{16}{9}\pi\beta\varepsilon\rho\sigma^3. \tag{46}$$

Although the combined MFMT-MFA-WDA approximation of the excess Helmholtz free energy functional is physically less significant, it has been found that the combined approximation generates very good predictions which are comparable to that from FMSA, and at high concentration the predictions from the combined approximation are even more accurate than those from FMSA.

Note that several variants of the above functional (i.e., Eq. 41) exist. Instead of using WDA for treating $F_{\mathrm{cor}}^{\mathrm{ex}}$ as shown in Eq. (43), the LDA can also be applied which provides comparable accuracy as discussed by Fu et al. (2015a). In addition, if we simply ignore this correlation term $F_{\mathrm{cor}}^{\mathrm{ex}}$ and apply WCA scheme for the decomposition of repulsion and attraction in the LJ potential, the above combined functional recovers to the so-called nonlocal density functional theory (NLDFT) initiated by Balbuena and Gubbins (1993) and extensively applied by Neimark et al. (Landers et al., 2013; Olivier et al., 1994).

Because the formulation is relatively simple, NLDFT can be easily extended to investigate immiscible fluid mixture in which a liquid–liquid interface exists. A simple liquid–liquid interface system can be modeled with nonadditive LJ potentials, i.e., the LJ interaction applies for the same species, while for the cross-interaction between the unlike pair, the interaction follows $u_{12}(r) = 4\varepsilon_{12}\kappa\left[\left(\frac{\sigma_{12}}{r}\right)^{12} - \left(\frac{\sigma_{12}}{r}\right)^{6}\right]$. Here ε_{12} and σ_{12} are the crossed LJ parameters determined by the conventional Lorentz–Berthelot combining rule. The dimensionless constant κ is less than one accounting for the immiscibility between species 1 and 2. Figure 10 plots our theoretical prediction of the liquid density profiles at the interfacial zone from NLDFT. Comparing to simulation results, the DFT predictions are satisfactory. The accurate produce of the liquid densities at the interface allows the reliable descriptions of the thermodynamic interfacial properties.

3.2.3 Simple Charged Systems

As in the primitive model of electrolyte solutions, the macroions and small ions are usually represented by charged HSs and the solvent is modeled as a continuous dielectric medium. The pair potential $u_{ij}(r)$ between two species i and j is given by

$$u_{ij}(r) = \begin{cases} \infty & r < d_{ij} \\ \dfrac{z_i z_j e^2}{\varepsilon r} & r \geq d_{ij} \end{cases}, \tag{47}$$

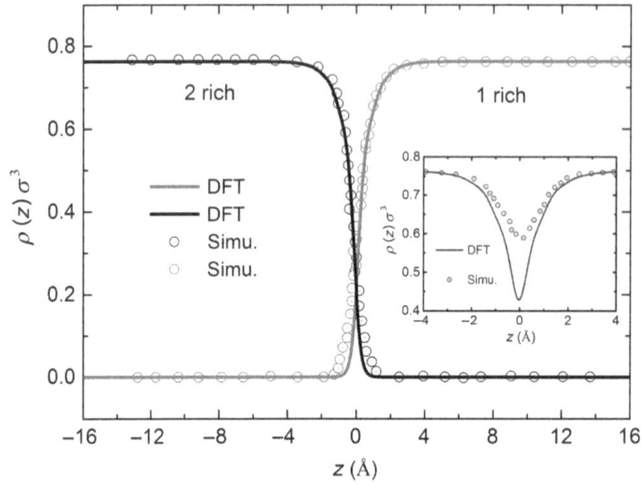

Figure 10 Density profiles of two species modeled with nonadditive LJ fluids with immiscible constant $\kappa = 0.25$ at the interfacial region. The solid lines represent our DFT predictions, and the symbols are simulation results from Galliero (2010). The figure inset plots the total number density across the interface.

where e is the elementary charge, z_i and d_i are the charge valence and diameter of particle i, respectively, $d_{ij} = \frac{(d_i + d_j)}{2}$, and again r is the center-to-center distance between two particles. For a neutral particle, the charge valence is zero and the relevant pair potential reduces HS potential. In above equation, $\varepsilon = 4\pi\varepsilon_r\varepsilon_0$ is the dielectric constant with ε_0 being the vacuum permittivity and ε_r the solvent relative permittivity.

For a mixture of charged and neutral HSs, the grand potential functional takes the similar expression as Eq. (2), and the ideal Helmholtz free energy functional is the summation of the contributions from each species which has the analytic expression as Eq. (12), and the excess Helmholtz free energy functional can be constructed as (Li and Wu, 2004; Yu et al., 2004)

$$F^{\mathrm{ex}}[\rho_i(\mathbf{r})] = F^{\mathrm{ex}}_{\mathrm{hs}}[\rho_i(\mathbf{r})] + F^{\mathrm{ex}}_{C}[\rho_i(\mathbf{r})] + F^{\mathrm{ex}}_{\mathrm{el}}[\rho_i(\mathbf{r})] \quad (48)$$

The first term is the contribution from HS repulsion, and again it can be addressed by MFMT. F^{ex}_{C} represents the direct Coulomb interaction contribution

$$F^{\mathrm{ex}}_{C}[\rho_i(\mathbf{r})] = \frac{1}{2}\int d\mathbf{r}d\mathbf{r}' \sum_{i,j} \frac{z_i z_j e^2 \rho_i(\mathbf{r})\rho_j(\mathbf{r}')}{\varepsilon|\mathbf{r}-\mathbf{r}'|}. \quad (49)$$

The residual part $F_{\mathrm{el}}^{\mathrm{ex}}$ arises from the correlations due to the Coulomb and HS interactions (Li and Wu, 2004), and it can be formulated by using functional expansion method and truncating the expansion at second order, i.e.,

$$\beta F_{\mathrm{el}}^{\mathrm{ex}}[\rho_i(\mathbf{r})] = \beta F_{\mathrm{el}}^{\mathrm{ex}}(\rho_i^{\,b}) + \int d\mathbf{r} \sum_i \Delta\mu_{i-b}^{\mathrm{ex}} \Delta\rho_i(\mathbf{r}) \\ -\frac{1}{2}\int d\mathbf{r}d\mathbf{r}' \sum_i \sum_j \Delta c_{ij}^{\mathrm{el}}(|\mathbf{r}-\mathbf{r}'|)\Delta\rho_i(\mathbf{r})\Delta\rho_j(\mathbf{r}') \tag{50}$$

The first term on the right-hand side in above equation corresponds to the electrostatic part of excess Helmholtz free energy for a bulk electrolyte solution and it can be estimated with EOS derived from MSA (Palmer and Weeks, 1973; Waisman and Lebowitz, 1972), and similar to Eqs. (31) and (32), the residual excess chemical potential $\Delta\mu_{i-b}^{\mathrm{ex}}$ and residual two-body DCF $\Delta c_{ij}^{\mathrm{el}}(|\mathbf{r}-\mathbf{r}'|)$ can be obtained by taking functional derivatives with respect to the local density.

The minimization of the grand potential functional yields the Euler–Lagrange equation, which follows

$$\rho_i(\mathbf{r}) = \\ \rho_i^b \exp\begin{bmatrix} -\beta V_i^{\mathrm{ext}}(\mathbf{r}) - \beta z_i e\left(\psi_C(\mathbf{r}) - \psi_C^b\right) + \int d\mathbf{r}' \sum_j \Delta c_{ij}^{\mathrm{el}}(|\mathbf{r}-\mathbf{r}'|)\Delta\rho_j(\mathbf{r}') \\ -\beta\left(\mu_{i,\mathrm{hs}}^{\mathrm{ex}}(\mathbf{r}) - \mu_{i,\mathrm{hs}-b}^{\mathrm{ex}}\right) \end{bmatrix} \tag{51}$$

where $\psi_C(\mathbf{r})$ stands for the intrinsic mean-electrostatic potential, and its expression can be derived from Eq. (49) by taking derivative with respect to $\rho_i(\mathbf{r})$. With the help of Fourier transform method it can be proven that $\psi_C(\mathbf{r})$ satisfies the following conventional Poisson equation (Li and Wu, 2004; Liu et al., 2013a)

$$\varepsilon\nabla^2\psi_C(\mathbf{r}) = -4\pi\sum_j z_j e\rho_j(\mathbf{r}). \tag{52}$$

ψ_C^b is the homogeneous limit of $\psi_C(\mathbf{r})$, and mathematically it is the integral constant when solving the Poisson equation. By taking the limit of $d_i \to 0$, the charged HSs become charged points, and then the second line in Eq. (51) vanishes. If we further ignore the correlation contribution and replace $\Psi(\mathbf{r}) = \psi_C(\mathbf{r}) - \psi_C^b$ which still satisfies the Poisson equation, we immediately recover the original Poisson–Boltzmann (PB) equation (Lamm, 2003)

$$\rho_i(\mathbf{r}) = \rho_i^b \exp\left[-\beta V_i^{\text{ext}}(\mathbf{r}) - \beta z_i e \Psi(\mathbf{r})\right]. \tag{53}$$

The comparison of Eq. (51) from DFT and Eq. (53) from PB clearly indicates the physical essence of the crude approximation involved in PB theory, i.e., both the ion size effect and correlation contribution are completely ignored, and therefore PB theory is applicable only for dilute electrolyte systems in which the direct Coulombic interaction dominates and the size effect is negligible.

In contrast, the DFT goes beyond the PB theory, and it faithfully captures both the short-range correlation effect and long-range correlation effect. Note that in the extended versions of PB theory, the dielectric constants are position-dependent or coupled with the ion distributions (Wang, 2010), and a dielectric mismatch may lead the PB theory fail even for dilute charged systems (Wang and Wang, 2013). As shown by Xu and coworkers (Ma and Xu, 2014; Xu et al., 2014), a self-consistent field model can be developed by properly accounting the coupling between the local dielectric constant and local charge density. Such coupling can, in principle, be considered in DFT by properly modifying the intermolecular interaction as stated in Eq. (47), or by considering the solvent explicitly. Toward this direction, however, no progress has been reported yet.

Recently, the DFT for electrolyte solution presented in Eqs. (48)–(50) is further extended by Jiang et al. (2014) by incorporating the contact value theorem. It has been demonstrated that the so-called contact-corrected DFT conformed to the exact statistical–mechanical sum rule for the contact ionic densities, and thus gave very accurate predictions for the density profiles of ions near solid surfaces. Figure 11 plots the predicted structures of ionic components near by a neutral surface.

3.3 Molecular DFT and Its Applications

3.3.1 Molecular Picture and Site Picture

In practical, most chemical molecules are not spherical and contain more than one sites, and for those complex molecules a conventional method for describing location of entire molecule is to identify the positions of all sites contained in this molecule, namely, for a molecule composed of m sites, its configuration can be characterized by $\mathbf{R} = \{\mathbf{r}_1, \mathbf{r}_2, \ldots, \mathbf{r}_m\}$ with $\mathbf{r}_i$ being the position of ith site.

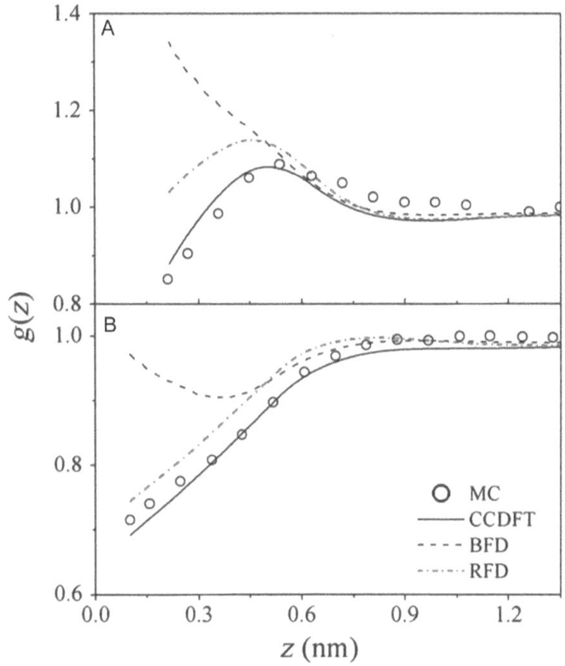

Figure 11 The ionic density profiles of a 1:2 electrolytes near a neutral surface predicted by different version of DFT and by MC simulation: (A) cations with diameter $d_+ = 4.25$ Å and valence $Z_+ = 1$, and (B) anions with diameter $d_- = 2.0$ Å and valence $Z_- = -2$. The reduced temperature $T^* = 0.28$ and reduced bulk densities are $\rho_+^* = 4.82 \times 10^{-3}$ and $\rho_-^* = 9.64 \times 10^{-3}$. *After Jiang et al. (2014).*

When the molecule is rigid (i.e., all bond lengths and bond angles of this molecule are fixed), the location of this molecule can be alternatively described by using both the position of the mass center and the orientation of this molecule. The orientation can be characterized by Euler angles in space. The Euler angles (θ, ϕ, ψ) are a particular sequence of three rotations around the reference frame axes, and detailed definitions of these Euler angles can be found elsewhere (Gray and Gubbins, 1985). Figure 12 depicts the locations of a three-site water molecule in space. As illustrated in Fig. 12A, the location of water molecule can be characterized with $(\mathbf{r}, \theta, \phi, \psi)$ with $\mathbf{r}$ being the position of oxygen site and (θ, ϕ, ψ) are the Euler angles of water molecule in space within the given coordinate system. This characterization is referred as molecular picture. Alternatively, as depicted in Fig. 12B, the location of the same water molecule can be described with

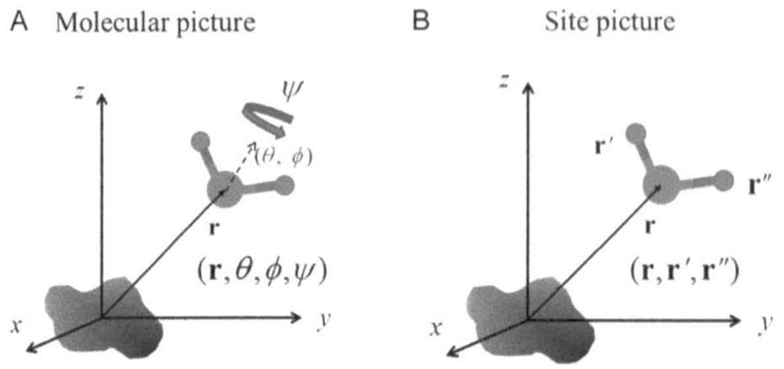

Figure 12 Illustration of two sets of descriptions for the location of a water molecule at given coordinate system.

$(\mathbf{r}, \mathbf{r}', \mathbf{r}'')$, and here $\mathbf{r}$, $\mathbf{r}'$, and $\mathbf{r}''$ are the positions of oxygen and two hydrogen sites. This characterization is referred as site picture.

Two sets of characterizations are rigorously equivalent. We briefly prove this equivalence with the water system in Fig. 12 as an example. In both descriptions, the position of oxygen is the same. For the two hydrogen sites in Fig. 12B, without loss of generality, we denote one hydrogen site locates at $\mathbf{r}_{H_1}$ and the other at $\mathbf{r}_{H_2}$, and then $(\mathbf{r}_{H_1}, \mathbf{r}_{H_2})$ can be uniquely determined by (θ, ϕ, ψ) with the help of following equation (Gray and Gubbins, 1985; Liu et al., 2013a),

$$\begin{cases} \mathbf{r}_{H_1} = \mathbf{r} + d_{OH}(\sin\theta\cos\phi, \sin\theta\sin\phi, \cos\theta) \\ \mathbf{r}_{H_1} = \mathbf{r} + d_{OH}\mathbf{M}_{\theta,\phi}(\sin\Theta\cos\phi', \sin\Theta\sin\phi', \cos\Theta)^T \end{cases}, \tag{54}$$

where d_{OH} is the OH bond length, and Θ is the bond angle for $\angle$HOH, and ϕ' is the rotation angle of the second OH bond (connecting oxygen site with the second hydrogen site) around the first one. Superscript T denotes the transformation of a matrix. $\mathbf{M}_{\theta,\phi}$ is the rotation matrix which is used to convert the coordinates within one Cartesian frame to those within another one, and the subscript (θ, ϕ) represents the set of Euler angles by rotating which the second frame can be obtained from the first one. Following the right-hand clock rule for the rotation, the rotation matrix follows (Gray and Gubbins, 1985)

$$\mathbf{M}_{\theta,\phi} = \begin{pmatrix} \cos\theta\cos\phi & -\sin\phi & \sin\theta\cos\phi \\ \cos\theta\sin\phi & \cos\phi & \sin\theta\sin\phi \\ -\sin\theta & 0 & \cos\theta \end{pmatrix} \tag{55}$$

On the other hand, it is known that (θ, ϕ, ψ) can be uniquely determined from $(\mathbf{r}_{H_1}, \mathbf{r}_{H_2})$ supposing we know the position of oxygen site $\mathbf{r}$. The arguments above suggest that variable $(\mathbf{r}, \theta, \phi, \psi)$ has one-to-one correspondence with $(\mathbf{r}, \mathbf{r}', \mathbf{r}'')$. In other word molecular picture and site picture are mathematically equivalent.

Both molecular picture and site picture have been employed in Monte Carlo simulation, in which the configuration of each molecule is generated randomly. For molecular picture, the random numbers are employed to generate the Euler angles, while for site picture the random numbers are used to generate site positions.

Both pictures are also adopted in integral equation theories. Over the past decades, within the framework of integral equation theories (Beglov and Roux, 1996; Cortis et al., 1997; Du et al., 2000; Hansen and McDonald, 2006) two subbranches have been developed, and historically they are molecular Ornstein–Zernike (MOZ) theories (Beglov and Roux, 1996; Cortis et al., 1997; Ratkova et al., 2010) and reference interacting site models (RISM) (Chandler, 1976; Hirata, 2003; Hirata and Rossky, 1981; Kovalenko and Hirata, 1998; Kovalenko et al., 1999, 2000). MOZ deals with the direct and the total molecular–molecular correlation functions, and those correlation functions are both position and orientation-dependent quantities. Along this line, Fries (Richardi et al., 1997, 1998, 1999), Patey, and Belloni (Belloni and Chikina, 2014; Puibasset and Belloni, 2012) made many interesting contributions. On the other hand, RISM presents the integral relations among the intra- and intermolecular pair correlations between atomic sites. Those site–site correlation functions are position-dependent only. RISM is first proposed by Chandler et al. as an approximated equation (Chandler, 1976; Ladanyi and Chandler, 1975) and thereafter greatly extended by Hirata, Fedrov, and others (Hirata and Rossky, 1981; Kovalenko and Hirata, 1998; Kovalenko et al., 1999; Ratkova et al., 2015). While both versions of the integral equation theories have their individual merits, and it is generally believed that RISM (particularly 3D-RISM) contains less information but is more robust in comparison with MOZ theory.

In analogy with integral equation theories, two versions of MDFTs exist. One follows the molecular picture wherein the position- and orientation-dependent local density is involved, and the other one follows the site picture, in which the local densities of all different sites are involved. In the following, two versions are introduced separately.

3.3.2 Three-Dimensional Molecular DFT

Similar to that for simple systems, the grand potential functional Ω for rigid molecule system at temperature T, volume V, and chemical potential μ can generally be expressed as

$$\Omega[\rho(\mathbf{r},\boldsymbol{\omega})]=F[\rho(\mathbf{r},\boldsymbol{\omega})]+\int \mathrm{d}\mathbf{r}\mathrm{d}\boldsymbol{\omega}\rho(\mathbf{r},\boldsymbol{\omega})[V^{\mathrm{ext}}(\mathbf{r},\boldsymbol{\omega})-\mu]. \tag{56}$$

Here $\rho(\mathbf{r},\boldsymbol{\omega})$ is local molecule density with $\boldsymbol{\omega}$ standing for Euler angle. The Euler angle refers to (θ,ϕ) for linear molecule and to (θ,ϕ,ψ) for nonlinear molecule. Note that generally the external potential V^{ext} also depends on the orientation of the molecule.

The Helmholtz free energy is composed of ideal gas part and excess part, and the ideal part takes the same functional as Eq. (12) with the variable $\rho(\mathbf{r})$ replaced by $\rho(\mathbf{r},\boldsymbol{\omega})$. For the excess part, it can be formulated with functional expansion method. By truncating the functional expansion at the second order, we have (Ramirez and Borgis, 2005; Ramirez et al., 2002),

$$\begin{aligned}\Omega[\rho(\mathbf{r},\boldsymbol{\omega})]-\Omega_0=&\,k_{\mathrm{B}}T\int \mathrm{d}\mathbf{r}\mathrm{d}\boldsymbol{\omega}\left(\ln\frac{\rho(\mathbf{r},\boldsymbol{\omega})}{\rho_0}-\rho(\mathbf{r},\boldsymbol{\omega})+\rho_0\right)\\&+\int \mathrm{d}\mathbf{r}\mathrm{d}\boldsymbol{\omega}\rho(\mathbf{r},\boldsymbol{\omega})V^{\mathrm{ext}}(\mathbf{r},\boldsymbol{\omega})\\&-\frac{k_{\mathrm{B}}T}{2}\int \mathrm{d}\mathbf{r}\mathrm{d}\boldsymbol{\omega}\mathrm{d}\mathbf{r}'\mathrm{d}\boldsymbol{\omega}'\Delta\rho(\mathbf{r},\boldsymbol{\omega})c(|\mathbf{r}-\mathbf{r}'|,\boldsymbol{\omega},\boldsymbol{\omega}')\Delta\rho(\mathbf{r}',\boldsymbol{\omega}').\end{aligned} \tag{57}$$

In above equation, Ω_0 is the counterpart of $\Omega[\rho(\mathbf{r},\boldsymbol{\omega})]$ at homogeneous condition, and it is a constant depending on the thermodynamic condition. $\rho_0=\rho^b/\Pi$ with $\Pi=\int \mathrm{d}\boldsymbol{\omega}$. Specifically, Π equals to 4π for linear molecule and $8\pi^2$ for nonlinear molecule. Again, $c(|\mathbf{r}-\mathbf{r}'|,\boldsymbol{\omega},\boldsymbol{\omega}')$ is the multi-dimensional DCFs of bulk molecule fluid, and as illustrated by Zhao et al. (2013a) it can be elaborately prepared through the MOZ equation by injecting the total correlation function which is extracted in prior from simulation.

Minimization of the functional in Eq. (57) gives the local structure in the homogeneous reference fluid approximation (HRFA) (Ramirez and Borgis, 2005)

$$\rho(\mathbf{r},\boldsymbol{\omega}) = \rho_0 \exp\left[-\beta V^{\text{ext}}(\mathbf{r},\boldsymbol{\omega}) + \int \mathrm{d}\mathbf{r}'\mathrm{d}\boldsymbol{\omega}' c(|\mathbf{r}-\mathbf{r}'|,\boldsymbol{\omega},\boldsymbol{\omega}')\Delta\rho(\mathbf{r}',\boldsymbol{\omega}')\right]. \tag{58}$$

Borgis and his coworker demonstrated (Ramirez et al., 2002, 2005; Zhao et al., 2011a) that the HRFA approximation yields very good predictions of the structure and thermodynamic properties for Stockmayer fluids and linear polar fluids like acetonitrile. The thermodynamic property they addressed mainly refers to the solvation free energy. The determination of solvation free energy of a solute dissolving in a specific solvent at given temperature and pressure has been all through a challenge task in physical chemistry. Theoretically, the solvation free energy is defined as the reversible work to transfer a solute molecule from a hypothetical ideal gas state into the pure solvent at a fixed temperature and pressure. From a thermodynamic point of view, it corresponds to the excess chemical potential of the solute and can be calculated by the change in grand potentials of the solvent system with and without the presence of solute molecule (Chong and Ham, 2015; Hirata, 2003), i.e., $F_{\text{sol}} = \Omega[\rho(\mathbf{r},\boldsymbol{\omega})] - \Omega_0$. Note that the solvation free energy calculation is irrelevant with the choice of thermodynamics ensemble (Chong and Ham, 2015), and it can be directly calculated with the help of Eq. (57). In this circumstance, the solute–solvent interaction gives rise to the external potential. With the help of angular grid implementation Gendre et al. (2009) showed that the minimization of Eq. (57) generated the molecular solvent structure around the solute and the solvation free energy simultaneously.

Figure 13 depicts the solvent structure of the water-like Stockmayer fluid around three individual spherical neutral or charged solutes. The Stockmayer molecule is a single LJ particle centered with a dipole and presents similar bulk density and molecular dipole as water. Due to the spherical symmetry of solute molecule, the solvent structure depends only on the distance to the center of solute after integrating over the solvent orientation, i.e., $\rho(r)/\rho^b = \int \rho(\mathbf{r},\boldsymbol{\omega})\mathrm{d}\boldsymbol{\omega}$. The comparisons in Fig. 13 show that the predictions from MDFT with HRFA approximation agree very well with the simulation results. Similar excellent agreements can be achieved for the solvation properties in acetonitrile wherein the acetonitrile are modeled with three-site linear molecule. Figure 14 plots the solvation structures of acetonitrile around alkaline and halide ions predicted from MDFT in comparison with simulation, and a satisfactory agreement can be found.

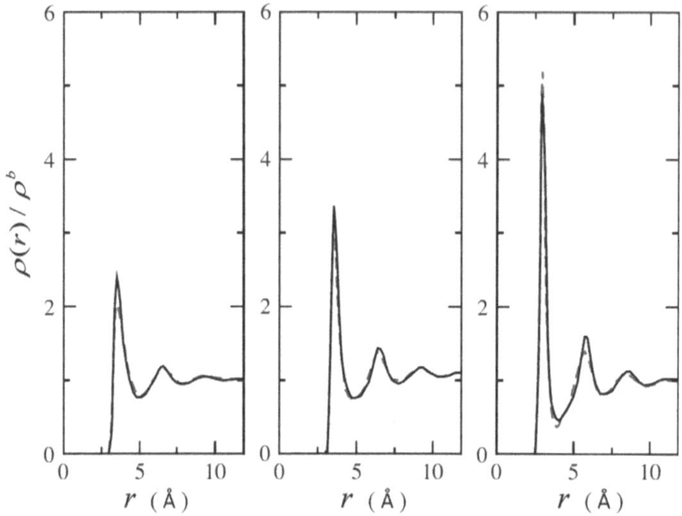

Figure 13 Reduced density of the Stockmayer solvent around various solutes from MDFT prediction (solid black lines) and MD simulation results (dashed red lines; gray color in the print version). From left to right, the solutes are: CH_4, Cl^-, K^+. *After Zhao et al. (2011a).*

Whereas HRFA proves to be a very good approximation for investigating the solvation in linear solvents like acetonitrile, it generates poor solvation structure for the ion dissolved in water as demonstrated by Zhao et al. (2011a). By incorporating an empirical three-body bridge function, the improvement of prediction accuracy is significant (Jeanmairet et al., 2013; Zhao et al., 2011a). Figure 15 plots the predicted solvation free energies of halide ions in acetonitrile and in water. In comparison with experimental data, those predictions are overall satisfactory.

3.3.3 Three-Dimensional Site DFT

Following site description for molecular densities and pair correlation functions, the site DFT (SDFT) has been first proposed by Chandler, McCoy, and Singer originally for bulk rigid molecular fluids (Chandler et al., 1986) and then extended to inhomogeneous molecular fluids by Liu et al. (2013a). The modified SDFT involves two extensions. First, the site–site DCFs of bulk solvent are prepared by combining RISM and simulation instead of using RISM/HNC theory. Second, the universality hypothesis of the bridge functional is incorporated. The extended SDFT allows for quantitative predictions of 3D solvent structure near a solute with moderate molecular size and of the corresponding solvation free energy (Fu et al., 2015b; Liu et al., 2013a).

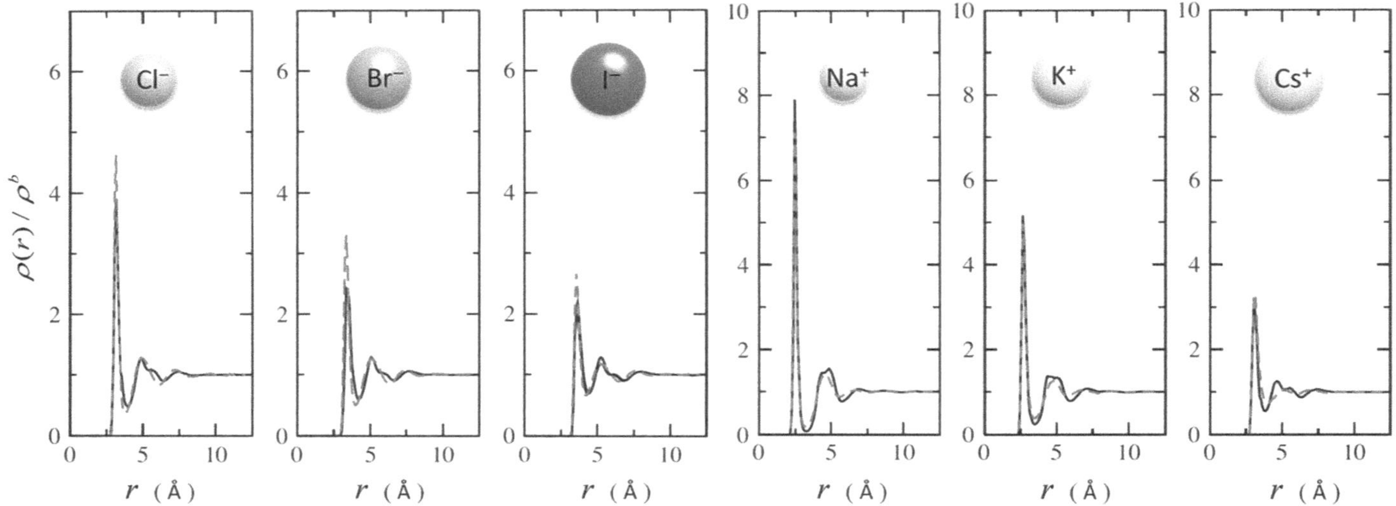

Figure 14 Reduced water density distributions around various solutes from MDFT predictions (solid black lines) and MD simulation results (dashed red lines; gray color in the print version). *After Zhao et al. (2011a).*

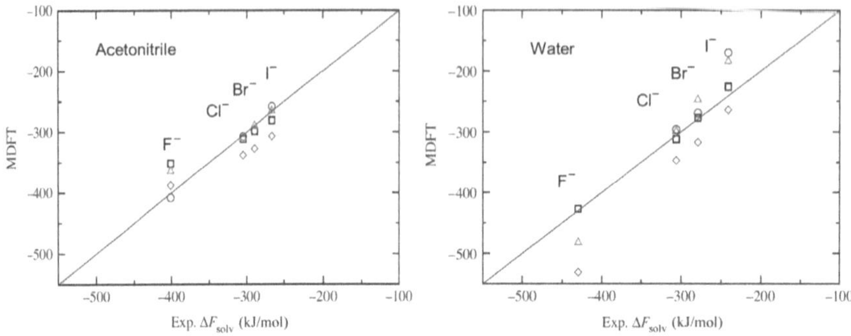

Figure 15 The predicted solvation free energies for various alkane ions in acetonitrile and in water correlated with experimental measurements. Different sets of symbols represent different LJ force fields. *After Zhao et al. (2011a).*

Within the framework of 3D-SDFT (Liu et al., 2013a), the local structure of solvent and the solvation free energy are calculated separately. Considering an arbitrary rigid solute composed of M sites dissolving in water, the site density profile of water, which can be modeled with SPC, SPC/E, or TIP3P, follows (Liu et al., 2013a)

$$\rho_i(\mathbf{r}) = \rho_i^b S_i(\mathbf{r}) \exp\left[-\beta V_i^{\text{ext}}(\mathbf{r}) + \sum_j \int \Delta\rho_j(\mathbf{r}') c_{ij}^{(2)}(|\mathbf{r}-\mathbf{r}'|)\mathrm{d}\mathbf{r}' + B_i(\mathbf{r})\right], \quad (59)$$

where $S_i(\mathbf{r})$ represents the normalized intramolecular correlation function. $V_i^{\text{ext}}(\mathbf{r})$ is the external potential for solvent site i at position $\mathbf{r}$ originating from the solute–solvent interaction. $c_{ij}^{(2)}(|\mathbf{r}-\mathbf{r}'|)$ is the site–site DCF of bulk water between site i and j in two different water molecules. The DCFs are prepared as inputs, and they are extracted from simulation and supposedly exact (Chuev et al., 2013; Zhao and Wu, 2011a; Zhao et al., 2013b). $B_i(\mathbf{r})$ are the site–site bridge terms.

The normalized intramolecular correlation function is given by

$$S_\alpha(\mathbf{r}) = \frac{1}{8\pi^2}\int_{-1}^{1} \mathrm{d}\cos\theta \int_0^{2\pi} \mathrm{d}\phi \int_0^{2\pi} \mathrm{d}\phi' \exp\left[-\lambda_\beta(\mathbf{r}_\beta) - \lambda_\gamma(\mathbf{r}_\gamma)\right], \quad (60)$$

where the subscripts α, β, and γ stand for three types of interacting sites from the same water molecule, i.e., one oxygen site and two hydrogen sites. $\mathbf{r}_\beta$ and $\mathbf{r}_\gamma$ represent the positions of the other two sites from the same water molecule, given the first site α located at position $\mathbf{r}$. In Eq. (60), $\lambda_i(\mathbf{r})$ is given by

$$\lambda_i(\mathbf{r}) = \beta V_i^{\text{ext}}(\mathbf{r}) - \sum_j \int \Delta\rho_j(\mathbf{r}')c_{ij}^{(2)}(|\mathbf{r}-\mathbf{r}'|)\mathrm{d}\mathbf{r}' + B_i(\mathbf{r}). \tag{61}$$

Following the universality hypothesis of bridge functional, the bridge contribution to free energy mainly rises from the repulsion between water molecules. On the other hand, in most popular molecular models of water including SPC, SPC/E, and TIP3P, the hydrogen sites are sizeless, and the excluded volume effect of water molecule is dominated by the oxygen site size. As a result, the bridge functional $F^B[\rho_i(\mathbf{r})]$ is approximated by the counterpart of a HS system which shares the same density profile as the oxygen site, i.e.,

$$F^B[\rho_i(\mathbf{r})] \approx F_{\text{hs}}^B[\rho_{\text{O}}(\mathbf{r})]. \tag{62}$$

Similar to Eq. (37), the bridge functional of the corresponding HS system can be calculated from MFMT. The functional derivative of $F^B[\rho_i(\mathbf{r})]$ in Eq. (62) with respect to site local density gives rise to the bridge term involved in Eq. (59),

$$\begin{cases} B_{\text{O}}(\mathbf{r}) = B_{\text{HS}}(\mathbf{r}) \\ B_{\text{H}}(\mathbf{r}) = 0 \end{cases}. \tag{63}$$

The treatment of HS bridge term follows Eq. (40). To implement such calculation the effective HS diameter is needed. This parameter is determined by equating the predicted solvation free energy of a particular solute with its experimental measurement. Following this criterion the effective HS diameters are determined as 2.91, 2.96, and 2.86 Å for SPC/E, SPC, and TIP3P waters, respectively. Conventionally, above approximation is referred as 3D-SDFT/HNCB, and it recovers to 3D-SDFT/HNC if the bridge contributions to both the structure and the free energy are omitted, and in that circumstance the Euler–Lagrange equation for the solvation structure presents the same expression as the HNC closure in RISM theories.

As discussed, the solvation free energy can be expressed as the grand potential change of a solvent system at given temperature and chemical potential after introducing a single solute molecule. The derivation of the grand potential change in SDFT is different to that in MDFT, and the difference mainly stems from the treatment of the ideal Helmholtz free energy in both versions. Within the framework of MDFT the ideal gas part accounts for the intrinsic Helmholtz free energy of ideal-molecular-gas free of intermolecular interaction, while within the framework of SDFT the ideal gas part addresses the intrinsic Helmholtz free energy of the system without both the intra- and intermolecular interactions. In other word, the ideal gas term in SDFT corresponds to the ideal system which composed of non-bonded sites instead of the ideal molecular gas system (note that a water molecule is composed of three bonded sites). The contribution from the intramolecular interaction to Helmholtz free energy is additionally accounted with the help of a hypothetical reference ideal molecule gas which shares the same site density profiles as the target system. With this treatment, the solvation free energy functional can be derived, reading

$$\begin{aligned}\beta F[\rho_i(\mathbf{r})] = & -\Delta N + \frac{1}{2}\sum_{i,j}\iint c_{ij}^{(2)}(|\mathbf{r}-\mathbf{r}'|)\left[\rho_i(\mathbf{r})\rho_j(\mathbf{r}) - \rho_i^b\rho_j^b\right]\mathrm{d}\mathbf{r}\mathrm{d}\mathbf{r}' \\ & + \beta F^B[\rho_i(\mathbf{r})] + \sum_i \int \rho_i(\mathbf{r})B_i(\mathbf{r})\mathrm{d}\mathbf{r}.\end{aligned} \tag{64}$$

Here ΔN is the difference in water number, and it can be computed by integrating the deviation of the oxygen site local density to the bulk one, i.e., $\rho_{\mathrm{O}}(\mathbf{r}) - \rho_{\mathrm{O}}^b$, over the system volume. Note that the system volume should be sufficiently large so that the local density of water at the edge of system recovers to bulk water density. The solvation free energy can finally be calculated by injecting the local site density profile determined by Eq. (59) into above equation.

With the help of 3D-SDFT, Liu et al. predicted the solvation structures and free energies of 15 molecular analogs of amino-acid side chains in water (Liu et al., 2013a). They demonstrated that the incorporation of the reference HS bridge functional significantly improves the numerical accuracy for the predictions of the solvation free energy in comparison with the simulation results from Shirts et al. (Shirts and Pande, 2005). Those simulation results were generated by using the thermodynamic integration method through the usage of folding@home program (Shirts et al., 2003). Simulation has been popularly employed to compute the solvation free energy with different simulation methods with the advent of modern computers and rapid developments for semiempirical force fields (Chang et al., 2007; Deng and Roux, 2004; Hess and van der Vegt, 2006; Maccallum and Tieleman, 2003). Notwithstanding the widespread usage, simulation of solvation free energy is time consuming. For each side chain, it may take up to thousands of CPU hours (Shirts et al., 2003), and by contrast, the 3D-SDFT

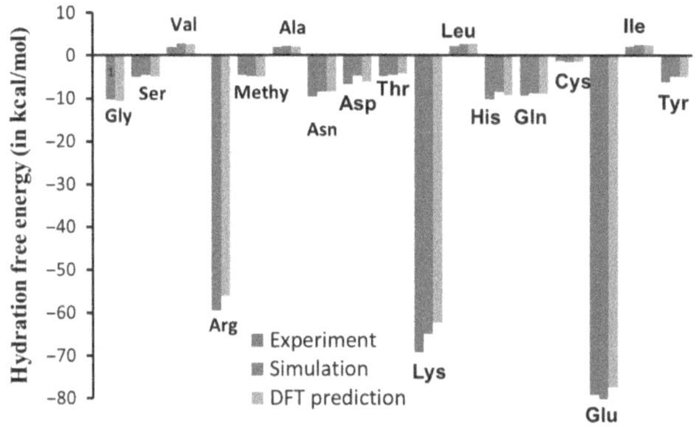

Figure 16 Comparisons of the hydration free energies of amino-acid side chain analogs predicted from 3D-SDFT to both the simulation results and the available experimental measurements. *After Liu et al. (2013a).*

calculation costs only tens of minutes with the help of FFT calculation method. The comparison of the 3D-SDFT prediction to both the simulation results and available experimental results are displayed in Fig. 16. The agreements are impressively excellent considering that the deviations between the simulation and experimental results are about 1–3 kcal/mol.

The predicted 3D site density profiles are also quantitatively good but less impressive compared to the solvation free energy predictions. It has been argued that 3D-SDFT/HNC significantly overestimates the first peak of oxygen site density profile, which is similar to the case in MDFT/HRFA (Zhao et al., 2011a). While with the inclusion of the reference HS bridge SDFT underestimates the first peak. Such overcorrection of solvent structure is clearly visible when comparing to the simulation result. Similar situation has been reported for the 3D-RISM/HNCB theory (Perkyns et al., 2010). It is generally believed that the only path to improve the accuracy of the predicted thermodynamic quantity is to improve the prediction of local structure (Chong and Ham, 2015). The overcorrection of the oxygen site density profile in 3D-SDFT/HNCB most likely arises from the approximation of the solute–solvent bridge functions, which completely ignores the bridge contribution from hydrogen sites. In short distance, the bridge terms involving hydrogen sites dominated by coulomb interaction may be qualitatively different from that involving oxygen site (see Fig. 17 for the site–site bridge functions in bulk water as an illustration). However, the density profiles of oxygen and hydrogen sites are iterative.

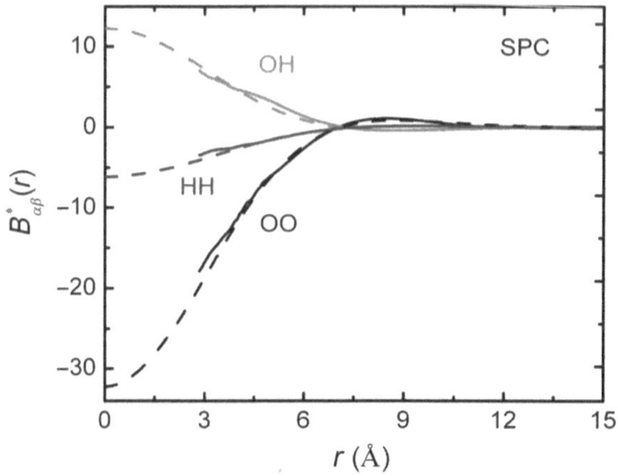

Figure 17 Site–site bridge functions of bulk SPC water systems under ambient condition. The solid lines represent the calculated results from modified 1D-RISM with total correlation functions extracted from simulation, and the dashed lines are fitting curves. *After Zhao et al. (2013b).*

In other words, the 3D-SDFT/HNCB fails to capture the charge-dipole multibody correlation as discussed in the framework of 1D integral equation theories (Kusalik and Patey, 1988; Lombardero et al., 1999), and it underestimates the local structure of oxygen site and thus overestimates that of hydrogen site. However, likely due to error cancellation, the calculated solvation free energies for neutral solute molecules turn to be very accurate. Nevertheless, 3D-SDFT/HNCB fails to predict the solvation properties of charged single solutes including alkane ions. In this spirit, inclusion of a proper solute–hydrogen bridge term or a three-body empirical bridge term (Zhao et al., 2011a) may be promising toward further improvement of SDFT.

Finally, it is worthwhile to mention that the computational efficiency and numerical accuracy of 3D-SDFT in principle allows us to investigate the solvation properties of large solutes like protein, and in that circumstance the internal flexibility of the solute, however, should be considered properly. In this direction, no progress has been reported yet.

3.4 Polymeric DFT and Its Applications

Generally, we consider a complex thermodynamic system composed of polyatomic molecules and simple atomic molecules. Similarly, the Helmholtz free energy can be divided into the summation of ideal gas term F^{id}, which account for the system free of nonbonded interaction, and the excess term. The ideal gas term can be formulated analytically,

$$F^{\text{id}} = k_\text{B}T\int d\mathbf{R}\rho_M(\mathbf{R})[\ln\rho_M(\mathbf{R})-1] \\ +\int d\mathbf{R}\rho_M(\mathbf{R})V_M(\mathbf{R}) + k_\text{B}T\sum_\alpha\int d\mathbf{r}\rho_\alpha(\mathbf{r})[\ln\rho_\alpha(\mathbf{r})-1]. \tag{65}$$

Here $\mathbf{R}\equiv(\mathbf{r}_1, \mathbf{r}_2, \ldots, \mathbf{r}_M)$ represents the positions of M segments of the polymeric molecule. Accordingly, $\rho_M(\mathbf{R})$ is the local density of the polymeric molecule, and $\rho_\alpha(\mathbf{r})$ is the local density for the monomeric fluid (also referred as simple atomic fluid) of species α. $V_M(\mathbf{R})$ specifies the chain connectivity of the segments in a polymeric molecule, and it takes different expressions for different models of a linear chain molecule. Specifically, for a freely jointed chain model with bond length l_0, it satisfies

$$\exp[-\beta V_M(\mathbf{R})] = \prod_{i=1}^{m-1}\frac{\delta(|\mathbf{r}_{i+1}-\mathbf{r}_i|-l_0)}{4\pi l_0{}^2}. \tag{66}$$

To a certain degree, the contribution from the chain connectivity into F^{id} can be interpreted as a constraint, which accounts for the configurations of a polymeric molecule. The excess Helmholtz free energy can be decomposed into different contributions,

$$F^{\text{ex}} = F^{\text{ex}}_{\text{hs}} + F^{\text{ex}}_{\text{C}} + F^{\text{ex}}_{\text{el}} + F^{\text{ex}}_{\text{ch}}. \tag{67}$$

The first three terms on the right-hand side of above equation correspond to HS repulsion and Coulombic interaction, of which the expressions are discussed above (see. Eq. 48). The last term, i.e., $F^{\text{ex}}_{\text{ch}}$, corresponds to the intrachain correlations, and following the thermodynamic circle proposed by Zhou and Stell (1992) and illustrated in Fig. 18, this contribution accounting for the difference in the free energy of polymerization in fluid or in vacuum can be determined through the cavity correlation function $y(\mathbf{R})$.

A systematic procedure to approximate $F^{\text{ex}}_{\text{ch}}$ has been developed following TPT developed by Wertheim (1984, 1986a, 1986b, 1988). The key idea is to represent the cavity correlation function $y(\mathbf{R})$ of the polymeric molecule by that of a reference monomeric system. For example, in the first-order TPT (TPT1), the multibody cavity correlation function $y(\mathbf{R})$ is approximated as a superposition of two-body cavity correlation functions in the corresponding monomeric fluid. The derivations of those approximations

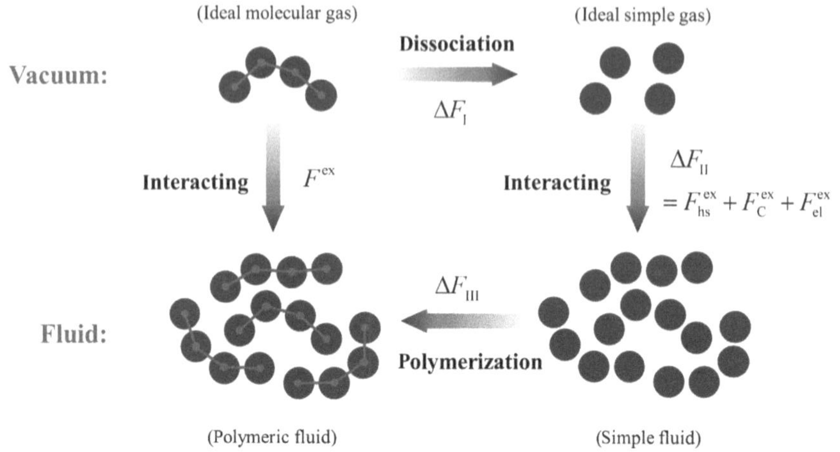

Figure 18 Illustration of the thermodynamic circle: (I) dissociation of polymeric molecule at the ideal-molecular-gas state, and the free energy change $\Delta F_{\mathrm{I}} = -V_M(\mathbf{R})$; (II) switching on the interactions between monomers, and the free energy change is ΔF_{II} wherein the final segment density profiles are identical to those in the polymeric fluid; (III) polymerization of the simple fluid system, and the free energy change $\Delta F_{\mathrm{III}} = -k_{\mathrm{B}}T \ln g(\mathbf{R})$. With the help of this thermodynamic circle, we have $F_{\mathrm{ch}}^{\mathrm{ex}} = \Delta F_{\mathrm{I}} + \Delta F_{\mathrm{III}} = -k_{\mathrm{B}}T \ln y(\mathbf{R})$.

for $F_{\mathrm{ch}}^{\mathrm{ex}}$ are nontrivial, and the readers who are interested to those details should refer to the original work (Wertheim, 1984, 1986a, 1986b, 1988).

Development of polymeric DFT provides us a versatile tool to address complex molecule systems quantitatively. Compared to the other theories including scaling analysis (De Gennes, 1979), self-consistent-field theory (Grosberg and Khokhlov, 1994), PRISM theory (Curro et al., 1989; Schweizer and Curro, 1989), etc., the merits of polymeric DFT are threefold: first, it faithfully capture the major characteristics of polymeric molecule, i.e., connectivity, segment excluded volume effect, and segment–segment attraction, and these enable the descriptions of the physical phenomena of a wide range. Second, the formulation is physically rigorous and thus one can improve the description accuracy in a systematic way by including various contributions. Third, the involved mathematics is relatively simple for implementation. The polymeric DFT has been extensively employed to investigate the properties of bulk polymer systems and the polymer systems near substrates, and many interesting physical phenomena have been explained or predicted (Wu and Li, 2007).

The accurate description of the thermodynamic behaviors of polymeric systems provides benchmarks to phenomenological theories. Scaling analysis is a phenomenological theory pioneered by De Gennes (1979), and it provides an adequate way to describe the thermodynamic properties of polymer systems in particular for those the polymer contour length is long. Scaling analysis incorporates repulsion between monomers and thus is believed to be beyond the conventional theories basing on MFA method (e.g., Flory–Huggins theory). Following the scaling analysis, the osmotic pressure p for the confined polymer system have a simple power-law dependence on the packing fraction, namely $p \sim \eta^{3\upsilon/(3\upsilon-1)}$. Here η is the packing fraction of the confined system defined as the ratio of total segment volumes to the cavity volume, and υ is the Flory exponent and typically it takes 0.5 to 0.588 at different polymer concentrations. Figure 19 depicts the reduced segment density profiles of a HS chain confined in a cavity and the osmotic pressure in function of packing fraction predicted from polymeric DFT by Jin et al. (2010). The cavity radius is fixed at $R_c = 10\sigma$ with σ being the diameter of the HS segment. When the chain length is larger, the packing fraction is higher. Figure 19A shows the excellent agreement of the reduced density profiles with Monte Carlo simulation results at three typical chain lengths, and this validates the numerical accuracy of predictions from polymeric DFT and suggests that the predicted osmosis pressure from DFT is reliable and accurate. The osmosis pressure distribution in terms of packing fraction is thereafter compared with that from scaling analysis upon a best fit. Clearly, this comparison demonstrates that the power law in scaling analysis brings larger and larger deviation as the packing fraction increases.

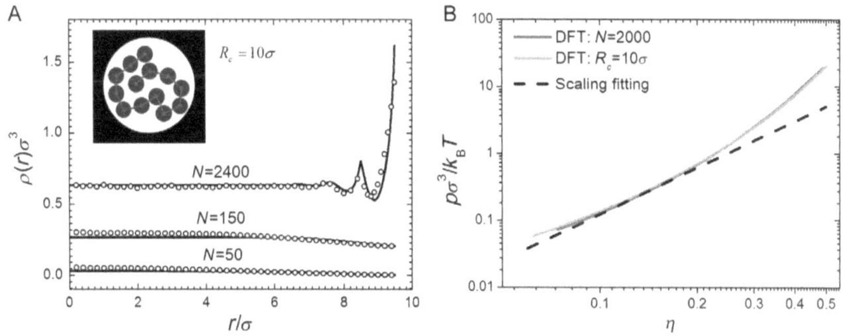

Figure 19 Predicted thermodynamic behaviors of a single hard-sphere polymeric chain confined in a rigid cage: (A) segment density profile from DFT (solid lines) in comparison with MC simulation results (symbols); (B) the reduced osmotic pressure in function of polymer packing fraction comparing with the best fit by using scaling analysis. *After Jin et al. (2010).*

By employing polymeric DFT, we also investigated the properties of various homopolymer/copolymer systems in bulk solution or in confinements (Chen et al., 2006a, 2006b, 2008, 2011; Liu et al., 2010a; Xu et al., 2012, 2013; Ye et al., 2005, 2006, 2007; Zhang et al., 2004). For instance, mixed polymer brushes have been introduced as switchable interfaces for many promising applications. These brushes are usually composed of two kinds of distinct grafted polymers, which have different interactions with solvent, and the surface wetting properties can be switched by appropriate selection of solvent. Recently, Ionov et al. (Ionov and Minko, 2012) reported an interesting phenomenon that wetting behavior of these mixed polymer brushes could be locked in the hydrophobic state, and after dipping the brushes into a special solvent this lock state could be unlocked (see the inset in Fig. 20). Whereas this exotic property enables the useful applications for the development of functional materials, a mechanistic understanding is necessitated. By using polymeric DFT, we investigated the structural distributions of mixed brushes in solvents with different sets of particle–particle interactions (Xu et al., 2012), and identified the condition at which a barrier in grand potential exists when transferring the system from "A-outer lock state" to "B-outer lock state" (see Fig. 20) (Xu et al., 2013). This free energy barrier results in the locked hydrophobic state. In addition, we revealed that with the treatment of nonselective solvents for both kinds of polymers the free energy barrier disappeared and the lock state could be unlocked.

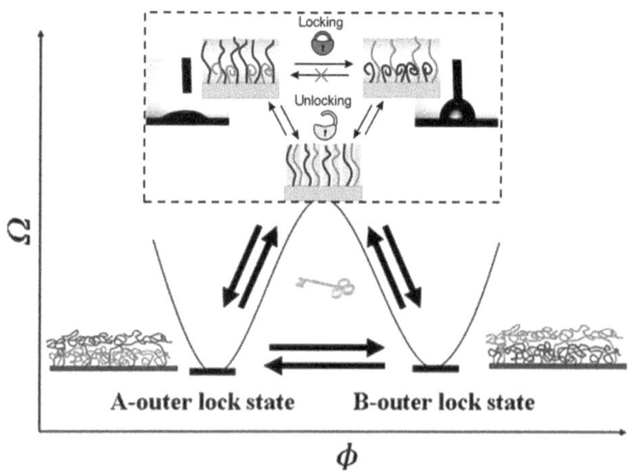

Figure 20 After elaborative design mixed brushes present nonswitchable wetting property (inset), and this lock state was explained by the barrier in grand potential upon a DFT study. *The figure inset from Ionov and Minko (2012).*

The extension of polymeric DFT into more complex polymer systems, e.g., with different nonlinear polymer architectures, has been proposed by Cao and Xu (Xu and Cao, 2009; Xu et al., 2009b), and this extension allows for the investigation of polymeric systems with complex polymer configurations including star-shaped polymers and dendrimers. Apart from this, a special branch of polymeric DFT, which is formulated on 3D lattice, should be mentioned, and this branch refers to Lattice Polymeric DFT, also-called Lattice DFT (LDFT) (Chen et al., 2009). Whereas the conventional density functionals are formulated in the continuous Hilbert space, the LDFT is developed on 3D cubic lattice, in which each grid is occupied either by one segment of polymer chain or by one solvent particle. Like the Ising model, the particle in each grid interacts only with other ones in the nearest grids. The grand advantage of LDFT is that it does not require the explicit interactions for different pairs of particles. Instead, it only needs the reduced exchange energy $\tilde{\varepsilon}=\beta(\varepsilon_{pp}+\varepsilon_{ss}-2\varepsilon_{ps})$ with ε_{pp}, ε_{ps}, and ε_{ss} being the energy for describing the segment–segment, segment–solvent, and solvent–solvent interactions. This engineering characteristic of LDFT enables it to be easily employed by nonspecialist. With the help of LDFT, we investigated the swelling behaviors of thermos-responsive polymer brushes (Lian et al., 2014a) and copolymer hydrogels (Lian et al., 2015) and identified the substrate effect on the phase behavior of polymer brushes (Lian et al., 2014b).

4. DEVELOPMENT OF DFT METHODS

DFTs are continuous approaches which somehow are similar to the phenomenological methods, and however, the local structure of the system is accounted. On the other hand, DFTs deal with the microscopic structure information but do not handle with the individual particles as simulation does. From this regard, DFT represents a compromised approach between simulations and phenomenological methods. Different approaches have their individual merits at different length scales. Simulation essentially is an exact theoretical approach, and although the calculation is time consuming it provides abundant information about the microscopic details including the kinetics. While EOS does not supply any piece of information about the microscopic details, it provides analytical description for the macroscopic properties upon a few parameters. DFT connects the macroscopic properties with the microscopic structure. In the following, we illustrate that two or more approaches can be combined to address various mesoscale problems.

4.1 Combination of Quantum DFT and Statistical DFT

4.1.1 Large-Scale Screening of Hypothetical MOF Materials

A wide variety of porous materials such as zeolites, metal-organic frameworks (MOFs), and silica nanopores are utilized for gas storage and membrane separations and catalysis. Compared to the traditional porous materials, the MOF family has attracted more and more attentions in gas storage field due to its excellent performance. In addition, the structure of MOF can be easily tailored and customized to enhance its functions. Although significant progress has been made toward the fabrications of excellent MOF materials, for instance, the NU-100 (Farha et al., 2010) is reported for CO_2 adsorption ability up to 2315 mg/g, it is reveal that, due to the large account of combinations of metal ions and organic linkers, the discovered MOFs are still negligible (less than 1%) compared to the total number of hypothetic MOFs that can be synthesized. In other word, it is very likely that new materials with extremely excellent performance on gas adsorption still hide in the shadow to be explored. To speed up the process of such exploration, large-scale screening method is urgent desiderata.

The implementation of large-scale screening of new materials demands both the efficiency and accuracy of the applied method. Given the differences in properties between the gases to be separated are relatively small, thus the computation method aiming at accurate prediction should stand on molecular level of control. Consequently, tons of molecular simulations have been reported. Along this line, Snurr's and Zhong's groups (Yang et al., 2013) and other simulation groups (Xiang et al., 2009, 2010) have made significant contributions in developing high-efficient simulation methods, by which Snurr et al. (Colon and Snurr, 2014; Gomez-Gualdron et al., 2014; Ryan et al., 2011; Snurr et al., 2004; Walton et al., 2008; Wilmer et al., 2012) have, for example, examined methane adsorption in 137,000 MOFs and identified over 300 structures that have a higher uptake than any known materials at a specified set of temperature and pressure.

The classical DFTs have proven to be an excellent alternative approach to the simulation method. The gas confined in MOF materials physically represents nothing but a highly inhomogeneous fluid system, in which the MOF materials exert external potential to the fluid system. As demonstrated above, statistical DFTs present the same level of accuracy with but superior efficiency than computer simulations for the predictions of the physicochemical properties of inhomogeneous fluid systems. Similar to classical simulation, the practical implementation of DFT calculations relies on a semiempirical force field that describes the gas-material interaction.

In most works, the gas–material interaction is characterized with the summation of pairwise-additive interaction between a gas (guest) molecule and each atom (host) of the material framework, and pair interaction is determined by fitting the distance-dependent energy calculated from QDFT (Dzubak et al., 2012) or by fitting the thermodynamic properties obtained from experiments. For example, by using QDFT, Vydrov determined the size and energy parameters of LJ interaction in weakly bound systems (Vydrov and Van Voorhis, 2010), and Dzubak et al. (2012) developed the MP2-derived force-field interaction between CO_2 and MOF materials (see Fig. 21A). Unsurprisingly, because of the differences in those quantum calculations and in the methods for date fitting, a few sets of force fields have been developed for the same interaction between gas molecules and MOF materials. The most popular ones include DREIDING (Mayo et al., 1990) and UFF (Rappe et al., 1992), and they provide LJ description for the guest–host interaction. This simple description proves to be very accuracy for H_2 gas, while for CO_2, as argued (Dzubak et al., 2012; Krishna and van Baten, 2011; Yang et al., 2013) the UFF fails to describe correctly the adsorption of CO_2. Dzubak et al. (2012) showed that a three-site model for CO_2 involving electrostatic interaction with MOF framework should be adopted in order to achieve accurate description in comparing to experimental measurement. Figure 21B plots the calculated Henry coefficients of CO_2 confined in Mg-MOF-74 by using simulations with three-site model described with MP2-drived force field and with a single atom model described with UFF. The comparisons to experimental results show that the deviations in the predicted Henry coefficients of CO_2 upon two different models are significant. Besides the guest–host interaction, the manually constructed

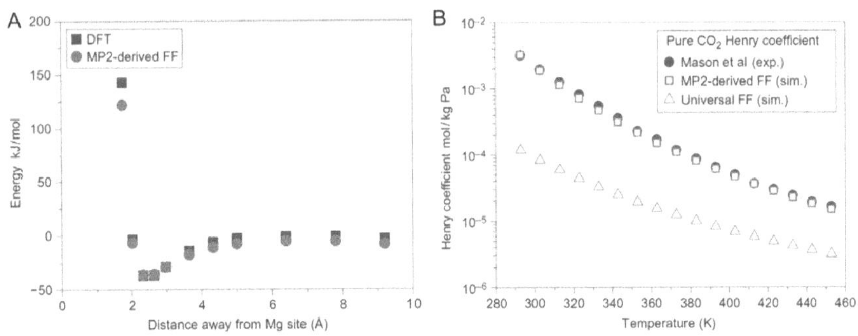

Figure 21 (A) interaction energy comparison of force field with quantum DFT; (B) comparison of simulated and experimental Henry coefficients. *After Dzubak et al. (2012).*

crystalline structure of the MOF materials should be corrected through energy minimizations by using QDFT. The detailed information can be found in relevant reviews (Keskin et al., 2009; Odoh et al., 2015).

The uptake (namely, loading) is the essential concern for gas adsorption and storage. For an adsorption system, the uptake N_{ads} of the guest molecule can be obtained by integrating the local density $\rho(\mathbf{r})$

$$N_{ads} = \int \rho(\mathbf{r}) d\mathbf{r}. \tag{68}$$

Because H_2 can be modeled with spherical LJ particle, for those confined H_2 systems in different MOFs materials, the local density $\rho(\mathbf{r})$ can be efficiently and accurately predicted by using atomic DFT as stated above (Liu et al., 2009). Besides, we also investigated the adsorptions of H_2 in Zeolitic imidazolate frameworks (ZIFs) (Liu, 2011). As shown in Fig. 22, the DFT predictions at three representative temperatures display very good agreement with simulation results.

Although the employment of spherical LJ models for CO_2, CH_4, and N_2 is not sufficiently accurate to quantify the absolute uptakes in a specific MOF framework, the separation behaviors of those gases can be well described. With this argument, we have studied (Liu et al., 2010b) the separation behavior of CO_2/CH_4 and CO_2/N_2 mixtures in Zn_2 $(BDC)_2$ (ted) and in ZIF-8 at room temperature by using atomic DFT. We predicted that

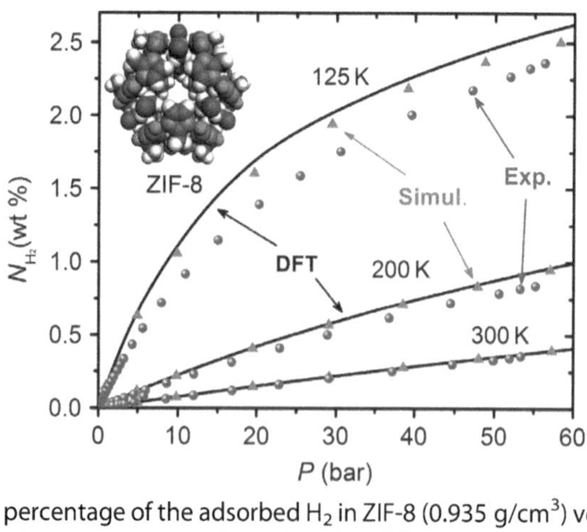

Figure 22 Mass percentage of the adsorbed H_2 in ZIF-8 (0.935 g/cm^3) versus pressure at three representative temperatures from DFT (solid lines), simulation (triangles), and experiments (dots). *After Liu (2011).*

Table 1 Top 10 MOF Structures for H_2 Adsorption Under Different Thermodynamic Conditions Predicted by Atomic DFT (Uptake Unit: mol/L)

	298 K, 0.1 MPa		298 K, 10 MPa		77 K, 0.1 MPa		77 K, 10 MPa	
Rank	**Code**	**Uptake**	**Code**	**Uptake**	**Code**	**Uptake**	**Code**	**Uptake**
1	254	0.1451	254	6.818	216	21.33	401	37.94
2	226	0.1432	226	6.727	227	21.27	399	37.89
3	231	0.1419	231	6.613	252	19.98	373	36.64
4	258	0.1273	207	6.515	215	19.80	409	36.27
5	298	0.1270	258	6.413	254	19.79	21	33.12
6	342	0.1241	298	6.391	207	19.74	17	33.08
7	207	0.1225	232	6.333	311	19.46	18	33.08
8	232	0.1210	222	6.263	222	19.37	11	33.07
9	222	0.1181	292	6.140	258	19.30	20	33.04
10	292	0.1140	227	6.136	360	19.18	1	33.03

The material code corresponds to the northwestern database.

the selectivities of CO_2 over N_2 are larger than those of CO_2 over CH_4 in both materials, and this generally accords with the experimental observations. Recently, Wu and coworker performed an impressive screening analysis of the H_2 storage abilities of 1200 MOF materials under two specific thermodynamic conditions (Fu et al., 2015a), and we have extended these high-throughput predictions of H_2 adsorption in 712 MOF materials to the whole temperature–pressure space, and at high and low temperatures the best candidatures out of those MOF materials are separately identified (Liu et al., 2015a). Table 1 presents the top 10 MOF structures for H_2 adsorption under different thermodynamic conditions. The identification of those excellent structures is helpful for the optimal energy storages at different practical environments. Most recently, we have extended this fast screening strategy to the identification of excellent desulfurization adsorbents (Liu et al., 2015b).

It should be mentioned atomic DFT has been widely applied to characterize the pore size distribution of mesoporous and microporous materials. Such an interesting application is initialized by Quirke and his coworkers in 1989 by using local DFT (Seaton et al., 1989), and then carried out independently by Lastoskie et al. (1993a, 1993b) and by Olivier (1995) with the help of NLDFT. In recent years, Neimark and coworkers (Landers et al., 2013) have extended this application to abundant of mesoporous and microporous materials. Especially, they developed (Cychosz et al., 2012; Neimark et al., 1998, 2009; Ravikovitch and Neimark, 2001, 2002, 2006; Ravikovitch et al., 2000; Thommes et al., 2006) very skillful methods to calculate the pore size distributions in traditional and newly discovered nanoporous solids. This impressive application of atomic DFT provides a feasible method to conveniently characterize the pore size distribution with the help of adsorption experiment (Miyasaka et al., 2009).

4.1.2 Properties of Ion–Electron Mixtures

By using "ion-sphere" type approximation, the system of dense plasma can be considered as a two-component fluid (Dharma-wardana and Perrot, 1982; Hong et al., 1994), i.e., ions carrying a charge Ze with Z being the charge valance, and electrons carrying a charge $-e$. The ion–electron interaction is purely Coulombic, i.e., $u_{\mathrm{ie}}(r) = -Ze^2/r$. The Coulomb interaction strength between the ions is characterized by the coupling constant $\Gamma = Z^2e^2/ak_{\mathrm{B}}T$ with a being the ion-sphere radius defined by $\rho_i^b a^3 4\pi/3 = 1$. Conventionally, the electron bulk density is described with the average reduced distance r_s through $\rho_e^b = 3/\left(4\pi r_s^3 a_B{}^3\right)$ with a_B being the Bohr radius. Following Hohenberg–Kohn–Mermin theorem, the grand potential the plasma system under the external potential $V_\alpha^{\mathrm{ext}}(\mathbf{r})$ ($\alpha = i$, e for ion and electron, respectively) follows

$$\Omega[\rho_\alpha(\mathbf{r})] = F[\rho_\alpha(\mathbf{r})] + \sum_\alpha \int \rho_\alpha(\mathbf{r})\left(V_\alpha^{\mathrm{ext}}(\mathbf{r}) - \mu_\alpha\right)\mathrm{d}\mathbf{r}. \tag{69}$$

Here $\rho_\alpha(\mathbf{r})$ is the density profile of species α, and F is the free energy functional. As stated above, it can be split into the noninteraction part (F^{id}) and the excess part (F^{ex}). F^{id} summarizes the contributions from ion species and electron species, i.e., $F^{\mathrm{id}} = F_i^{\mathrm{id}}[\rho_i(\mathbf{r})] + F_e^{\mathrm{id}}[\rho_e(\mathbf{r})]$, and $F^{\mathrm{ex}} = F_{\mathrm{ii}}^{\mathrm{ex}} + F_{\mathrm{ie}}^{\mathrm{ex}} + F_{\mathrm{ee}}^{\mathrm{ex}}$ accounts for the contribution from ion–ion interaction, ion–electron interaction, and electron–electron interaction.

For temperature $T \ll T_F$, we have $F_e^{\mathrm{id}}[\rho_e(\mathbf{r})] = E_e^{\mathrm{id}}[\rho_e(\mathbf{r})]$, and this noninteracting term takes the same functional form as Eq. (6). The treatment of $F_{\mathrm{ee}}^{\mathrm{ex}}$ is similar as Eq. (7). Unlike in the pure electron systems, the local density of electron here is determined by not only the external potential but also the coupling with the ion local density via $F_{\mathrm{ie}}^{\mathrm{ex}}$, which, similar to $F_{\mathrm{ii}}^{\mathrm{ex}}$, can be formulated within the framework of statistical DFT for charge particles as described in Eq. (48). It is to be noted that during the minimization of the grand potential, two additional constraints should be considered: particle reservation and the overall charge neutrality of the entire system. Physically, the neutrality condition should include the external potential source (Dharma-wardana and Perrot, 1982).

By using this combined DFT, Dharma-wardana and his coworker (Dharma-wardana and Perrot, 1982) investigated the hydrogen plasma or equivalently the proton–electron mixture, and with the help of quantum many-body theory for the exchange-correlation potential and the HNC approximation for the ion–ion correlation contribution, they demonstrated that the local structure of hydrogen plasma can be well predicted. Xu and Hansen used the procedure of Percus (1962), i.e., fixing one ion at the origin providing the external potential, and then applied the HNC approximation for ion–ion correlation and LDA approximation for the exchange energy and neglected the correlation energy, they investigated the structure of hydrogen plasma and the phase transition (Hong, 2002). Because in their calculation the kinetic energy functional was treated with the Thomas–Fermi plus square-gradient (Weizsacker') correction at different λ value (see Eq. 6), they called this approach as HNC-TFW method. The predictions of the radial distribution by HNC-TFW/DFT are plotted in Fig. 23 in comparison with *ab initio* calculations and overall good agreements can be found.

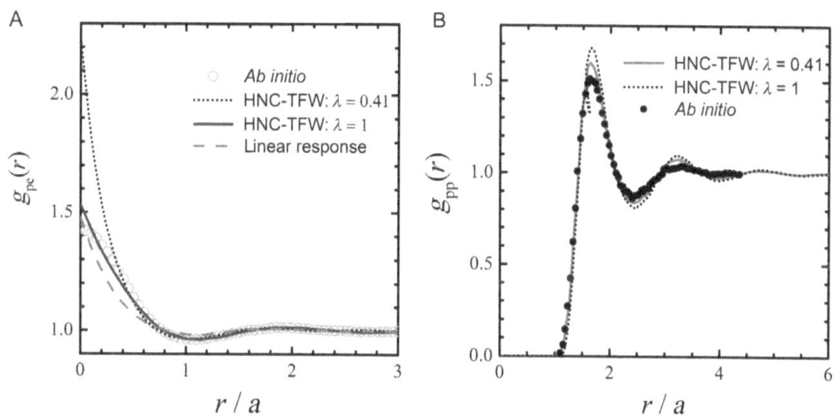

Figure 23 Predicted structure of hydrogen plasma by HNC-TFW method in comparison with *ab initio* calculations: (A) proton–electron pair distribution function at electron density $r_s = 1$ and temperature $T = 10{,}000$ K; (B) the proton–proton pair distribution function at electron density $r_s = 1$ and temperature $T = 3000$ K. *After Hong (2002).*

Interestingly, Dharma-wardana et al. (Dharma-Wardana, 2012; Dharma-wardana and Perrot, 2000) developed a classical mapping method which enables a classical description for the interaction between two electrons. By incorporating the HS bridge term into this classical mapping approach, Zhao et al. (2013c) showed that the structure and the correlation energy of bulk electron gas can be accurately predicted comparing to quantum simulation results. This method has been recently extended to bulk electron gas with finite temperature (Liu and Wu, 2014a, 2014b). With the help of this classical mapping, a new exchange-correlation functional is proposed by Liu and Wu (2014c), displaying promising features.

4.2 Combination of DFT with Simulation

The combination of DFT with simulation has been developed from various aspects. The famous combination is the Car–Parrinello QM/MM approach (Car and Parrinello, 1985) in which the QDFT is incorporated into MD simulation. Since its foundation, there is tremendous desire to perform mixed QM/MM (Burke, 2012). Regarding combination of polymeric DFT with simulation, it has been demonstrated that a single-chain simulation can be performed for collecting the intrachain correlation information, which is utilized in polymeric DFT for the investigation of polymer systems with finite chain concentration (Cao et al., 2006; Chen et al., 2008). In the following, we demonstrate two other examples on the combinations of atomic DFT and MDFT with simulation.

The first example is concerned with the calculation of the solvent-mediated interaction between lock and key. Solvent-mediated interaction plays pivotal role in molecular recognition, association, and binding in solvent environment, and this interaction is usually characterized with the potential of mean force (PMF). By definition, the PMF is associated with the total correlation function (Hirata, 2003)

$$\begin{aligned} W(r, \boldsymbol{\omega}_1, \boldsymbol{\omega}_2) &= -k_B T \ln g_{ss}(r, \boldsymbol{\omega}_1, \boldsymbol{\omega}_2) \\ &= V(r, \boldsymbol{\omega}_1, \boldsymbol{\omega}_2) - k_B T \ln y_{ss}(r, \boldsymbol{\omega}_1, \boldsymbol{\omega}_2). \end{aligned} \tag{70}$$

Here r is the center-to-center distance, and $\boldsymbol{\omega}_1$ and $\boldsymbol{\omega}_2$ are the orientations of two involved molecules in solvent. $g_{ss}(r, \boldsymbol{\omega}_1, \boldsymbol{\omega}_2)$ is the total correlation function of the solute species (two solutes could be in different species but for simplicity we assume they are identical molecules). In the second line in Eq. (70), the first term is the direct interaction of the solute molecules, and the second term is the solvent-induced interaction with $y_{ss}(r, \boldsymbol{\omega}_1, \boldsymbol{\omega}_2)$ being the cavity correlation function. Unsurprisingly, two attractive objects could present repulsive PMF due to the presence of solvent, of which the local structure rather than a specific microscopic energy gives rise to the solvent-induced interaction. It is generally believed that if the involved two particles are hydrophilic, they are individually warped tightly by water molecules and thus present repulsive solvent-mediated interaction due to the excluded volume effect from the surrounding water molecule. Basing on this interpretation, quite a few coarse-grained implicit solvent models have been developed involving hydrophobic/hydrophilic interactions for the investigation of polyelectrolyte and gene delivery (Liu et al., 2012, 2015c; Zhan et al., 2015).

If the density of solute species is finite, the total correlation functions in Eq. (70) can be easily sampled from simulations and thereafter the PMF can be computed straightforwardly. However, in the infinitely dilute limit of solute density, e.g., only two solute molecules are immersed in solvent, it is almost impossible to directly sample the solute–solute total correlation functions from simulation, and the solute–solute correlation function can then be calculated through the integral equation theories with or without the incorporation of simulation for computing the solute–solvent correlation functions.

On the other hand, from a thermodynamic point of view the PMF between two objects corresponds to the reversible work to separate them far apart or equivalently the free energy difference between the systems wherein the involved two objects stay in a given distance in solvent or infinitely apart. In the case of two spherical particles, the PMF is simply distance-dependent

$$W(r) = \Omega(r) - \Omega(\infty). \tag{71}$$

Here r is the center-to-center distance between two solutes. $\Omega(r)$ and $\Omega(\infty)$ are the grand potentials of systems in which two particles are fixed in distance r or sufficiently far away. Both systems have identical volume, temperature, and solvent chemical potential. Again, the volume in principle should be sufficiently large so that at the edge of system the solvent density recovers to bulk density. Note that $\Omega(r)$ includes not only the contribution from solvent system with the presence of solutes but also the contribution from solute–solute interaction $V(r)$. By adding and subtracting the same grand potential of the relevant bulk solvent system, the PMF in Eq. (71) can be easily rewritten as

$$W(r) = V(r) + F_{\text{solv}}(r) - F_{\text{solv}}(\infty) \tag{72}$$

Here $F_{\text{solv}}(r)$ is the solvation free energy (or equivalently the excess chemical potential) of the solute pair with fixed distance r. In some work, above equation refers to potential distribution theorem (PDT) wherein the one-body direct correlations (namely, the excess chemical potentials) are present. As demonstrated above, the grand potential for an inhomogeneous system, or equivalently the solvation free energy for a small individual solute molecule can be calculated accurately by using DFT. While for the calculation of PMF between a particle and a macroscopic obstacle such as a wall, a MC-DFT or MD-DFT algorithm can be used.

This hybrid method has been developed basing on the idea that DFTs connect the microscopic structure directly with the thermodynamic property. Unlike the thermodynamic integration method which has been widely used in computer simulation for the determination of the free energies of fluid systems, the hybrid method avoids the lengthy thermodynamic integration, and after a one-shot calculation it gives straightforwardly the free energy as long as the final equilibrium structural information is substituted into the functional. With the advent of modern computers and the developments of efficient simulation packages, the structural information of a moderate fluid system can be easily collected.

With the help of this hybrid method, Jin and Wu (2011) investigated the PMFs between a lock and different keys in HS solvents. As demonstrated in Fig. 24, the lock is represented with a substrate with a hemispherical pocket with a given diameter, and the keys are represented by big spherical particles with different diameters. Jin and Wu extracted the local density of solvent around the lock and key with fixed separation r from simulation, and then injected it into the grand potential functional for determining the PMF through Eq. (71). They showed that the PMFs are closely associated with the match degree between the geometry of the lock pocket and that of the key. In particular, the potential barrier occurred in the path of key going straight into the pocket is sensitive to the key size.

By employing the similar algorithm, Zhao et al. developed MD-MDFT method for the fast and accurate predictions of hydration free energies of alkane and halide ions by incorporating the accurate density profiles of water, which are extracted from MD simulations, and the universality hypothesis of bridge functional (Zhao et al., 2011b, 2011c).

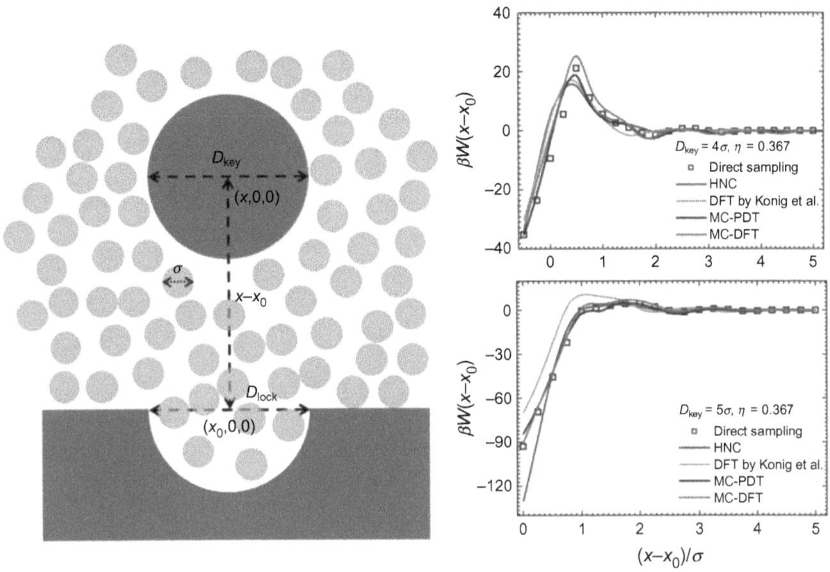

Figure 24 A simple model (the left panel) for the lock–key system represented by a substrate with a hemispherical pocket and big spherical particle, respectively, in HS solvent. The right panel plots the corresponding PMF under different conditions from direct sampling, HNC integral equation, MC-PDT, and MC-DFT. *After Jin and Wu (2011).*

4.3 Combination of DFT with Equation of State

4.3.1 Incorporation of EOS into DFT

The successful applications of DFTs rely on the formulations of proper Helmholtz free energy functional, in which the accurate treatment of excess free energy functional is the key task. Generally, there are two typical routes to construct the excess Helmholtz free energy functional. The first one is based on the functional expansion method, and the second one relies on the incorporation of equations of state.

By incorporating the DCF of bulk reference system and the approximation for bridge functional upon the universality ansatz or empirical interpretation, an accurate excess free energy functional can be devised within the framework of functional expansion method. In most cases, the DCFs do not have analytical expressions, and moreover it has been recognized that the accuracy of the prepared DCFs has large influence on the final accuracy of DFT calculations (Mi et al., 2006). In other word, the availability of exact DCFs is prerequisite for the construction of satisfactory functional. For this reason, the elaborative calculations of exact DCFs of the reference bulk systems turn to be very important (Zhao and Wu, 2011a; Zhao et al., 2013a). As introduced above, the exact DCFs can be in principle obtained by combining the computer simulation for generating the total correlation functions and the integral equations for solving the DCFs. Although this procedure seems simple, the calculation is nontrivial at all. The difficulty mainly comes from the finite size effect of simulation which destroys the values of DCFs inside hard cores. Whereas several methods (Chuev et al., 2013; Zhao and Wu, 2011a; Zhao et al., 2013a) have been developed for solving this numerical problem, the difficulty in preparing exact DCFs generally hinders the wide applications of functional expansion approach and thereafter the spread of DFTs into chemistry lab.

The incorporations of EOS into DFTs by using LDA and/or WDA(s) provide an excellent alternative route for the construction of free energy functional. As illustrated in Eq. (41), one can generally decompose the excess free energy into a HS repulsion contribution, a mean field contribution from the attraction contribution and the residual part. The formulations of the first two parts are standard and kind of universal, and the residual part can be complemented by using the corresponding EOS upon the LDA or WDA method. The development of MFMT is based on BMCSL EOS (Boublik, 1970; Boublík, 1970; Mansoori et al., 1971). With the help of SAFT EOS for associating fluids (Chapman et al., 1989, 1990; Muller and Gubbins, 2001), a premier equation in polymer science (Zhou and Solana, 2009), Chapman et al. developed a polyatomic density functional based on the first-order thermodynamic perturbation approach (Bymaster and Chapman, 2010; Tripathi and Chapman, 2005). By using EOS, Yu developed a density functional for LJ inhomogeneous system (Yu, 2009). Liu et al. developed a lattice-based density functional with the incorporation of the EOS for linear and branched polymer system (Yang et al., 2006). Mier-y-Teran et al. developed a density functional (Mier-y-Teran et al., 1990) accounting for Coulomb interaction with the EOS for charged HS (Palmer and Weeks, 1973; Waisman and Lebowitz, 1972).

Apparently, this approach is consistent with the homogeneous fluid system limit. Namely, when the local density becomes uniform, the free energy functional recovers to the free energy of the according bulk system described by the EOS. In addition, for highly inhomogeneous fluid systems such as gas adsorptions in MOF materials near the critical condition, the numerical evidences demonstrate that the EOS-based DFT gives even more accurate predictions than, e.g., FMSA, which is developed by using functional expansion method. This comparison is illustrated in Fig. 25, which shows the

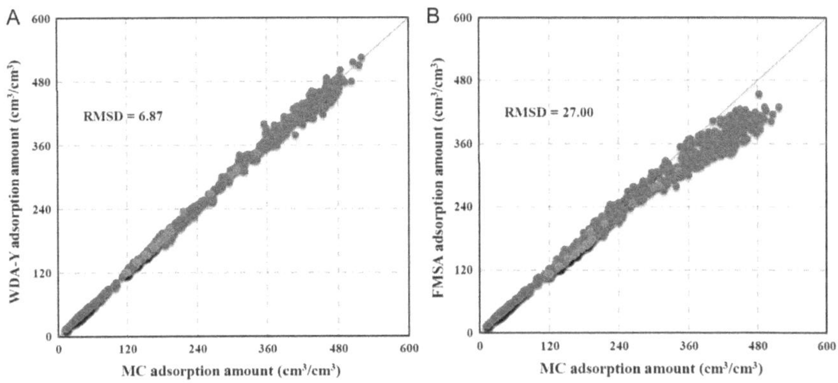

Figure 25 Adsorption amount of H_2 in 1200 MOF materials at temperature $T = 77$ K and pressure $P = 1$ bar predicted by two different versions of DFT in comparison with GCMC simulation results. The versions are (A) MBWR-EOS-based DFTand (B) FMSA. *After Fu et al. (2015a).*

adsorption amount of H_2 in 1200 MOF materials predicted by EOS-based DFT and by FMSA in comparison with grand canonical Monte Carlo (GCMC) simulation results at 1 ATM. The EOS used therein refers to MBWR for LJ system, and the WDA originally proposed by Yu (2009) (denoted as WDA-Y) was employed for the residual part. The overall deviation of the WDA-Y DFT is smaller than that of FMSA, indicating that the EOS-based DFT can generate more satisfactory predictions than FMSA. This trend is more evident at high pressures (Fu et al., 2015a).

Another merit of the EOS-based DFT approach is that plenty of EOS has been developed for different simple or complex fluid systems (Chen et al., 2010; Eliezer et al., 2002; Holovko and Dong, 2009; Lemmon and Jacobsen, 2005; Span, 2000). Besides the EOS discussed above, Gross and Sadowski developed an EOS for chain molecules (Gross and Sadowski, 2001); Maeso and Solana developed an EOS for hard-convex body fluids (Maeso and Solana, 1994); Slattery et al. developed an EOS for one-component plasmas (Slattery et al., 1980); Lemmon and Span developed short fundamental equations of state for 20 industrial fluids by fitting the simulation data (Lemmon and Span, 2006); A few equations of state have been proposed at a wide range of temperature and pressure for the most important industrial fluids, namely water (Fuentevilla and Anisimov, 2006; Harrington et al., 1997; Hill et al., 1982; Holten et al., 2014; Jeffery and Austin, 1999; Mitchell and Nellis, 1982), and its mixture with other fluids (Croft et al., 1988; Kabadl and Danner, 1985) as well as sea water (Friedrich and

Levitus, 1972). In addition, the EOS for bulk fluid system is essentially accessible by experiment and/or by performing simulation as noted below. Hopefully, those merits will finally enable statistical DFTs to become a laboratory tool.

4.3.2 Incorporation of DFT into the Development of EOS

EOS gives the analytical relation among the complete set of thermodynamic functions for a given substance over a wide range of temperature and pressure, and it is the center equation in thermodynamics. A sound knowledge of the EOS is a very helpful for understanding the processes under extreme conditions. The EOS was originally developed for ideal gas, and from then on increasingly sophisticated equations of state have been developed either empirically by fitting the experimental measurements (Span, 2000; Valderrama, 2003) or through statistical–mechanical approaches by taking into account molecular interactions (Lemmon and Span, 2006; Song and Mason, 1989). Here we show that the development of EOS may be implemented with the incorporation of DFT.

First, at homogeneous limit the local density becomes a constant, and then DFT naturally gives rise to the EOS in Helmholtz type (Chapman et al., 1989; Lemmon and Jacobsen, 2005). Second, EOS can be developed with the knowledge of local structure through Virial equation. Specifically, for the fluid composed of spherically symmetric molecules of m components, the pressure is determined by

$$\frac{P}{\rho k_{\mathrm{B}} T}=1-\frac{2\pi\rho}{3k_{\mathrm{B}} T}\sum_{i=1}^{m}\sum_{j=1}^{m} x_i x_j \int_0^{\infty} g_{ij}(r)\frac{\mathrm{d}u_{ij}(r)}{\mathrm{d}r} r^3 \mathrm{d}r \tag{73}$$

where ρ is the total number density defined with $\rho=\sum_i^m \rho_i^b$, and x_i is the molar fraction of component i, and $g_{ij}(r)$ is the RDF between one molecule of component i and the other of component j. $u_{ij}(r)$ is the pair intermolecular potential. For ideal gas, we have $g_{ij}(r)=1$ and $u_{ij}(r)=0$, and above equation recovers to the famous EOS for ideal gas.

There are three typical methods to obtain the RDF: radiation scattering experiments, computer simulation, and liquid state theories such as DFTs and integral equation methods. Whereas the first two methods allow us to construct the EOS of the highest possible accuracy, numerous efforts are needed to collect the measured results at different particular thermodynamic state (namely, different sets of composition and temperature). On the other hand, the DFT is a theoretical method with predictive ability. As demonstrated, statistical DFT can accurately predict the microscopic structures for both homogeneous (along with Percus' trick) and inhomogeneous fluids. By substituting the microscopic structure, i.e., density profiles, into Eq. (73), the EOS can be obtained straightforwardly. This algorithm displays great advantage for the developments of equations of state for inhomogeneous systems such as confined fluid (Nojiri and Odintsov, 2005; Travalloni et al., 2010; Zarragoicoechea and Kuz, 2002).

As demonstrated by Kalyuzhnyi and Cummings (2000), the equations of state for various systems, including adhesive HS fluid, hard-core Yukawa fluid, and some associating fluids could be derived by using integral equation theories. Similarly, those equations of state can be obtained by using DFTs due to the close correspondence between DFTs and integral equation theories. In addition, for confined simple fluids the DFT calculation is capable of generating the fluid densities in confinement at a wide range of pressure and temperature, which is equivalent to giving the EOS for this confined fluid systems. Along this line, numerous examples can be found in the literature, and just for illustration, Fig. 26 plots the mass percentage of the stored H_2 gas in MOF-5, which presents a density of 0.593 g/cm^3, in terms of pressure at two representative temperatures calculated by DFT in comparison with experimental and simulation results.

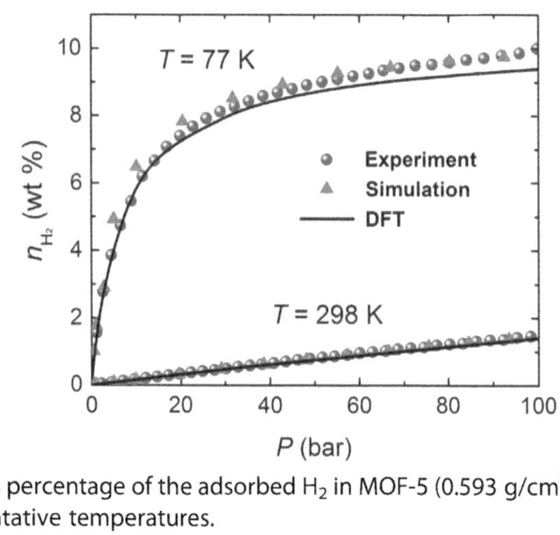

Figure 26 Mass percentage of the adsorbed H_2 in MOF-5 (0.593 g/cm^3) versus pressure at two representative temperatures.

5. DISCUSSIONS AND EXPECTATION

Before closing this chapter, it will be beneficial to have a general discussion on the up-to-date progress and weakness of DFT and the expectations on their future development. As discussed above, DFT links the microscopic structure directly with the macroscopic property, and since its foundation significant progress and developments have been achieved. Nowadays, DFT not only provides an excellent alternative to a large variety of conventional theoretical methods and molecular simulations but also acts as a powerful and versatile tool to address the structural and thermodynamic properties of wide ranges of complex systems. Those applications are exhaustively reviewed in literature. Lowen gave an impressive schematic figure for presenting the diverse applications of statistical DFT in different model systems with increased complexity (as x-axis) for different problems (as y-axis) (Lowen, 2002). Wu tabulated the representative applications of statistical DFT in various practical systems including colloids, polymer, protein, etc. (Wu and Li, 2007), while in his another excellent review, the applications to different physical problems ranging from capillarity to soft materials were discussed item by item (Wu, 2006). In a very recent review given by Fedorov and coworkers focused on molecular integral equation theories, they also presented a comprehensive introduction on the developments of MDFT in the past 10 years (Ratkova et al., 2015). Reviews on the applications of statistical DFT on special topics can also be found. Neimark and coworkers gave an extensive review of the classical atomic DFT studies on the characterization of porous materials (Landers et al., 2013). Regarding QDFT, even more excellent reviews, book chapters, and monographs are available, and just for example, see Geerlings et al. (2003), Koch and Holthausen (2001), Becke (2014), Burke (2012), and the references therein. We believe the future will certainly see many more applications of DFTs.

The key challenge in the applications of DFT is to properly formulate the free energy functional, to which a number of generic strategies have been established. As discussed above, the formulation of the free energy functional is closely associated with the description of the involved two-body interactions. For quantum systems, besides the quantum effects, the interaction between two particles follows the coulomb potential. We speculate that the simplicity of this interaction leads to the consensus of different strategies in formulating the nonideal free energy functional in QDFT. Although quite a few approximations have been developed, several ones have been widely accepted (Burke, 2012). This partially explains that the application of QDFT is very wide and its popularity goes much beyond the statistical one introduced above. Nowadays, many efficient computational packages and softwares have been developed, which allows the QDFT to act as a robust theoretical tool accessible to nonspecialists.

Unlike those in typical quantum systems, the two-body interactions involved in classical model systems are completely empirical and very diverse. Numerous types of pairwise-additive interactions have been proposed and employed to describe the effective interaction between two particles. The diversity of those molecular interactions unfortunately makes it nearly impossible to develop a universal Helmholtz free energy functional applicable for all classical systems. In practical, the formulations of Helmholtz free energy functional are quite different for the systems with different types of molecular interactions, and no systematical procedure for those formulations has been set up. As we discussed above, three major versions of statistical DFTs exist which are atomic DFT applicable for atomic fluid systems and MDFT for rigid molecular systems, and polymeric DFT for polymer systems. Generally, the difficulty in formulating a universal density functionals for classical complex systems creates a barrier of knowledge for the beginners and also for the programmers, and this barrier leads the statistical DFT nearly inaccessible to nonspecialists.

Interestingly, the atomic DFT has been recently applied for the high-throughput prediction of the H_2 adsorption in a large number of MOF materials (Fu et al., 2015a; Liu et al., 2015a), and for the screening of desulfurization adsorbent (Liu et al., 2015b). The success of those large screenings relies not only on the accuracy of the classical modeling with LJ potentials but also on the generality of atomic DFT. We optimistically expect the appearance of efficient computational packages of atomic DFT in the near further.

The fast screening practice has also been carried out concerning the hydration free energy predictions. By using 3D-SDFT, Wu and coworker predicted the hydration free energies of 500 neutral molecules with average unsigned errors of 0.96 and 1.04 kcal/mol in comparison with the experimental and simulation data, respectively (Liu et al., 2013b). By taking into account the correct pressure for liquid–vapor coexistence by means of incorporating an empirical bridge contribution in MDFT/HRFA, Borgis and his coworkers predicted the hydration free energies of small molecules with a small systematic error of the order of k_BT (Jeanmairet et al., 2015). However, those hydration free energy predictions from Wu's and Borgis's groups both relies on the elaborative preparation of accurate DCFs of bulk water with site description and with molecular description, respectively. The dependence of these additional inputs together with the difficulty in their preparation, as we noted above, greatly hinder the extensive applications of MDFT to diverse molecular systems. Considering the numerous combinations of possible solvent compositions and the relevant thermodynamic conditions, countless solvent systems exist in reality. It is practically impossible to prepare the DCFs for all those solvent systems not to mention the difficulty of calculating them. In addition, the solvation system represents only a special inhomogeneous system in which one solo solute molecule is present and thus the inhomogeneity of the solvent systems is generally less significant comparing to that of the confined fluids in porous materials. In this sense, much extension beyond the predictions of solvation free energies is highly expected for MDFT. For example, the accurate predictions of isothermal adsorptions of molecular gas, like three-site CO_2 molecule carrying partial charges on each site, in MOF materials are of great challenging because (i) the DCF cannot be prepared once for all in this circumstance since it varies with the pressure; (ii) the significant inhomogeneity of confined fluid systems likely causes the failure of MDFT/HNCB approach as revealed in the relevant applications of FMSA (Fu et al., 2015a). Despite of those challenges, we note that MDFT has been involving from a qualitative approach to a quantitatively one.

From a fundamental point of view, one imperative task for future work is to develop combined DFTs for describing mesoscale phenomena in complex fluids, among which, an interesting example is how to incorporate the solvent effects into the calculation of activation free energies in liquid-phase chemical reactions. As demonstrated in many works (Hirata, 2003; Hirata et al., 2001), the solvent effects have large influence on reaction barriers, and without properly taking account of the solvent effects, the calculated activation and reaction free energies from QDFT display quantitative or even qualitative deviations. The solvent effect in principle can be well treated using the Car–Parrinello method (Car and Parrinello, 1985) of combining DFT with MD, in which the chemically active part is treated by QDFT and the rest part by classical molecular dynamics simulation. This approach is thriving, but limited to a few hundred atoms per simulation (Guner et al., 2003), and in addition the free energy calculation from simulation is time consuming. With the help of continuum medium model for solvent a few attempts have been made in quantum calculation which gave improved calculation accuracy. On the other hand, Hirata and coworkers (Hirata et al., 2001) showed that the combination of QDFT with the liquid state theory, i.e., RISM theory, is promising to take account of the solvent effects into chemical reactions. Because of the intrinsic correspondence between RISM and statistical DFT, we expect that the combined DFTs, i.e., QDFT/statistical DFT, can provide an alternative tool for resolving this problem. By decoupling the solvent effect and the chemical reaction within the framework of transition state theory and treating them separately by using MDFT and QDFT, Borgis et al. have demonstrated the electron transfer in solute (and thus isomerization) could be well described (Borgis et al., 2012). Although the isomerization does not represent a typical chemical reaction, this work laid the first stone to this encouraging direction. Future application of combined DFTs depends on continuing progress toward more suitable treatments of the coupling between the information of the subsystems at two length scales, and we expect that the combined DFTs will be fruitful for application to mesoscale problems.

Finally, we should mention the extension of static DFT to dynamic DFT. In chemical engineering processes, most of the involved systems are in nonequilibrium state, and the local structure of fluids not only varies with spatial position but also evolves with time. Whereas thermodynamics can predict the direction of the evolution, it fails to provide the evolution details. To characterize the dynamical properties, nonequilibrium statistical mechanism should be developed and applied. In the past decades, impressive progress has been made including the unfolding of dynamical DFT (Goddard et al., 2012; Xu et al., 2007a, 2007b, 2008). However, the development of dynamical DFT relies on the near-equilibrium assumption and it follows the principle that the Helmholtz free energy of nonequilibrium system always decreases in time until arriving its minimum. This principle does not apply for the dynamical systems far from equilibrium wherein the structural factors change significantly in time. By combining Liouville theorem and information theory, Zhao and Wu (2011b) derived a self-consistent theory for colloid dynamics including the dynamic equation for the two-body correlation function. They showed that this theory recovers to dynamical DFT at the near-equilibrium approximation. In general, the framework of nonequilibrium statistical mechanics is not yet well established, and toward the development of such a framework the generic knowledge and understanding accumulated from the investigations of thermodynamic properties are surely beneficial.

6. CONCLUSIONS

In this chapter, we have introduced a unified framework of DFTs and show that this unified framework provides a promising theoretical tool for the investigation of mesoscale problems. Within the same framework, four typical versions of DFTs including QDFT, atomic DFT, MDFT, and polymeric DFT are presented one by one followed by their individual representative applications including those from our own contributions. Afterward, we demonstrate that the combinations of these DFTs with each other or with the conventional theoretical methods and simulations are capable of treating the coupling between the systems of two different length scales. Finally, a general discussion on the recent progress of those DFTs and their challenges during practical applications is given followed by the expectation on several future directions for extensions. These "bottom-up" and "top-down" strategies for combining different versions of DFTs, which share the same physical concept and mathematical framework, provide an effectual methodology for the investigations of different mesoscale problems.

ACKNOWLEDGMENTS

This work is supported by the National Basic Research Program of China (2014CB748500), the National Natural Science Foundation of China (Nos. 21206036, 91334203, and 91434110), the Open Project of State Key Laboratory of Chemical Engineering of China (SKL-ChE-13C04), and the 111 Project of China (No. B08021). S.Z. acknowledges the support of the Shanghai Science and Technology Committee Rising-Star Program (Grant No. 14QA1401300).

REFERENCES

Accelrys, Materials Studio. Accelrys Software Inc.

Allen MR, Frame DJ, Huntingford C, et al.: Warming caused by cumulative carbon emissions towards the trillionth tonne, *Nature* 458(7242):1163–1166, 2009.

Ashbaugh HS, Pratt LR: Colloquium: scaled particle theory and the length scales of hydrophobicity, *Rev Mod Phys* 78(1):159–178, 2006.

Ayadim A, Amokrane S: Generalization of Rosenfeld's functional to non-additive hard-spheres: pair structure and test-particle consistency, *J Phys Condens Matter* 22(3): 035103, 2010.

Balbuena PB, Gubbins KE: Theoretical interpretation of adsorption behavior of simple fluids in slit pores, *Langmuir* 9(7):1801–1814, 1993.

Barker JA, Henderson D: Perturbation theory and equation of state for fluids. 2. A successful theory of liquids, *J Chem Phys* 47(11):4714–4721, 1967.

Becke AD: Perspective: fifty years of density-functional theory in chemical physics, *J Chem Phys* 140(18):18A301, 2014.

Beglov D, Roux B: Solvation of complex molecules in a polar liquid: an integral equation theory, *J Chem Phys* 104(21):8678–8689, 1996.

Belloni L, Chikina I: Efficient full Newton-Raphson technique for the solution of molecular integral equations—example of the SPC/E water-like system, *Mol Phys* 112(9–10):1246–1256, 2014.

Berendsen HJC, Grigera JR, Straatsma TP: The missing term in effective pair potentials, *J Phys Chem* 91(24):6269–6271, 1987.

Borgis D, Gendre L, Ramirez R: Molecular density functional theory: application to solvation and electron-transfer thermodynamics in polar solvents, *J Phys Chem B* 116(8):2504–2512, 2012.

Boublík T: Hard-sphere equation of state, *J Chem Phys* 53(1):471–472, 1970.

Burke K: Perspective on density functional theory, *J Chem Phys* 136(15):150901, 2012.

Bymaster A, Chapman WG: An ISAFT density functional theory for associating polyatomic molecules, *J Phys Chem B* 114(38):12298–12307, 2010.

Cai Q-X, Wang J-G, Wang Y-G, Mei D: Mechanistic insights into the structure-dependent selectivity of catalytic furfural conversion on platinum catalysts, *AIChE J* 61(11): 3812–3824, 2015.

Cao D, Jiang T, Wu J: A hybrid method for predicting the microstructure of polymers with complex architecture: combination of single-chain simulation with density functional theory, *J Chem Phys* 124(16):164904, 2006.

Car R, Parrinello M: Unified approach for molecular dynamics and density-functional theory, *Phys Rev Lett* 55(22):2471–2474, 1985.

Carnahan NF, Starling KE: Equation of state for nonattracting rigid spheres, *J Chem Phys* 51:635–636, 1969.

Chai JD, Weeks JD: Orbital-free density functional theory: kinetic potentials and ab initio local pseudopotentials, *Phys Rev B* 75(20), 2007. Artn 205122.

Chandler D: Cluster diagrammatic analysis of the RISM equation, *Mol Phys* 31(4):1213–1223, 1976.

Chandler D, McCoy JD, Singer SJ: Density functional theory of nonuniform polyatomic systems. I. General formulation, *J Chem Phys* 85(10):5971–5976, 1986.

Chang J, Lenhoff AM, Sandler SI: Solvation free energy of amino acids and side-chain analogues, *J Phys Chem B* 111(8):2098–2106, 2007.

Chapman WG, Gubbins KE, Jackson G, Radosz M: SAFT: equation-of-state solution model for associating fluids, *Fluid Phase Equilib* 52:31–38, 1989.

Chapman WG, Gubbins KE, Jackson G, Radosz M: New reference equation of state for associating liquids, *Ind Eng Chem Res* 29(8):1709–1721, 1990.

Charpentier J-C: Modern chemical engineering in the framework of globalization, sustainability, and technical innovation, *Ind Eng Chem Res* 46(11):3465–3485, 2007.

Charpentier JC: Among the trends for a modern chemical engineering, the third paradigm: the time and length multiscale approach as an efficient tool for process intensification and product design and engineering, *Chem Eng Res Des* 88(3A):248–254, 2010.

Chen H, Ye Z, Peng C, Liu H, Hu Y: Density functional theory for the recognition of polymer at nanopatterned surface, *J Chem Phys* 125(20):204708, 2006a.

Chen X, Cai J, Liu H, Hu Y: Depletion interaction in colloid/polymer mixtures: application of density functional theory, *Mol Simul* 32(10–11):877–885, 2006b.

Chen H, Cai J, Ye Z, et al.: Adsorption of copolymers in a selective nanoslit: a hybrid density functional theory, *J Phys Chem B* 112(32):9568–9573, 2008.

Chen X, Sun L, Liu H, Hu Y, Jiang J: A new lattice density functional theory for polymer adsorption at solid-liquid interface, *J Chem Phys* 131(4):044710, 2009.

Chen W, Dong W, Holovko M, Chen XS: Comment on "a highly accurate and analytic equation of state for a hard sphere fluid in random porous media", *J Phys Chem B* 114(2):1225, 2010.

Chen X, Chen H, Liu H, Hu Y: A free-space density functional theory for polymer adsorption: influence of packing effect on conformations of polymer, *J Chem Phys* 134(4):044713, 2011.
Chiappe C, Pomelli CS: Computational studies on organic reactivity in ionic liquids, *Phys Chem Chem Phys* 15(2):412–423, 2013.
Chong S-H, Ham S: Thermodynamic-ensemble independence of solvation free energy, *J Chem Theory Comput* 11(2):378–380, 2015.
Chremos A, Nikoubashman A, Panagiotopoulos AZ: Flory-Huggins parameter X, from binary mixtures of Lennard-Jones particles to block copolymer melts, *J Chem Phys* 140(5):054909, 2014.
Chu JW, Ayton GS, Izvekov S, Voth GA: Emerging methods for multiscale simulation of biomolecular systems, *Mol Phys* 105(2–3):167–175, 2007.
Chuev GN, Vyalov I, Georgi N: Extraction of atom-atom bridge and direct correlation functions from molecular simulations: a test for ambient water, *Chem Phys Lett* 561:175–178, 2013.
Cohen AJ, Mori-Sánchez P, Yang W: Challenges for density functional theory, *Chem Rev* 112(1):289–320, 2012.
Colon YJ, Snurr RQ: High-throughput computational screening of metal-organic frameworks, *Chem Soc Rev* 43(16):5735–5749, 2014.
Cortis CM, Rossky PJ, Friesner RA: A three-dimensional reduction of the Ornstein-Zernicke equation for molecular liquids, *J Chem Phys* 107(16):6400–6414, 1997.
Croft SK, Lunine JI, Kargel J: Equation of state of ammonia-water liquid: derivation and planetological applications, *Icarus* 73(2):279–293, 1988.
Curro JG, Schweizer KS, Grest GS, Kremer K: A comparison between integral equation theory and molecular dynamics simulations of dense, flexible polymer liquids, *J Chem Phys* 91(2):1357–1364, 1989.
Cychosz KA, Guo X, Fan W, et al.: Characterization of the pore structure of three-dimensionally ordered mesoporous carbons using high resolution gas sorption, *Langmuir* 28(34):12647–12654, 2012.
De Gennes P-G: *Scaling concepts in polymer physics,* Ithaca, 1979, Cornell University Press.
Deng Y, Roux B: Hydration of amino acid side chains: nonpolar and electrostatic contributions calculated from staged molecular dynamics free energy simulations with explicit water molecules, *J Phys Chem B* 108(42):16567–16576, 2004.
Dharma-Wardana MWC: The classical-Map hyper-netted-chain (CHNC) method and associated novel density-functional techniques for warm dense matter, *Int J Quant Chem* 112(1):53–64, 2012.
Dharma-wardana MWC, Perrot F: Density-functional theory of hydrogen plasmas, *Phys Rev A* 26(4):2096–2104, 1982.
Dharma-wardana MWC, Perrot F: Simple classical mapping of the spin-polarized quantum electron gas: distribution functions and local-field corrections, *Phys Rev Lett* 84(5):959–962, 2000.
Douglas Frink LJ, Salinger AG: Two- and three-dimensional nonlocal density functional theory for inhomogeneous fluids: I. Algorithms and parallelization, *J Comput Phys* 159(2):407–424, 2000.
Du QH, Beglov D, Roux B: Solvation free energy of polar and nonpolar molecules in water: an extended interaction site integral equation theory in three dimensions, *J Phys Chem B* 104(4):796–805, 2000.
Dzubak AL, Lin L-C, Kim J, et al.: Ab initio carbon capture in open-site metal–organic frameworks, *Nat Chem* 4(10):810–816, 2012.
Ebner C, Saam WF, Stroud D: Density-functional theory of simple classical fluids. I. Surfaces, *Phys Rev A* 14(6):2264–2273, 1976.

Eliezer S, Ghatak A, Hora H: *Fundamentals of equations of state,* Singapore, 2002, World Scientific Publishing Co. Pte. Ltd.

Errington JR, Debenedetti PG: Relationship between structural order and the anomalies of liquid water, *Nature* 409(6818):318–321, 2001.

Estrada-Tejedor R, Ros-Blanco L, Teixido Closa J: Multiscale modeling for complex chemical systems: highlights about the Nobel prize in chemistry 2013, *Afinidad* 71(566):89–94, 2014.

Evans R: The nature of the liquid-vapour interface and other topics in the statistical mechanics of non-uniform, classical fluids, *Adv Phys* 28(2):143–200, 1979.

Evans R: Density functionals in the theory of nonuniform fluids. In Henderson D, editor: *Fundamentals of inhomogeneous fluids*, New York, 1992, Marcel Dekker, pp 85–175.

Farha OK, Yazaydin AO, Eryazici I, et al.: De novo synthesis of a metal-organic framework material featuring ultrahigh surface area and gas storage capacities, *Nat Chem* 2(11):944–948, 2010.

Foley M, Madden PA: Further orbital-free kinetic-energy functionals for ab initio molecular dynamics, *Phys Rev B* 53(16):10589–10598, 1996.

Friedrich H, Levitus S: An approximation to the equation of state for sea water suitable for numerical ocean models, *J Phys Oceanogr* 2(4):514–517, 1972.

Frisch MJ, Trucks GW, Schlegel HB, et al.: *Gaussian 09,* Wallingford, CT, USA, 2009, Gaussian, Inc.

Fu J, Liu Y, Tian Y, Wu J: Density functional methods for fast screening of metal–organic frameworks for hydrogen storage, *J Phys Chem C* 119(10):5374–5385, 2015a.

Fu J, Liu Y, Wu J: Molecular density functional theory for multiscale modeling of hydration free energy, *Chem Eng Sci* 126:370–382, 2015b.

Fuentevilla DA, Anisimov MA: Scaled equation of state for supercooled water near the liquid-liquid critical point, *Phys Rev Lett* 97:19, 2006.

Galliero G: Lennard-Jones fluid-fluid interfaces under shear, *Phys Rev E* 81(5):056306, 2010.

Gao DF, Zhou H, Wang J, et al.: Size-dependent electrocatalytic reduction of CO_2 over Pd nanoparticles, *J Am Chem Soc* 137(13):4288–4291, 2015.

Geerlings P, De Proft F, Langenaeker W: Conceptual density functional theory, *Chem Rev* 103(5):1793–1873, 2003.

Gendre L, Ramirez R, Borgis D: Classical density functional theory of solvation in molecular solvents: angular grid implementation, *Chem Phys Lett* 474(4–6):366–370, 2009.

Ghatee MH, Moosavi F: Physisorption of hydrophobic and hydrophilic 1-alkyl-3-methylimidazolium ionic liquids on the graphenes, *J Phys Chem C* 115(13): 5626–5636, 2011.

Goddard BD, Nold A, Savva N, Pavliotis GA, Kalliadasis S: General dynamical density functional theory for classical fluids, *Phys Rev Lett* 109(12):120603, 2012.

Gomez-Gualdron DA, Wilmer CE, Farha OK, Hupp JT, Snurr RQ: Exploring the limits of methane storage and delivery in nanoporous materials, *J Phys Chem C* 118(13):6941–6951, 2014.

Goodpaster JD, Ananth N, Manby FR, Miller TF: Exact nonadditive kinetic potentials for embedded density functional theory, *J Chem Phys* 133(8), 2010. Artn 084103.

Gray CG, Gubbins KE: Theory of molecular fluids: fundamentals. In *International series of monographs on chemistry*, vol. 1, New York, 1985, Oxford University Press.

Gray CG, Gubbins KE, Joslin CG: Theory of molecular fluids: applications. In Rowlinson JS, editor: *International series of monographs on chemistry*, vol. 2, New York, 2011, Oxford University Press.

Groot RD, Warren PB: Dissipative particle dynamics: bridging the gap between atomistic and mesoscopic simulation, *J Chem Phys* 107(11):4423–4435, 1997.

Grosberg AY, Khokhlov AR: *Statistical physics of macromolecules,* New York, 1994, American Institute of Physics.

Gross J, Sadowski G: Perturbed-chain SAFT: an equation of state based on a perturbation theory for chain molecules, *Ind Eng Chem Res* 40(4):1244–1260, 2001.

Guillot B: A reappraisal of what we have learnt during three decades of computer simulations on water, *J Mol Liq* 101(1–3):219–260, 2002.

Guner V, Khuong KS, Leach AG, Lee PS, Bartberger MD, Houk KN: A standard set of pericyclic reactions of hydrocarbons for the benchmarking of computational methods: the performance of ab initio, density functional, CASSCF, CASPT2, and CBS-QB3 methods for the prediction of activation barriers, reaction energetics, and transition state geometries, *J Phys Chem A* 107(51):11445–11459, 2003.

Hafner J: Ab-initio simulations of materials using VASP: density-functional theory and beyond, *J Comput Chem* 29(13):2044–2078, 2008.

Hansen JP, McDonald IR: *Theory of simple liquids,* ed 3, London, 2006, Academic Press.

Hansen JP, McDonald IR: *Theory of simple liquids,* ed 4, London, 2013, Academic Press.

Hansen-Goos H, Mecke K: Fundamental measure theory for inhomogeneous fluids of nonspherical hard particles, *Phys Rev Lett* 102(1):018302, 2009.

Hansen-Goos H, Wettlaufer JS: A fundamental measure theory for the sticky hard sphere fluid, *J Chem Phys* 134(1):014506, 2011.

Harrington S, Poole PH, Sciortino F, Stanley HE: Equation of state of supercooled water simulated using the extended simple point charge intermolecular potential, *J Chem Phys* 107(18):7443–7450, 1997.

Henderson D, Sokolowski S: Hard-sphere bridge function calculated from a 2nd-order Percus-Yevick approximation, *J Chem Phys* 103(17):7541–7544, 1995.

Hendrik H-G, Klaus M: Tensorial density functional theory for non-spherical hard-body fluids, *J Phys Condens Matter* 22(36):364107, 2010.

Hess B, van der Vegt NFA: Hydration thermodynamic properties of amino acid analogues: a systematic comparison of biomolecular force fields and water models, *J Phys Chem B* 110(35):17616–17626, 2006.

Hill PG, MacMillan RDC, Lee V: A fundamental equation of state for heavy water, *J Phys Chem Ref Data* 11(1):1–14, 1982.

Hirata F, editor: *Understanding chemical reactivity, Molecular theory of solvation*, Dordrecht, 2003, Kluwer Academic Publishers.

Hirata F, Rossky PJ: An extended RISM equation for molecular polar fluids, *Chem Phys Lett* 83(2):329–334, 1981.

Hirata F, Sato H, Ten-no S, Kato S: The RISM-SCF/MCSCF approach for the chemical processes in solutions. In Becker OM, MacKerell JAD, Benoit R, Masakatsu W, editors: *Computational biochemistry and biophysics*, New York, 2001, Marcel Dekker Inc.

Ho GS, Ligneres VL, Carter EA: Analytic form for a nonlocal kinetic energy functional with a density-dependent kernel for orbital-free density functional theory under periodic and Dirichlet boundary conditions, *Phys Rev B* 78(4):045105, 2008.

Hodak H: The Nobel prize in chemistry 2013 for the development of multiscale models of complex chemical systems: a tribute to martin karplus, Michael Levitt and arieh warshel, *J Mol Biol* 426(1):1–3, 2014.

Hoffert MI, Caldeira K, Benford G, et al.: Advanced technology paths to global climate stability: energy for a greenhouse planet, *Science* 298(5595):981–987, 2002.

Hohenberg P, Kohn W: Inhomogeneous electron gas, *Phys Rev* 136(3B):B864–B871, 1964.

Holovko M, Dong W: A highly accurate and analytic equation of state for a hard sphere fluid in random porous media, *J Phys Chem B* 113(18):6360–6365, 2009.

Holten V, Sengers JV, Anisimov MA: Equation of state for supercooled water at pressures up to 400 Mpa, *J Phys Chem Ref Data* 43(4):23, 2014.

Hong X: Density functional theory applied to metallic hydrogen: pair correlations and phase transitions, *J Phys Condens Matter* 14(40):9109, 2002.
Hong X, Hansen JP, Chandler D: Density functional theory and freezing of an ion-electron plasma, *EPL (Europhys Lett)* 26(6):419, 1994.
Hu Y, Liu H, Prausnitz JM: Equation of state for fluids containing chainlike molecules, *J Chem Phys* 104(1):396–404, 1996.
Hu Y-F, Lv W-J, Zhao S, Shang Y-Z, Wang H-L, Liu H-L: Effect of surfactant SDS on DMSO transport across water/hexane interface by molecular dynamics simulation, *Chem Eng Sci* 134:813–822, 2015.
Ionov L, Minko S: Mixed polymer brushes with locking switching, *ACS Appl Mater Interfaces* 4(1):483–489, 2012.
IPCC: Climate change 2014: synthesis report. In Core Writing Team , Pachauri RK, Meyer LA, editors: *The fifth assessment report of the intergovernmental panel on climate change*, Geneva, Switzerland, 2014, IPCC, p 151.
Janesko B: Density functional theory beyond the generalized gradient approximation for surface chemistry. In Johnson ER, editor: *Density functionals*, Berlin, 2015, Springer International Publishing, pp 25–51.
Jeanmairet G, Levesque M, Vuilleumier R, Borgis D: Molecular density functional theory of water, *J Phys Chem Lett* 4(4):619–624, 2013.
Jeanmairet G, Levesque M, Sergiievskyi V, Borgis D: Molecular density functional theory for water with liquid-gas coexistence and correct pressure, *J Chem Phys* 142(15):154112, 2015.
Jeffery CA, Austin PH: A new analytic equation of state for liquid water, *J Chem Phys* 110(1):484–496, 1999.
Jiang J, Cao D, Henderson D, Wu J: A contact-corrected density functional theory for electrolytes at an interface, *Phys Chem Chem Phys* 16(9):3934–3938, 2014.
Jin Z, Wu J: Hybrid MC – DFT method for studying multidimensional entropic forces, *J Phys Chem B* 115(6):1450–1460, 2011.
Jin Z, Zhao S, Wu J: Entropic forces of single-chain confinement in spherical cavities, *Phys Rev E* 82(4):041805, 2010.
Johnson JK, Zollweg JA, Gubbins KE: The Lennard-Jones equation of state revisited, *Mol Phys* 78(3):591–618, 1993.
Jorgensen WL, Chandrasekhar J, Madura JD, Impey RW, Klein ML: Comparison of simple potential functions for simulating liquid water, *J Chem Phys* 79(2):926–935, 1983.
Kabadl VN, Danner RP: A modified Soave-Redlich-Kwong equation of state for water-hydrocarbon phase equilibria, *Ind Eng Chem Process Des Dev* 24(3):537–541, 1985.
Kahl G, Bildstein B, Rosenfeld Y: Structure and thermodynamics of binary liquid mixtures: universality of the bridge functional, *Phys Rev E* 54(5):5391–5406, 1996.
Kalyuzhnyi YV, Cummings PT: Equations of state from analytically solvable intergral equation approximations. In Sengers JV, Kayser RF, Peters CJ, White HJ Jr, editors: *Equations of state for fluids and fluid mixtures*, New York, 2000, International Union of Pure and Applied Chemistry, Elsevier, p 455.
Karplus M: Development of multiscale models for complex chemical systems: from H + H-2 to biomolecules (Nobel lecture), *Angew Chem Int Ed* 53(38):9992–10005, 2014.
Keskin S, Liu J, Rankin RB, Johnson JK, Sholl DS: Progress, opportunities, and challenges for applying atomically detailed modeling to molecular adsorption and transport in metal – organic framework materials, *Ind Eng Chem Res* 48(5):2355–2371, 2009.
Kirchner B: Ionic liquids from theoretical investigations. In Kirchner B, editor: *Ionic liquids*, Berlin, Heidelberg, 2010, Springer, pp 213–262.
Knepley MG, Karpeev DA, Davidovits S, Eisenberg RS, Gillespie D: An efficient algorithm for classical density functional theory in three dimensions: ionic solutions, *J Chem Phys* 132(12):124101, 2010.

Koch W, Holthausen MC: *A chemist's guide to density functional theory,* ed 2, New York, 2001, Wiley-VCH.
Kohn W, Sham LJ: Self-consistent equations including exchange and correlation effects, *Phys Rev* 140(4A):A1133–A1138, 1965.
Kovalenko A, Hirata F: Three-dimensional density profiles of water in contact with a solute of arbitrary shape: a RISM approach, *Chem Phys Lett* 290(1–3):237–244244, 1998.
Kovalenko A, Ten-No S, Hirata F: Solution of three-dimensional reference interaction site model and hypernetted chain equations for simple point charge water by modified method of direct inversion in iterative subspace, *J Comput Chem* 20(9): 928–936, 1999.
Kovalenko A, Hirata F, Kinoshita M: Hydration structure and stability of Met-enkephalin studied by a three-dimensional reference interaction site model with a repulsive bridge correction and a thermodynamic perturbation method, *J Chem Phys* 113(21):9830–9836, 2000.
Krishna R, van Baten JM: In silico screening of metal-organic frameworks in separation applications, *Phys Chem Chem Phys* 13(22):10593–10616, 2011.
Kusalik PG, Patey GN: The solution of the reference hypernetted-chain approximation for water-like models, *Mol Phys* 65(5):1105–1119, 1988.
Ladanyi BM, Chandler D: New type of cluster theory for molecular fluids: interaction site cluster expansion, *J Chem Phys* 62(11):4308–4324, 1975.
Lamm G: The Poisson–Boltzmann equation. In Lipkowitz KB, Larter R, Cundari TR, Boyd DB, editors: *Reviews in computational chemistry*, Hoboken, USA, 2003, John Wiley & Sons, Inc, pp 147–365.
Landers J, Gor GY, Neimark AV: Density functional theory methods for characterization of porous materials, *Colloids Surf A Physicochem Eng Asp* 437:3–32, 2013.
Lastoskie C, Gubbins KE, Quirke N: Pore size distribution analysis of microporous carbons: a density functional theory approach, *J Phys Chem* 97(18):4786–4796, 1993a.
Lastoskie C, Gubbins KE, Quirke N: Pore size heterogeneity and the carbon slit pore: a density functional theory model, *Langmuir* 9(10):2693–2702, 1993b.
Leach AR: *Molecular modelling: principles and applications,* ed 2, Harlow, England; New York, 2001, Prentice Hall.
Lemmon EW, Jacobsen RT: A new functional form and new fitting techniques for equations of state with application to pentafluoroethane (HFC-125), *J Phys Chem Ref Data* 34(1):69–108, 2005.
Lemmon EW, Span R: Short fundamental equations of state for 20 industrial fluids, *J Chem Eng Data* 51(3):785–850, 2006.
Levesque M, Vuilleumier R, Borgis D: Scalar fundamental measure theory for hard spheres in three dimensions: application to hydrophobic solvation, *J Chem Phys* 137(3):034115, 2012.
Li ZD, Wu JZ: Density-functional theory for the structures and thermodynamic properties of highly asymmetric electrolyte and neutral component mixtures, *Phys Rev E* 70(3):031109, 2004.
Li H, Lu Y, Zhu X, et al.: CO_2 capture through halogen bonding: a theoretical perspective, *Sci China Chem* 55(8):1566–1572, 2012.
Li H, Lu Y, Wu W, et al.: Noncovalent interactions in halogenated ionic liquids: theoretical study and crystallographic implications, *Phys Chem Chem Phys* 15(12):4405–4414, 2013.
Li J, Ge W, Kwauk M: *Meso-scale phenomena from compromise—a common challenge, not only for chemical engineering*, 2015. I.o.P.E. State Key Laboratory of Multi-phase Complex Systems, Chinese Academy of Sciences, Editor. 2015, http://arxiv.org/ftp/arxiv/papers/0912/0912.5407.pdf.
Lian C, Wang L, Chen X, et al.: Modeling swelling behavior of thermoresponsive polymer brush with lattice density functional theory, *Langmuir* 30(14):4040–4048, 2014a.

Lian C, Chen X, Zhao S, et al.: Substrate effect on the phase behavior of polymer brushes with lattice density functional theory, *Macromol Theory Simul* 23(9):575–582, 2014b.
Lian C, Zhi D, Han X, Zhao S, Xu S, Liu H: A lattice molecular thermodynamic model for thermo-sensitive random copolymer hydrogels, *Colloid Polym Sci* 293(2):433–439, 2015.
Liu Y: *Application of density functional theory in MOF material and DNA denaturation,* Shanghai, 2011, School of Chemistry and Molecular Engineering, East China University of Science and Technology, 140.
Liu H, Hu Y: Equation of state for systems containing chainlike molecules, *Ind Eng Chem Res* 37(8):3058–3066, 1998.
Liu Y, Wu J: An improved classical mapping method for homogeneous electron gases at finite temperature, *J Chem Phys* 141(6):064115, 2014a.
Liu Y, Wu J: A bridge-functional-based classical mapping method for predicting the correlation functions of uniform electron gases at finite temperature, *J Chem Phys* 140(8):084103, 2014b.
Liu Y, Wu JZ: A new exchange-correlation functional free of delocalization and static correlation errors, *Phys Chem Chem Phys* 16(31):16373–16377, 2014c.
Liu Y, Liu H, Hu Y, Jiang J: Development of a density functional theory in three-dimensional nanoconfined space: H_2 storage in metal – organic frameworks, *J Phys Chem B* 113(36):12326–12331, 2009.
Liu Y, Chen X, Liu H, Hu Y, Jiang J: A density functional theory for Yukawa chain fluids in a nanoslit, *Mol Simul* 36(4):291–301, 2010a.
Liu Y, Liu H, Hu Y, Jiang J: Density functional theory for adsorption of gas mixtures in metal – organic frameworks, *J Phys Chem B* 114(8):2820–2827, 2010b.
Liu Z, Shang Y, Feng J, Peng C, Liu H, Hu Y: Effect of hydrophilicity or hydrophobicity of polyelectrolyte on the interaction between polyelectrolyte and surfactants: molecular dynamics simulations, *J Phys Chem B* 116(18):5516–5526, 2012.
Liu Y, Zhao S, Wu J: A site density functional theory for water: application to solvation of amino acid side chains, *J Chem Theory Comput* 9(4):1896–1908, 2013a.
Liu Y, Fu J, Wu J: High-throughput prediction of the hydration free energies of small molecules from a classical density functional theory, *J Phys Chem Lett* 4(21):3687–3691, 2013b.
Liu Y, Zhao S, Liu H, Hu Y: High-throughput and comprehensive prediction of H_2 adsorption in metal-organic frameworks under various conditions, *AIChE J* 61(9):2951–2957, 2015a.
Liu Y, Guo F, Hu J, Zhao S, Liu H, Hu Y: Screening of desulfurization adsorbent in metal–organic frameworks: a classical density functional approach, *Chem Eng Sci* 137:170–177, 2015b.
Liu ZH, Lv WJ, Zhao SL, et al.: Effects of the hydrophilicity or hydrophobicity of the neutral block on the structural formation of a block polyelectrolyte/surfactant complex: a molecular dynamics simulation study, *Comput Condens Matter* 2:16–24, 2015c.
Llanorestrepo M, Chapman WG: Bridge function and cavity correlation-function for the Lennard-Jones fluid from simulation, *J Chem Phys* 97(3):2046–2054, 1992.
Loehnert S, Belytschko T: A multiscale projection method for macro/microcrack simulations, *Int J Numer Methods Eng* 71(12):1466–1482, 2007.
Lombardero M, Martin G, Jorge S, Lado F, Lomba E: An integral equation study of a simple point charge model of water, *J Chem Phys* 110(2):1148–1153, 1999.
Lowen H: Density functional theory of inhomogeneous classical fluids: recent developments and new perspectives, *J Phys Condens Matter* 14(46):11897–11905, 2002.
Lowen H: Density functional theory: from statics to dynamics, *J Phys Condens Matter* 15(6): V1–V3, 2003.
Lyu JH, Wang JG, Lu CS, et al.: Size-dependent halogenated nitrobenzene hydrogenation selectivity of Pd nanoparticles, *J Phys Chem C* 118(5):2594–2601, 2014.

Ma M, Xu Z: Self-consistent field model for strong electrostatic correlations and inhomogeneous dielectric media, *J Chem Phys* 141(24):244903, 2014.
Maccallum JL, Tieleman DP: Calculation of the water-cyclohexane transfer free energies of neutral amino acid side-chain analogs using the OPLS all-atom force field, *J Comput Chem* 24(15):1930–1935, 2003.
Macdonald RW, Harner T, Fyfe J: Recent climate change in the arctic and Its impact on contaminant pathways and interpretation of temporal trend data, *Sci Total Environ* 342(1–3):5–86, 2005.
Maeso MJ, Solana JR: Equation of state for hard convex body fluids from the equation of state of the hard sphere fluid, *J Chem Phys* 100(4):3142–3148, 1994.
Mansoori GA, Carnahan NF, Starling KE, Leland TW: Equilibrium thermodynamic properties of the mixture of hard spheres, *J Chem Phys* 54(4):1523–1525, 1971.
Matthias S: Rosenfeld functional for non-additive hard spheres, *J Phys Condens Matter* 16(30): L351, 2004.
Matthias S: Density functional for ternary non-additive hard sphere mixtures, *J Phys Condens Matter* 23(41):415101, 2011.
Mayo SL, Olafson BD, Goddard WA: Dreiding: a generic force field for molecular simulations, *J Phys Chem* 94(26):8897–8909, 1990.
Mennucci B, Cammi R: *Continuum solvation models in chemical physics: from theory to applications,* Chichester, England; Hoboken NJ, 2007, John Wiley & Sons.
Mermin ND: Thermal properties of the inhomogeneous electron gas, *Phys Rev* 137(5A): A1441–A1443, 1965.
Mi J, Tang Y, Zhong C, Li Y-G: Prediction of phase behavior of nanoconfined Lennard-Jones fluids with density functional theory based on the first-order mean spherical approximation, *J Chem Phys* 124(14):144709, 2006.
Mier-y-Teran L, Suh SH, White HS, Davis HT: A nonlocal free-energy density-functional approximation for the electrical double layer, *J Chem Phys* 92(8):5087–5098, 1990.
Mitchell AC, Nellis WJ: Equation of state and electrical conductivity of water and ammonia shocked to the 100 Gpa (1 Mbar) pressure range, *J Chem Phys* 76(12): 6273–6281, 1982.
Miyasaka K, Neimark AV, Terasaki O: Density functional theory of in situ synchrotron powder X-Ray diffraction on mesoporous crystals: argon adsorption on MCM-41, *J Phys Chem C* 113(3):791–794, 2009.
Muller EA, Gubbins KE: Molecular-based equations of state for associating fluids: a review of SAFT and related approaches, *Ind Eng Chem Res* 40(10):2193–2211, 2001.
Neimark AV, Ravikovitch PI, Grün M, Schüth F, Unger KK: Pore size analysis of MCM-41 type adsorbents by means of nitrogen and argon adsorption, *J Colloid Interface Sci* 207(1):159–169, 1998.
Neimark AV, Lin Y, Ravikovitch PI, Thommes M: Quenched solid density functional theory and pore size analysis of micro-mesoporous carbons, *Carbon* 47(7):1617–1628, 2009.
Nojiri S, Odintsov SD: Inhomogeneous equation of state of the universe: phantom era, future singularity, and crossing the phantom barrier, *Phys Rev D* 72(2), 2005.
Odoh SO, Cramer CJ, Truhlar DG, Gagliardi L: Quantum-chemical characterization of the properties and reactivities of metal–organic frameworks, *Chem Rev* 115(12):6051–6111, 2015.
Olivier JP: Modeling physical adsorption on porous and nonporous solids using density functional theory, *J Porous Mater* 2(1):9–17, 1995.
Olivier JP, Conklin WB, Szombathely MV: Determination of pore size distribution from density functional theory: a comparison of nitrogen and argon results. In Rodriguez-Reinoso F, Sing KSW, Rouquerol J, Unger KK, editors: *Studies in surface science and catalysis,* Amsterdam, 1994, Elsevier, pp 81–89.

Palmer RG, Weeks JD: Exact solution of the mean spherical model for charged hard spheres in a uniform neutralizing background, *J Chem Phys* 58(10):4171–4174, 1973.
Parr RG, Yang W: Density-functional theory of atoms and molecules. In *International series of monographs on chemistry*, New York, 1994, Oxford University Press.
Percus JK: Approximation methods in classical statistical mechanics, *Phys Rev Lett* 8(11):462–463, 1962.
Perkyns JS, Lynch GC, Howard JJ, Pettitt BM: Protein solvation from theory and simulation: exact treatment of coulomb interactions in three-dimensional theories, *J Chem Phys* 132(6):064106, 2010.
Puibasset J, Belloni L: Bridge function for the dipolar fluid from simulation, *J Chem Phys* 136(15):154503, 2012.
Ramanathan V, Feng Y: Air pollution, greenhouse gases and climate change: global and regional perspectives, *Atmos Environ* 43(1):37–50, 2009.
Ramirez R, Borgis D: Density functional theory of solvation and its relation to implicit solvent models, *J Phys Chem B* 109(14):6754–6763, 2005.
Ramirez R, Gebauer R, Mareschal M, Borgis D: Density functional theory of solvation in a polar solvent: extracting the functional from homogeneous solvent simulations, *Phys Rev E* 66(3):031206, 2002.
Ramirez R, Mareschal M, Borgis D: Direct correlation functions and the density functional theory of polar solvents, *Chem Phys* 319(1-3):261–272, 2005.
Rappe AK, Casewit CJ, Colwell KS, Goddard WA, Skiff WM: Uff, a full periodic table force field for molecular mechanics and molecular dynamics simulations, *J Am Chem Soc* 114(25):10024–10035, 1992.
Ratkova EL, Chuev GN, Sergiievskyi VP, Fedorov MV: An accurate prediction of hydration free energies by combination of molecular integral equations theory with structural descriptors, *J Phys Chem B* 114(37):12068–12079, 2010.
Ratkova EL, Palmer DS, Fedorov MV: Solvation thermodynamics of organic molecules by the molecular integral equation theory: approaching chemical accuracy, *Chem Rev* 115:6312–6356, 2015.
Ravikovitch PI, Neimark AV: Characterization of micro- and mesoporosity in SBA-15 materials from adsorption data by the NLDFT method, *J Phys Chem B* 105(29):6817–6823, 2001.
Ravikovitch PI, Neimark AV: Density functional theory of adsorption in spherical cavities and pore size characterization of templated nanoporous silicas with cubic and three-dimensional hexagonal structures, *Langmuir* 18(5):1550–1560, 2002.
Ravikovitch PI, Neimark AV: Density functional theory model of adsorption on amorphous and microporous silica materials, *Langmuir* 22(26):11171–11179, 2006.
Ravikovitch PI, Vishnyakov A, Russo R, Neimark AV: Unified approach to pore size characterization of microporous carbonaceous materials from N_2, Ar, and CO_2 adsorption isotherms†, *Langmuir* 16(5):2311–2320, 2000.
Richardi J, Krienke H, Fries PH: Dielectric constants of liquid formamide, N-methylformamide and dimethylformamide via molecular Ornstein-Zernike theory, *Chem Phys Lett* 273(3–4):115–121, 1997.
Richardi J, Fries PH, Fischer R, Rast S, Krienke H: Liquid acetone and chloroform: a comparison between Monte Carlo simulation, molecular Ornstein-Zernike theory, and site-site Ornstein-Zernike theory, *Mol Phys* 93(6):925–938, 1998.
Richardi J, Millot C, Fries PH: A molecular Ornstein-Zernike study of popular models for water and methanol, *J Chem Phys* 110(2):1138–1147, 1999.
Roland R: Fundamental measure theory for hard-sphere mixtures: a review, *J Phys Condens Matter* 22(6):063102, 2010.
Rosenfeld Y: Free-energy model for the inhomogeneous hard-sphere fluid mixture and density functional theory of freezing, *Phys Rev Lett* 63(9):980–983, 1989.

Roth R, Evans R, Lang A, Kahl G: Fundamental measure theory for hard-sphere mixtures revisited: the white bear version, *J Phys Condens Matter* 14(46): 12063–12078, 2002.

Rozen AM, Kostanyan AE: Scaling-up effect in chemical engineering, *Theor Found Chem Eng* 36(4):307–313, 2002.

Ryan P, Farha OK, Broadbelt LJ, Snurr RQ: Computational screening of metal-organic frameworks for xenon/krypton separation, *AIChE J* 57(7):1759–1766, 2011.

Saam WF, Ebner C: Density-functional theory of classical systems, *Phys Rev A* 15(6):2566–2568, 1977.

Schweizer KS, Curro JG: Integral equation theory of the structure and thermodynamics of polymer blends, *J Chem Phys* 91(8):5059–5081, 1989.

Seaton NA, Walton JPRB, N. Quirke: A New analysis method for the determination of the pore size distribution of porous carbons from nitrogen adsorption measurements, *Carbon* 27(6):853–861, 1989.

Shakourian-Fard M, Kamath G, Jamshidi Z: Trends in physisorption of ionic liquids on boron-nitride sheets, *J Phys Chem C* 118(45):26003–26016, 2014.

Shakourian-Fard M, Jamshidi Z, Bayat A, Kamath G: Meta-hybrid density functional theory study of adsorption of imidazolium- and ammonium-based ionic liquids on graphene sheet, *J Phys Chem C* 119(13):7095–7108, 2015.

Sherwood P, Brooks BR, Sansom MSP: Multiscale methods for macromolecular simulations, *Curr Opin Struct Biol* 18(5):630–640, 2008.

Shi K, Lian C, Bai Z, Zhao S, Liu H: Dissipative particle dynamics study of the water/benzene/caprolactam system in the absence or presence of non-ionic surfactants, *Chem Eng Sci* 122:185–196, 2015.

Shirts MR, Pande VS: Solvation free energies of amino acid side chain analogs for common molecular mechanics water models, *J Chem Phys* 122(13):134508, 2005.

Shirts MR, Pitera JW, Swope WC, Pande VS: Extremely precise free energy calculations of amino acid side chain analogs: comparison of common molecular mechanics force fields for proteins, *J Chem Phys* 119(11):5740–5761, 2003.

Shukla KP: Thermodynamic properties of simple fluid mixtures from perturbation theory, *Mol Phys* 62(5):1143–1163, 1987.

Slattery WL, Doolen GD, DeWitt HE: Improved equation of state for the classical one-component plasma, *Phys Rev A (Gen Phys)* 21(6):2087–2095, 1980.

Snurr RQ, Hupp JT, Nguyen ST: Prospects for nanoporous metal-organic materials in advanced separations processes, *AIChE J* 50(6):1090–1095, 2004.

Song Y, Mason EA: Statistical-mechanical theory of a new analytical equation of state, *J Chem Phys* 91(12):7840–7853, 1989.

Span R: *Multiparameter equations of state: an accurate source of thermodynamic property data,* Berlin, 2000, Springer.

Steele WA: Interaction of gases with solid surfaces. In *International encyclopedia of physical chemistry and chemical physics: properties of interfaces*, Oxford, 1974, Pergamon Press.

Tang Y: On the first-order mean spherical approximation, *J Chem Phys* 118(9):4140–4148, 2003.

Tang Y: First-order mean spherical approximation for inhomogeneous fluids, *J Chem Phys* 121(21):10605–10610, 2004.

Tang Y, Lu BCY: Analytical description of the Lennard-Jones fluid and its application, *AIChE J* 43(9):2215–2226, 1997.

Tang Y, Wu J: Modeling inhomogeneous Van Der Waals fluids using an analytical direct correlation function, *Phys Rev E* 70(1):011201, 2004.

Thommes M, Smarsly B, Groenewolt M, Ravikovitch PI, Neimark AV: Adsorption hysteresis of nitrogen and argon in pore networks and characterization of novel micro- and mesoporous silicas, *Langmuir* 22(2):756–764, 2006.

Tomasi J, Persico M: Molecular interactions in solution: an overview of methods based on continuous distributions of the solvent, *Chem Rev* 94(7):2027–2094, 1994.

Tomasi J, Mennucci B, Cammi R: Quantum mechanical continuum solvation models, *Chem Rev* 105(8):2999–3094, 2005.

Travalloni L, Castier M, Tavares FW, Sandler SI: Critical behavior of pure confined fluids from an extension of the Van Der Waals equation of state, *J Supercrit Fluids* 55(2):455–461, 2010.

Tripathi S, Chapman WG: Microstructure and thermodynamics of inhomogeneous polymer blends and solutions, *Phys Rev Lett* 94(8):087801, 2005.

Turner BL II, Lambin EF, Reenberg A: The emergence of land change science for global environmental change and sustainability, *Proc Natl Acad Sci U S A* 104(52): 20666–20671, 2007.

Valderrama JO: The state of the cubic equations of state, *Ind Eng Chem Res* 42(8):1603–1618, 2003.

Van Der Waals JD: *On the continuity of the gaseous and liquid state,* Netherlands, 1873, Universiteit Leiden[translated and published in Studies in Statistical Mechanics, vol. 14, Rowlinson J.S. (Ed.), North-Holand, Amsterdam, 1988].

Vega C, Abascal JLF: Simulating water with rigid non-polarizable models: a general perspective, *Phys Chem Chem Phys* 13(44):19663–19688, 2011.

Vydrov OA, Van Voorhis T: Nonlocal Van Der Waals density functional: the simpler the better, *J Chem Phys* 133(24):244103, 2010.

Waisman E, Lebowitz JL: Mean spherical model integral equation for charged hard spheres I. Method of solution, *J Chem Phys* 56(6):3086–3093, 1972.

Wallqvist A, Mountain RD: Molecular models of water: derivation and description. In Lipkowitz KB, Boyd DB, editors: *Reviews in computational chemistry*, vol. 13, London, 1999, John Wiley and Sons.

Walton KS, Millward AR, Dubbeldam D, et al.: Understanding inflections and steps in carbon dioxide adsorption isotherms in metal-organic frameworks, *J Am Chem Soc* 130(2):406–407, 2008.

Wang Z-G: Fluctuation in electrolyte solutions: the self energy, *Phys Rev E* 81(2):021501, 2010.

Wang R, Wang Z-G: Effects of image charges on double layer structure and forces, *J Chem Phys* 139(12):124702, 2013.

Weeks JD, Chandler D, Andersen HC: Role of repulsive forces in determining the equilibrium structure of simple liquids, *J Chem Phys* 54(12):5237–5247, 1971.

Wertheim MS: Fluids with highly directional attractive forces. I. Statistical thermodynamics, *J Stat Phys* 35(1–2):19–34, 1984.

Wertheim MS: Fluids with highly directional attractive forces. III. Multiple attraction sites, *J Stat Phys* 42(3–4):459–476, 1986a.

Wertheim MS: Fluids of dimerizing hard spheres, and fluid mixtures of hard spheres and dispheres, *J Chem Phys* 85(5):2929–2936, 1986b.

Wertheim MS: Integral equation for the Smith–Nezbeda model of associated fluids, *J Chem Phys* 88(2):1145–1155, 1988.

Wilmer CE, Leaf M, Lee CY, et al.: Large-scale screening of hypothetical metal-organic frameworks, *Nat Chem* 4(2):83–89, 2012.

Wu JZ: Density functional theory for chemical engineering: from capillarity to soft materials, *AIChE J* 52(3):1169–1193, 2006.

Wu JZ: Density functional theory for liquid structure and thermodynamics. In Lu XH, Hu Y, editors: *Molecular thermodynamics of complex systems*, Berlin, 2009, Springer, pp 1–73.

Wu JZ, Li ZD: Density functional theory for complex fluids, *Annu Rev Phys Chem* 58:85–112, 2007.

Wu W, Lu Y, Liu Y, et al.: Structures and electronic properties of transition metal-containing ionic liquids: insights from ion pairs, *J Phys Chem A* 118(13):2508–2518, 2014.

Wu W, Lu Y, Ding H, Peng C, Liu H: The acidity/basicity of metal-containing ionic liquids: insights from surface analysis and the Fukui function, *Phys Chem Chem Phys* 17(2):1339–1346, 2015.
Xiang Z, Lan J, Cao D, Shao X, Wang W, Broom DP: Hydrogen storage in mesoporous coordination frameworks: experiment and molecular simulation, *J Phys Chem C* 113(34):15106–15109, 2009.
Xiang Z, Cao D, Lan J, Wang W, Broom DP: Multiscale simulation and modelling of adsorptive processes for energy gas storage and carbon dioxide capture in porous coordination frameworks, *Energy Environ Sci* 3(10):1469–1487, 2010.
Xu X, Cao D: Density functional theory for adsorption of colloids on the polymer-tethered surfaces: effect of polymer chain architecture, *J Chem Phys* 130(16):164901, 2009.
Xu H, Hansen J-P: Density-functional theory of pair correlations in metallic hydrogen, *Phys Rev E* 57(1):211–223, 1998.
Xu H, Liu H, Hu Y: Dynamic density functional theory based on equation of state, *Chem Eng Sci* 62(13):3494–3501, 2007a.
Xu H, Liu H, Hu Y: The effect of pressure on the microphase separation of diblock copolymer melts studied by dynamic density functional theory based on equation of state, *Macromol Theory Simul* 16(3):262–268, 2007b.
Xu H, Wang T, Huang Y, Liu H, Hu Y: Microphase separation and morphology of the real polymer system by dynamic density functional theory, based on the equation of state, *Ind Eng Chem Res* 47(17):6368–6373, 2008.
Xu Z, Deng S, Cai W: Image charge approximations of reaction fields in solvents with arbitrary ionic strength, *J Comput Phys* 228(6):2092–2099, 2009a.
Xu X, Cao D, Zhang X, Wang W: Universal version of density-functional theory for polymers with complex architecture, *Phys Rev E* 79(2):021805, 2009b.
Xu Z, Cai W, Cheng X: Image charge method for reaction fields in a hybrid ion-channel model, *Commun Comput Phys* 9(4):1056–1070, 2011.
Xu Y, Chen X, Chen H, Xu S, Liu H, Hu Y: Density functional theory for the selective adsorption of small molecules on a surface modified with polymer brushes, *Mol Simul* 38(4):274–283, 2012.
Xu Y, Chen X, Han X, Xu S, Liu H, Hu Y: Lock/unlock mechanism of solvent-responsive binary polymer brushes: density functional theory approach, *Langmuir* 29(16):4988–4997, 2013.
Xu Z, Ma M, Liu P: Self-energy-modified Poisson-Nernst-Planck equations: WKB approximation and finite-difference approaches, *Phys Rev E* 90(1):013307, 2014.
Yang J, Peng C, Liu H, Hu Y, Jiang J: A generic molecular thermodynamic model for linear and branched polymer solutions in a lattice, *Fluid Phase Equilib* 244(2):188–192, 2006.
Yang Q, Liu D, Zhong C, Li J-R: Development of computational methodologies for metal–organic frameworks and their application in gas separations, *Chem Rev* 113(10):8261–8323, 2013.
Ye Z, Cai J, Liu H, Hu Y: Density and chain conformation profiles of square-well chains confined in a slit by density-functional theory, *J Chem Phys* 123(19):194902, 2005.
Ye Z, Chen H, Cai J, Liu H, Hu Y: Density functional theory of homopolymer mixtures confined in a slit, *J Chem Phys* 125(12):124705, 2006.
Ye Z, Chen H, Liu H, Hu Y, Jiang J: Density functional theory for copolymers confined in a nanoslit, *J Chem Phys* 126(13):134903, 2007.
Yethiraj A, van Blaaderen A: A colloidal model system with an interaction tunable from hard sphere to soft and dipolar, *Nature* 421(6922):513–517, 2003.
Yu Y-X: A novel weighted density functional theory for adsorption, fluid-solid interfacial tension, and disjoining properties of simple liquid films on planar solid surfaces, *J Chem Phys* 131(2):024704, 2009.

Yu YX, Wu JZ: Structures of hard-sphere fluids from a modified fundamental-measure theory, *J Chem Phys* 117(22):10156–10164, 2002.
Yu Y-X, Wu J, Gao G-H: Density-functional theory of spherical electric double layers and Z potentials of colloidal particles in restricted-primitive-model electrolyte solutions, *J Chem Phys* 120(15):7223–7233, 2004.
Zarragoicoechea GJ, Kuz VA: Van Der Waals equation of state for a fluid in a nanopore, *Phys Rev E* 65(2):021110, 2002.
Zeng M, Tang Y, Mi J, Zhong C: Improved direct correlation function for density functional theory analysis of pore size distributions, *J Phys Chem C* 113(40):17428–17436, 2009.
Zhan B, Shi K, Dong Z, et al.: Coarse-grained simulation of polycation/DNA-like complexes: role of neutral block, *Mol Pharm* 12:2834–2844, 2015.
Zhang SL, Cai J, Liu HL, Hu Y: Density functional theory of square-well chain mixtures near solid surface, *Mol Simul* 30(2-3):143–147, 2004.
Zhao Y, Truhlar DG: Density functionals with broad applicability in chemistry, *Acc Chem Res* 41(2):157–167, 2008.
Zhao SL, Wu JZ: An efficient method for accurate evaluation of the site-site direct correlation functions of molecular fluids, *Mol Phys* 109(21):2553–2564, 2011a.
Zhao S-L, Wu J: Self-consistent equations governing the dynamics of nonequilibrium colloidal systems, *J Chem Phys* 134(5):054514, 2011b.
Zhao S, Ramirez R, Vuilleumier R, Borgis D: Molecular density functional theory of solvation: from polar solvents to water, *J Chem Phys* 134(19):194102, 2011a.
Zhao S, Jin Z, Wu J: New theoretical method for rapid prediction of solvation free energy in water, *J Phys Chem B* 115(21):6971–6975, 2011b.
Zhao S, Jin Z, Wu J: Correction to "New theoretical method for rapid prediction of solvation free energy in water"*J Phys Chem B* 115(51):15445, 2011c.
Zhao S, Liu H, Ramirez R, Borgis D: Accurate evaluation of the angular-dependent direct correlation function of water, *J Chem Phys* 139(3):034503, 2013a.
Zhao S, Liu Y, Liu H, Wu J: Site-site direct correlation functions for three popular molecular models of liquid water, *J Chem Phys* 139(6):064509, 2013b.
Zhao S, Feng P, Wu J: A liquid-state theory for electron correlation functions and thermodynamics, *Chem Phys Lett* 556:336–340, 2013c.
Zhou SQ, Solana JR: Progress in the perturbation approach in fluid and fluid-related theories, *Chem Rev* 109(6):2829–2858, 2009.
Zhou Y, Stell G: Chemical association in simple models of molecular and ionic fluids. III. The cavity function, *J Chem Phys* 96(2):1507–1515, 1992.
Zhou BJ, Ligneres VL, Carter EA: Improving the orbital-free density functional theory description of covalent materials, *J Chem Phys* 122(4):44103, 2005.
Zhu X, Lu Y, Peng C, Hu J, Liu H, Hu Y: Halogen bonding interactions between brominated ion pairs and CO_2 molecules: implications for design of new and efficient ionic liquids for CO_2 absorption, *J Phys Chem B* 115(14):3949–3958, 2011.

CHAPTER TWO

Surface Structure and Interaction of Surface/Interface Probed by Mesoscale Simulations and Experiments

Linghong Lu*, Xuebo Quan†, Yihui Dong*, Gaobo Yu†, Wenlong Xie*, Jian Zhou†, Licheng Li‡, Xiaohua Lu*,[1], Yudan Zhu*

*State Key Laboratory of Materials-Oriented Chemical Engineering, College of Chemistry and Chemical Engineering, Nanjing Tech University, Nanjing, PR China

†School of Chemistry and Chemical Engineering, Guangdong Provincial Key Lab for Green Chemical Product Technology, South China University of Technology, Guangzhou, PR China

‡College of Chemical Engineering, Nanjing Forestry University, Nanjing, PR China

[1]Corresponding author: e-mail address: xhlu@njtech.edu.cn

Contents

Advances in Chemical Engineering, Volume 47
ISSN 0065-2377
http://dx.doi.org/10.1016/bs.ache.2015.10.011

Abstract

Heterogenous catalysis, fluid transportation, wetting, nanofriction, and protein adsorption are important phenomena in chemical and biological engineering. Mesoporous materials show great prospect in heterogenous catalysis. However, the reaction and transport in mesopores differ from that in the macroscale, so the complicated structure and intra-/interaction of the interface must be clarified to reveal the mechanism of regulation. Wetting is of key importance for many applications. Ionic liquids (ILs) are a new promising class of lubricants; its performance strongly depends on the wetting behavior at interface. The frictional properties of thin films depend on the film packing density, structure, and crystallinity. Protein adsorption onto surfaces is fundamental for the development of low-fouling surfaces, protein immobilization, and protein separations. The adsorbed proteins act as an intermediary between the implanted material and the surrounding tissues, subsequently influencing the host responses. Therefore, surface repellence to protein is critical for the development of advanced biomaterials. These microscopic phenomena cannot be accurately analyzed by only using simple experiment. Using molecular simulation, it has been found that the flow of polar liquid on the surface of TiO_2 can be improved significantly. Coupling different simulation methods is needed to consider the behavior of fluid on rough interface with reaction. We can investigate the reaction mechanism and the relationship between changes of the composition quantitatively, using ReaxFF method that considers reaction and transition simultaneously for a large system. Due to a greater impact of interfacial roughness on the stability of the catalytic, the new method "interface coarse-grained mesoscopic simulation" should be adopted. Because of the large size of the protein molecule, mesoscale approaches are needed. Researchers used coarse-grained simulation methods to study its adsorption and interfacial behavior and obtained a lot of results. Atomic force microscopy (AFM) can be used to measure the force of clusters or fluids on surfaces with different roughness in mesoscopic directly and then draw the law of interface behaviors. People can establish the relationship between AFM measurement results and coarse-grained force field; it is expected to become a powerful medium-scale method to promote the study of the current mesoscale phenomena and mechanisms.

1. INTRODUCTION

"Transport and reaction" will continue to be the core of chemical engineering in the twenty-first century, but research in chemical engineering must change in order to adapt to the economic demands and play an important role in the process of integration of cross disciplines (Li, 2008). The research of transport and reaction in mesoscale becomes the forefront of chemical engineering research in the twenty-first century, while the regulation of the structure of nanoscale materials to microstructure of molecules of the fluid and the effect of transport on the catalytic properties are the key of research. If we can understand its mechanisms and regulate it by designing materials, the current separation and reaction technology will have new revolutionary upgrades.

Mesoporous materials as the catalyst have good application prospects in the important multiphase reaction. The transport of media in the mesoporous materials is different from that in macro; the complex structure of interface affects the reaction and transport, which makes the transport of the same importance as the reaction. The activity of catalyst could be improved by mesoporous TiO_2. But various crystal facets make the recognizing of surface property difficult. The experimental phenomena are due to both complex structures and interactions in varied scales. It is necessary to care about not only catalysis mechanism but also transport for the design of catalyst with nano-/micro-/mesopore structures.

Wetting phenomena are an area where chemistry, physics, and engineering intersect (Bonn et al, 2009). It is resulting from intermolecular interactions when a liquid and a solid are brought together. The degree of wetting (wettability) is determined by a force balance between adhesive and cohesive forces. Much research has been devoted to modifying the surface chemistry of various solids in order to obtain specific wetting properties. One can modify the chemical properties of the surface of the material. Wetting phenomena are governed by the surface and interfacial interactions, acting usually at a few nanometers for van der Waals (vdW) or electrostatic interactions. These length scales are now being probed with relatively new experimental techniques, such as atomic force microscopy (AFM) and surface force apparatus, or theoretical tools, such as molecular simulation. While the roughness of the contact line leads to complicated dynamics, it is still poorly understood.

Ionic liquids (ILs) are a new promising class of lubricants that could significantly improve the wear life in micro-/nanoelectromechanical systems; the device performance strongly depends on the wetting behavior at interface between ILs and solid surfaces. The frictional properties of thin films composed of functional groups depend on the film packing density, structure, and crystallinity. Roughness-induced superhydrophobic surfaces show that low frictional and adhesive forces can be achieved by forming multiple roughness scales on surfaces (An et al, 2012, 2013).

Protein adsorption onto surfaces is fundamental for the development of low-fouling surfaces, protein immobilization, and protein separations. Shortly after the implantation of materials into the body, proteins adhere to the device's surfaces. The adsorbed proteins act as an intermediary between the implanted material and the surrounding tissues, subsequently influencing the host responses. Therefore, surface repellence to protein is critical for the development of advanced biomaterials (An et al, 2014).

Molecular simulation techniques can obtain the microscopic information that cannot be detected by current experimental conditions, but the conventional simulation methods still have inherent limitations with special mesoscopic scales of various complex forces and complex structure. It is necessary to establish a new mesoscale method that considers the chemical reaction and transport to the larger system at the same time. The roughness and chemical properties of catalyst supporting interface have great influence on chemical and physical adsorption stability of clusters. The problem is that the system is too large for traditional simulation in nano-/micro-/mesoscale. We need a new mesoscale method to study the effect of interface roughness on physical/chemistry phenomena.

The scale of heterogenous catalysis determines that the roughness of the interface cannot be ignored. Establishing a new mesoscale coarse-graining method is a good way to investigate the coeffect of physical and chemical interaction under roughness condition. How to develop accurate force fields of the new method is one of the difficulties. Based on AFM measuring, we found that the different TiO_2 with different surface structures have different nanofriction properties and different IL surface wettabilities (An et al, 2012, 2013). There is quantitative relationship between mesoporous TiO_2 surface structure and the adsorption of biological macromolecule clusters. This shows that the macroscopic surface roughness and surface phenomena are closely linked and can be detected by AFM. The roughness and chemical composition of the solid surface can affect the wettability of the surface. Large-scale coarse-grained simulations not only speed up the simulation time and make the study of the effect of roughness on hydrophilic/hydrophobic surface wetting properties more feasible but also make the comparison of simulation and experimental results more intuitive. AFM detection and coarse-grained force field simulation are in the same scale level. So using a combination of AFM technique and multiple integral atomic force fields to fit coarse-grained force field parameters is possible. This combination method will focus on coarse-grained solid interface and establish the coarse-grained force field in conjunction with AFM measurements.

Interfacial protein adsorption has stimulated a huge research interest in medicine, pharmaceutical sciences, analytic sciences, biotechnology, cell biology, and biophysics (Rabe et al, 2011). Although experimental techniques and theoretical models have been used to explore the protein adsorption, these explorations cannot directly obtain the molecular details of protein adsorption. Molecular dynamics (MD) simulations are well suited to the study of orientation and conformation change of proteins/peptides adsorbed on solid surfaces; the molecular-level details could also be provided by this method. However, the atomistic simulations are still limited to systems containing tens to hundreds of thousands of atoms or undergoing a submicrosecond timescale despite the advance in computer power and simulation algorithms. Many biologically interesting phenomena are beyond the capabilities of atomistic simulations, such as the whole process of protein adsorption, protein unfolding, and the formation of higher-order protein complexes. We need to simplify the model when simulating these motions. On the other hand, coarse-grained (CG) models (Thorpe et al, 2008; Zhou et al, 2007) allow simulations to be run on larger systems and longer timescales and can preserve basic structural detail of proteins. Consequently, coarse-grained simulation is an attractive alternative to atomistic models especially for interfacial protein adsorption and its interfacial behaviors.

In this chapter, we focus on the core issues of surface/interface structure and interaction under mesoscale based on the phenomena in mesoscale of heterogenous catalysis (mainly on mesoporous TiO_2), fluid transportation, wetting, nanofriction, and protein adsorption (Fig. 1). We propose two key factors, which are surface chemical property and surface roughness; these two are very critical for material and chemical–biological engineering. To research these objects, we need to combine simulation, AFM measurement, and adsorption experiment methods.

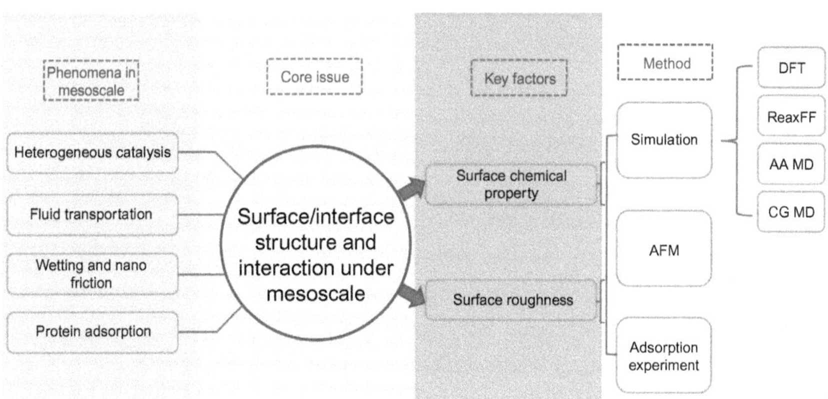

Figure 1 The content of this chapter.

2. CONTROLLING FACTORS IN HETEROGENOUS CATALYSIS AND THE MESOSCALES IN MESOPOROUS TIO_2

In the last century, scientists found that the micropore of molecular sieve played a very important role in the shape-selective catalysis in the oil refining industry. With the oil exhausting, scientists and engineers hoped to pyrolyze the hydrocarbons with large molecule size. However, the development of pyrolysis technologies was limited by the pore size of microporous molecular sieve. The researchers of Mobil Corporation first introduced the preparation procedures of mesoporous molecular sieve of MCM-41, which were published in the journal of *Nature* (Kresge et al, 1992). Till now, the total cited number of this paper has reached up to 13,000, which profoundly influenced the field of catalysis. Accordingly, Zhao synthesized the mesoporous molecule sieve SBA-15 with pore size of 5–30 nm using triblock copolymers as template. The relative researches were published in the top international journal *Science*, which was cited above 7000 times (Zhao et al, 1998). Mesoporous material, such as γ-Al_2O_3, TiO_2, and MCM-41, used as the catalytic support shows great application potentials in the heterogenous catalysis. The new requirement to catalytic process provides the nanoscale mesoporous material opportunity. However, the transport mechanism of substance in the mesopore is significantly different from that in the macroscopic scale. The substance may be concentrated and condensed in the mesopore. The transport mechanism plays an equal role to reaction mechanism in the catalytic process. To reveal the regulatory mechanism, we must clarify the complex structure and function at the interface and analyze the main factors affecting the reaction and transport.

2.1 Transport in Catalyst with Mesopores and Catalytic Mechanism

According to the conventional theories, the main role of catalytic support is to disperse the active species for increasing the reaction interface of active species with reactants. The interaction between active species and support is considered to be physical, but the interaction between active species and reactant is chemical. It is known that the strength of chemical interaction is more than two orders of magnitude of that of physical interaction. In fact, the influence of support on the chemical reaction cannot be ignored. But the conventional design of catalysts is first focused on the catalytically active site to improve the reaction activities of catalysts.

Little attentions are paid on the reaction transport in the catalysts unless the pore size of catalyst is too small to geometrically confine the diffusion of reactant molecules in the pore. However, if the catalytic support consisted of active species, rather than traditional support materials, the new catalytic support could also have chemical effect with reactants. The influence of chemical effect of such active support on the transport mechanism of reactants is unknown. In recent years, some investigators proposed the unsupported catalyst for designing the new catalyst. Instead of being loaded on the catalytic support, active species are made into the mesoporous material to increase the highly catalytic activity. Though it obtains some progress, the unit catalytic performance of unsupported catalyst cannot show a proportional improvement compared with that of supported catalysts (Eijsbouts et al, 2007).

As is known to all, TiO_2 as a catalytically active support can have synergistic effects with active species, which can enhance its intrinsic catalytic activity (Tauster et al, 1978). The effect of TiO_2 is similar to that of catalytic promoter, which can be considered as a part of active site. Therefore, using TiO_2 as the catalytic support, does the chemical interaction of TiO_2 with reactant influence the transport mechanism of reaction? In our previous work, we studied the transport behavior of polar molecules such as water molecules in TiO_2 slits and graphic carbon slits by MD. It was found that transport behavior of water can be significantly influenced by superficial chemical properties (Wei et al, 2011a). As shown in Fig. 2 (left), diffusion coefficient of water molecules in TiO_2 slits is obviously smaller than that in graphic carbon slits. The water molecules closed to TiO_2 slits show a low diffusion coefficient, which indicates that TiO_2 surface restrains the water molecules and inhibits their transport. In a contrast, water molecules in graphic carbon slits have different transport behaviors from that in TiO_2 slits that all the water molecules have similar diffusion coefficient, which indicates graphic carbon has no influence on the transport behavior of water molecules. Also, we further studied the state of water molecules on the TiO_2 surface by density functional theory (DFT) (Liu et al, 2010). The simulation results show that water molecules have chemical interaction with TiO_2 surface. The water molecules absorbed on TiO_2 surface are chemically dissociated into hydroxyls (Fig. 2, right), which can explain why the strong interaction between water molecules and TiO_2 surface occurs.

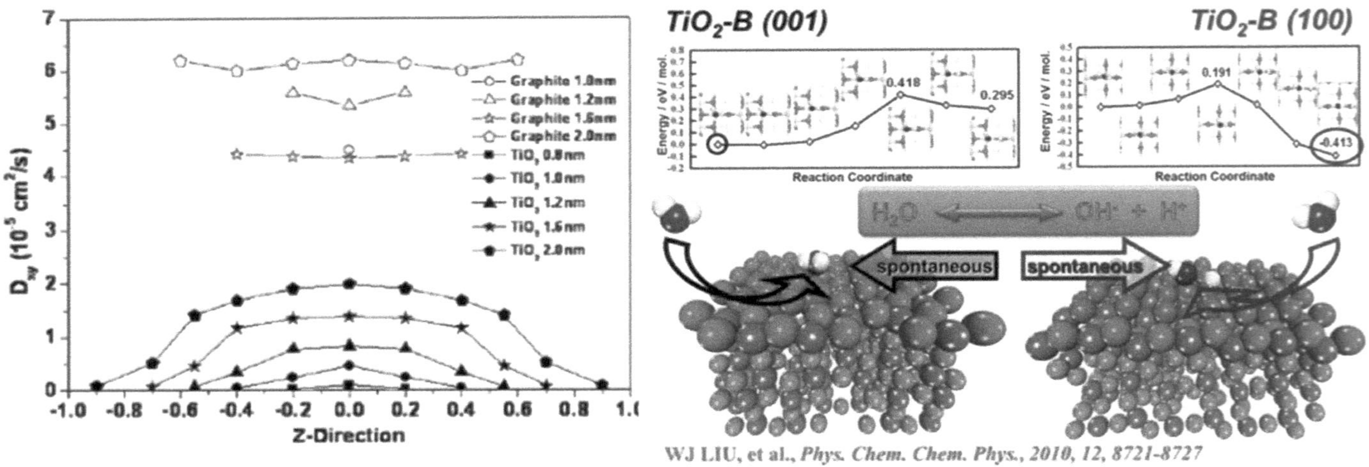

Figure 2 Diffusion coefficients of water molecules in slits (left) (Wei et al, 2011a); the scheme for the chemical dissociation process of water molecule on TiO_2 (right) (Liu et al, 2010).

The simulation results tell us that TiO_2 surface has a severe inhibition effect on the transport behavior of polar molecules. In mesopores, polar molecules closed to the pore wall of TiO_2 display obviously low transport rate. H_2S, hydrodesulfurization reaction resultant, is a kind of typical polar molecule. Provided that the transport rate of H_2S molecules in the pore of mesoporous TiO_2 was influenced, the whole hydrodesulfurization reaction could be inhibited.

2.2 Influence of Surface Chemistry Properties on Catalyst

Based on the analysis mentioned above, modification of surface chemical properties of TiO_2 to accelerate the transport of H_2S molecules can improve the desorption of H_2S from catalyst and further increase its hydrodesulfurization performance. The surface modification technology can be divided into two catalogs: blending and grafting. The grafting modification is fit for changing the surface properties of a specific structured substrate compared with blending. The silane coupling agent is one of the most conventional modifiers. The alkoxysilane group of modifiers can react with water to be silanol, which can be linked to the surface of substrate. Mutin et al pointed out that the silane coupling agent was very sensitive to the moisture that it can agglomerate with each other, which can break the modification effect (Mutin et al, 2005). In our previous work, the selected modifier is a kind of phenylphosphonic acid (PPA) that can be grafted onto the material surface by bonding its phosphonic-based group. The scheme for the influence of surface modification on the catalytic properties of TiO_2-supported catalyst is shown in Fig. 3 (Li et al, 2012). The modification procedure includes two steps: grafting of PPA onto TiO_2 and thermal treatment of PPA/TiO_2 to obtain a carbon surface-modified sample C/TiO_2.

This surface modification does not influence the pore structure of support. The modifier molecules do not cover all the surface of TiO_2. Only 29% of surface of TiO_2 is occupied by PPA, which is considered to be a heterogenous surface modification. Furthermore, this surface modification also does not change the dispersion of active species on the support due to the same morphology of active species and interaction between active species and support according to the results of transmission electron microscopy (TEM) and TPR analysis (Li et al, 2012). However, there is only one obvious influence by surface modification on the H_2S desorption from different supports. The H_2S-TPD patterns show that there are two absorption modes of H_2S on the TiO_2: The lower peak at 254 °C indicates weak adsorption

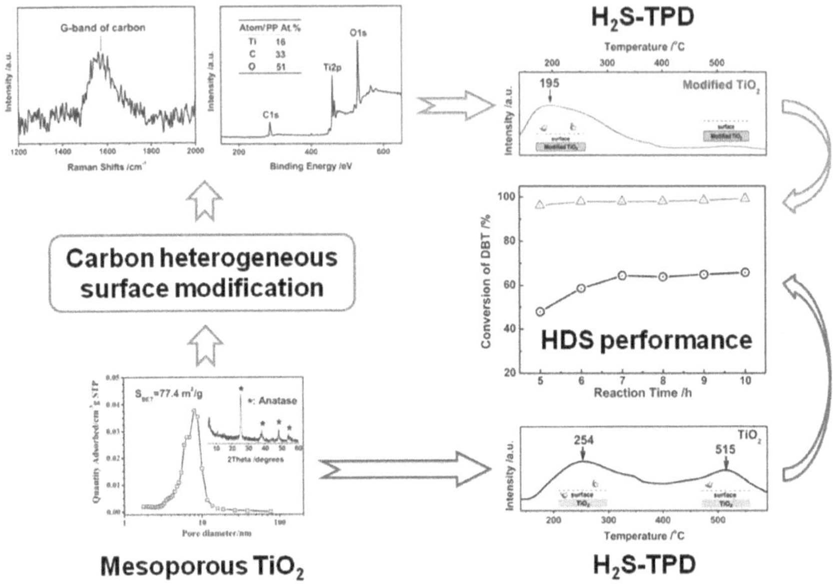

Figure 3 Scheme for the influence of surface modification on the catalytic performance of TiO_2-supported catalyst. *Figure taken from Li et al (2012).*

and the higher peak at 515 °C indicates strong adsorption. But the peak value of the lower peak of modified C/TiO_2 shifts to 195 °C, and the height of the higher peak decreases sharply. The total H_2S desorption amount of C/TiO_2 is 33% less than that of TiO_2. Especially, 95% of adsorbed H_2S of C/TiO_2 is ascribed to the lower peak. Moreover, the H_2S desorption amount of the higher peak is only about 11% of that of TiO_2. The reduction in desorption temperature of C/TiO_2 and the difference in the H_2S desorption amount between C/TiO_2 and TiO_2 indicate the decrease in absorption intensity of H_2S molecules on the surface of C/TiO_2 compared with that of TiO_2. That is to say, H_2S molecules are liable to be desorbed from C/TiO_2. Due to the same structure of TiO_2 and C/TiO_2, the improvement in H_2S desorption from C/TiO_2 can be attributed to the new surface properties brought by the carbon heterogenous surface modification.

As expected, the carbon heterogenous surface modification is indeed beneficial to the improvement in HDS performance of the TiO_2-supported catalyst. With the surface-modified catalyst, $MoNi/C/TiO_2$ exhibits an outstanding HDS performance, and the DBT conversion can reach up to 98%. By contrast, the conversion of $MoNi/TiO_2$ is only about 65%. Moreover, the DBT conversion of $MoNi/C/TiO_2$ remains constant throughout

1000 h of the continuous reaction, suggesting a good catalytic stability of the modified catalyst.

2.3 Complex Interactions and Complex Structures of Mesoporous TiO_2

Mesoporous TiO_2 surface is very complex; it has a mesoporous structure and a variety of crystal planes as an active support. The catalyst loading by mesoporous TiO_2 has multiple mesoscale structures and relative phenomena in heterogenous catalysis. Thus, unconventional mesoscale method is needed to study these structures and phenomena.

The mesostructure of TiO_2 makes the flow behavior of fluid in pore different from that in bulk. Especially, the phase state and wettability of fluid in the pore are other than that in the bulk. The phase state, wettability, and the other properties of fluid can influence the catalytic properties of TiO_2-supported catalyst. Some experimental characterization results can show that the properties of fluid in the pore can significantly influence the chemical reaction. However, we cannot obtain the detailed information about phase state or wettability of fluid from the experiment characterization results. Noteworthy is that understanding the effect of the flow behavior of fluid in the pore on the chemical reaction can significantly help us to improve the catalytic performance of TiO_2-supported catalysts. TiO_2 is an active catalytic support; the wall of TiO_2 pore can chemically absorb the fluid molecules or react with the fluid molecules. Therefore, it needs a united method for simultaneously studying the chemical absorption and physical absorption. Furthermore, the results from AFM study show that the topological structure of TiO_2 surface can also influence the properties of fluid, such as friction and adhesive force, which can further influence the catalytic properties of TiO_2-supported catalyst.

With the continuous development of nanoadvanced materials, chemical engineering has been extended to the nanoscale. Thus, the research object has changed from macrofluid to the solid/liquid interface that has complex structures and complex interactions. When introducing the materials into solid–liquid interface, it will generate some common problems that may influence the transport and other aspect properties. Molecular thermodynamics is a powerful weapon that can be used to solve these common problems, especially for heterogenous catalysis process that involved fluid transport behavior at the complex interface. Mesoporous TiO_2 as the catalyst carrier under the heterogenous reaction shows a series of experimental phenomena at the macroscale. Such as the conversion rate can be measured by the fixed bed and chromatography, the desorption temperature can be measured by the temperature-programmed desorption (TPD), the selectivity of the materials can be acquired by chromatograph, the TiO_2 facet can be distinguished by TEM, the wettability of TiO_2 for different fluids can be obtained by contact angle meter and AFM, the available surface area of clusters on the materials can be received by Brunauer–Emmett–Teller (BET) and AFM, and the friction and adhesion force of different material surfaces can be acquired by AFM. The macroscale experimental performances derive from the complex structures and complex interactions on the TiO_2 surface in micro-/nanoscale. For the complex interactions on the surfaces, first, DFT can be used to study the intramolecular interaction, which belongs to the electronic level (scale between 1 and 10 Å). Second, the MD simulation can acquire intermolecular interaction, which belongs to the molecular and atomic levels (scale between 10 and 100 Å). However, the force field used in MD is limited to particles with physical interactions, rather than chemical interactions. It cannot be used to simulate processes including chemical reaction. Quantum chemistry calculation might not be applied to deal with the large system because of the tremendous compute time. Reactive force field (ReaxFF) molecular simulation methods can handle reaction and transportation of large system at the same time. AFM can be used to acquire the cluster interaction (scale above 100 Å–1 μm). When dealing with the interaction between clusters and roughness surfaces, simulation methods with coarse-grained force field are a better way. On the other hand, complex structures of TiO_2 (scale between 1 Å and 1 μm) mainly reflect on solid–liquid interface structure, which includes the surface chemical structure (involving molecular, atomic, and electronic levels) and surface geometric structure (involving cluster level) and fluid microstructure (involving molecular and atomic levels).

3. MOLECULAR MODELING STUDIES OF COMPLEX SURFACE STRUCTURES

3.1 Study Surface Properties Using Molecular Simulation

Based on our long-term study conducted by molecular modeling interface delivery mechanism of small molecules under confining environment, we found that the surface properties of a material have important implications for transport of small molecules and the reaction behavior. The mobility of polar fluid increases significantly on the strongly hydrophobic surface of graphite comparing with titanium oxide; this effect is even larger than the influence of pore size. This indicates that the interfacial properties of titanium materials are an important factor affecting its performance.

The study of MD simulation shows that the existence of confined conditions can greatly restrict the mobility of water molecules in the channels. The mobility of water molecules in the slit space formed in different chemical properties is very different, and the surface carbon layer has a remarkable effect on the mobility of water molecules.

We performed MD simulations to study water diffusion in slits formed by rutile $TiO_2(110)$ and graphite (0001) with distance ranging from 0.8 to 2.0 nm at 300 K (Wei et al, 2011a). We found the residence time of water in TiO_2 slit is considerably longer than that in graphite slit. This is because the interaction between water and TiO_2 is stronger than that of water and graphite. The bound water at the TiO_2 surface has an identical orientation with two hydrogen atoms pointing out of surface. These hydrogen atoms may form hydrogen bond (HB) with free water molecules in the slits of TiO_2 (shown in Fig. 4). In contrast, there is no bound water, and accordingly, there is no opportunity to form HB between water molecule and graphite surface. We believe the molecules above the TiO_2 surfaces will contact the surface by the bound water layer, rather than contacting them directly. We can imagine for biological system the contact between cells and the TiO_2 surfaces is bridged by the interfacial water layer since the cell or protein is surrounded by water molecules. Protein can be adsorbed on a surface because of the interaction between the residue of the protein and the surface (Hayashi et al, 2006); the presence of water molecules bounded at surface could help the residue of proteins to recognize the surface

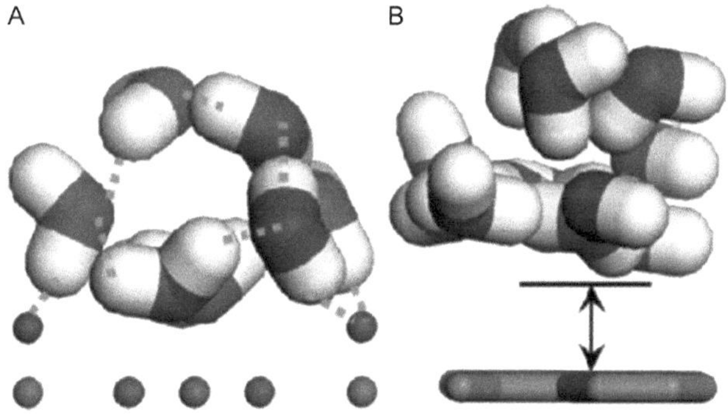

Figure 4 Sketch map of interfacial water near TiO_2 (A) and graphite (B) surface. Green (light gray in the print version) dashed lines stand for hydrogen bonds. *Figure taken from Wei et al (2011a).*

(Skelton et al, 2009). This is one of the reasons of good biocompatibility of TiO_2 as coatings in many implants (Koktysh et al, 2002; Popat et al, 2007).

We analyzed the distribution and orientation of the water molecules near the TiO_2 surfaces; the bound water molecules at the surfaces reduce the diffusivity of water in the TiO_2 slits because bound waters form HB network with other water molecules. It suggests a large majority of water molecules would be held in the hydrophilic nanopores. Surface chemistry is critical to the diffusion of water in nanoscale pores; in this case, it is more efficient to change surface chemistry, rather than to tune the size of nanopores (Alexiadis and Kassinos, 2008; Wei et al, 2011a).

From this work, we can find that the mobility of water molecules in the TiO_2 nanopores would be hindered because of the hydrophilicity of TiO_2, which will decrease the diffusion of reactant and product molecules in the nanopores. While the mobility of water molecules in hydrophobic nanopores (graphite slits) is larger than that in hydrophilic nanopores, it is interesting that the mobility of water molecules will increase or not on TiO_2 surfaces covered by carbon. Carbon nanotube or graphene-covered TiO_2 composites have excellent performance as photocatalysts (Ng et al, 2010; Xu et al, 2010); it can be used in Li-ion batteries and some other applications (Li et al, 2011; Williams et al, 2008).

To reveal the molecular behavior of water confined in slit pores with chemically heterogenous surfaces, we also presented a slit pore model with a carbon-modified TiO_2 surface with different coverages (Wei et al, 2012). We found the carbon covering could break the interfacial water structure on the carbon-covered region. But on the bare TiO_2 region, the interactions between water and TiO_2 molecules cannot be reduced, and also the mobility of interfacial water cannot increase. However, the mobility of water in the center region of the pores can be improved by the carbon covering, especially for wide pores, because the HB network between interfacial waters and that in the central region is broken. The calculated HBs are listed in Table 1 (upper portion); the average HB numbers of water molecules in the center of the pore (outside the interfacial regions) are very similar to each other in all models with various slit widths and carbon coverages. It seems that no general rule can be followed from these data. The total number of HBs between water molecules in the interface region and water molecules outside this region was calculated and also listed in Table 1 (lower portion). The total HB numbers decrease rapidly as the carbon coverage increases.

The diffusion of water molecules in the slit pore depends on the slit width. The diffusion is reduced in the narrow slit that provides limited room

Table 1 The Average Number of Hydrogen Bonds for Water Molecules in Slits

	The Average Number of HB					
Slit Width (nm)	**0%**	**7%**	**47%**	**53%**	**93%**	**100%**
	HBs of water molecules outside the interface region					
1.2	2.65	2.61	2.38	2.68	2.48	2.36
1.6	2.82	2.82	2.71	2.87	2.90	2.96
2.0	2.89	2.87	2.83	2.92	3.00	3.03
	Total numbers of HBs between interfacial water molecules and others					
1.2	–	132.28	69.51	68.45	6.11	0
1.6	–	164.58	41.58	64.54	4.69	0
2.0	–	107.75	31.71	45.40	2.31	0

Table taken from Wei et al (2012)

for water molecules to exchange with surrounding ones. In the 2.0 nm slits, the diffusion of water increased with the carbon coverage increase because more HBs were broken. In pores of width 1.2 nm, the pores are too narrow to let water molecule exchange on the carbon-covered regions; thus, water molecules could only move as a whole. On the other hand, water molecules on the bare TiO_2 region could exchange with others; larger bare TiO_2 region provides more opportunity for water to move in the 1.2 nm slits. In this case, the silts with 93% carbon coverage (smallest area of bare TiO_2 region) exhibit the lowest mobility. We could find when the position of water molecules approaches the center region, the diffusion coefficients of water become close to the value of bulk water. When the slit width is large enough (41.2 nm), the diffusion coefficients could be approaching that of bulk water, although the mobility of water molecules in hydrophilic slits will be hindered by the surface (Wei et al, 2011a). The diffusion of water could be enhanced in the fully hydrophobic slits (100% coverage); this phenomenon has been verified by the experimental work (Karan et al, 2012; Nair et al, 2012). Our experimental work showed that TiO_2-covered carbon material has good performance as an electrochemical capacitor because carbonization changed the hydrophilicity of TiO_2 and thus enhances the diffusion of aqueous electrolyte in the porous-electrode TiO_2 material (Lu et al, 2010). The experimental and simulation results confirm that surface chemistry is critical to the diffusion of molecules in nanoscale pores, and it is possible to enhance the performance of material by modifying the surfaces.

3.2 Study Surface Activity with ReaxFF Simulation

Conventional molecular simulation methods are inadequate to describe the control factor with mesoscale structures and phenomena in heterogenous catalysis. We need to couple simulation methods in different scales to establish new method.

Traditional MD simulations can investigate the different residence times due to the hydrophobic property of the surface of the pores and then obtain different surface fluid transport coefficients and provide a reliable basis for the experimental design. Quantum chemical calculations can consider reactive molecules with different reaction mechanisms for the active surface of pores; the reactivity will also affect the transport fluid to some extent. MD simulation cannot reflect the chemical interaction between the particles and cannot simulate a chemical reaction, while quantum chemical calculations cannot be applied to large systems. ReaxFF simulation method is a new simulation method rising in recent years; it can simulate not only the transport process but also chemical reactions and can simulate transport and reaction in a large system at the same time, which can fill the gap between quantum chemistry and classical force field (empirical force field). It simulates the process with more realistic method because we can not only obtain the transport properties of the fluid but also reflect a chemical reaction of fluid and the surface.

In 2001, van Duin (Van Duin et al, 2001) proposed a technique of reaction force field, this method was applied to carbohydrates, the geometry data of compounds from simulation are consistent with the literature well, and the bond parameters agree to the results of quantum chemistry, but the calculation time is much less than the time required for quantum chemical calculations. Subsequently, researchers come to realize the advantages of the reaction force field method compared with quantum chemistry and electrostatic force field methods. Reaction force field parameters that are suitable for other materials have gradually been developed, such as silicon oxide, platinum, and titanium. The reaction force field parameters are theoretically universal and general. When Kim et al (2013) studied the interaction between the titanium oxide and water, sodium ions, chloride ions, methanol, and formic acid and other substances, the interaction parameter of Cl/O/H and Na/O/H is from the literatures of different systems.

The reaction force field parameters for a specific crystal plane oxide can also be applied to other planes (Fang et al, 2011; Guo et al, 2012; Liu et al, 2010, 2011).

Take hydrodesulfurization catalysis, for example. We used an active TiO_2 as the catalytic support to load the active species for catalysis. Till now, there are many works devoted to research the hydrodesulfurization catalysis mechanism by DFT and experiment, and the reaction pathway of hydrodesulfurization has been figured out. DFT was almost focused on the simulation of the active site. For example, Raybaud et al investigated the interaction of the reactants and intermediate with the MoS_2 or NiMoS catalyst by DFT (Dupont et al, 2011). Also, they studied the interaction between active species and support (Costa et al, 2007). However, there was little work on the interaction of reactants or resultants with catalytic support. The ReaxFF method can be applied to investigate the reaction on TiO_2. It can combine the effect of surface and the effect of transport into reaction process so that the reaction time and the transport time can be separately studied to find out the key influence factor.

MD simulation study showed confining conditions will greatly restrict the mobility of water molecules in the channels. Water molecules in the slit space formed by the surface with different chemical properties exhibit very large differences in mobility (Wei et al, 2011a). The covering of the carbon layer on the TiO_2 surface affects the water molecules greatly on mobility (Wei et al, 2012). The ReaxFF method is introduced to this system; this can break through the limitations (only calculate the active site of the small system) and the limitations of traditional MD (cannot obtain the influence of reaction). The evolution of the system versus time can be calculated in the large system containing TiO_2, active metal clusters, and sulfur-containing material, as well as examining the role of the surface, the impact of reaction, and transport on reaction rate and conversion rate.

The surface structure, surface chemistry, surface reactivity, and hydrostatic and thermal stability are very important. The performance of TiO_2 materials in environmental and energy applications strongly depends on complex material factors. Using the ReaxFF and RxMD simulations, we researched theoretically the surface reactivity of TiO_2 materials to water dissociation (Huang et al, 2014). We studied the interactions between water and five different TiO_2 surfaces, which are anatase (001), rutile (011), rutile (110), TiO_2-B (100), and TiO_2-B (001). These five studied TiO_2 surfaces show different reactivities for water dissociation on four surfaces. And the reactivities are in the order of rutile (011) > TiO_2-B (100) > anatase

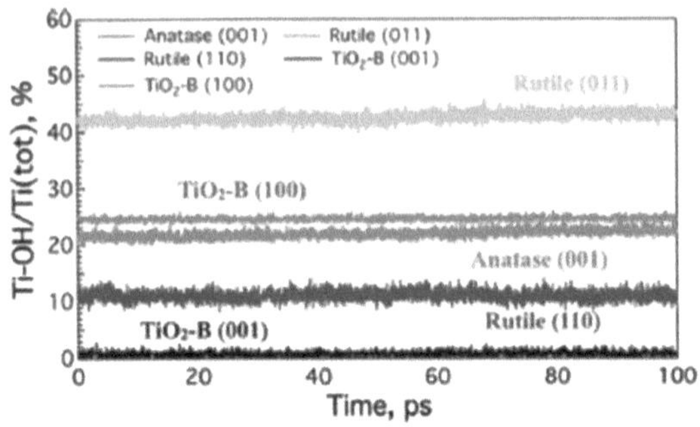

Figure 5 Surface hydroxyl group density evolution: Ti-OH is the number of titanium atoms that are bonded with the OH groups from water dissociation. Ti(tot) is the total number of surface Ti atoms. The ratio of Ti-OH/Ti(tot) is monitored as a function of simulation time. More water molecules will dissociate on a more reactive TiO_2 surface, which leads to a larger ratio of Ti-OH/Ti(tot). *Figure taken from Huang et al (2014).*

(001) > rutile (110). There is no water dissociation observed on the TiO_2-B (001) surface (Fig. 5).

Analysis of the near-surface water suggests that the water has dissociated. The TiO_2 surface chemistry changes to form the new surface with Ti-OH and O-H functional groups. These groups can affect the orientation of water molecules near surface. The first layer of water molecules dissociates on the rutile (011) surface, which is the reactive TiO_2 surface. At the same time, the surface changed, and the resulting new functional groups enhance the interaction between surface and near-surface waters. It also brings the second layer of water closer to the surface. On the TiO_2-B (001) surface, no molecular or dissociative water adsorption is observed; thus, the TiO_2-B (001) surface can represent nonreactive surface. The water near TiO_2-B (001) surface can also form an HB network, which includes the surface oxygen atoms of TiO_2, but the distance from first-layer water molecules to the surface is larger than that on reactive TiO_2 surfaces (Fig. 6).

From this work, we can understand the reactivity order of different TiO_2 surfaces; it also proved the surface property and structure of metal oxide surfaces are key factors to near-surface water behavior on surface. It will be important to understand how the property of water near surface changes quantitatively on different metal and metal oxide surfaces. How this change in the near surface affects the interaction between surface and adsorbents should be studied systematically. More efforts are needed from both ab initio quantum mechanics (QM) calculations. ReaxFF development for a fundamental understanding of complicated TiO_2 systems is important.

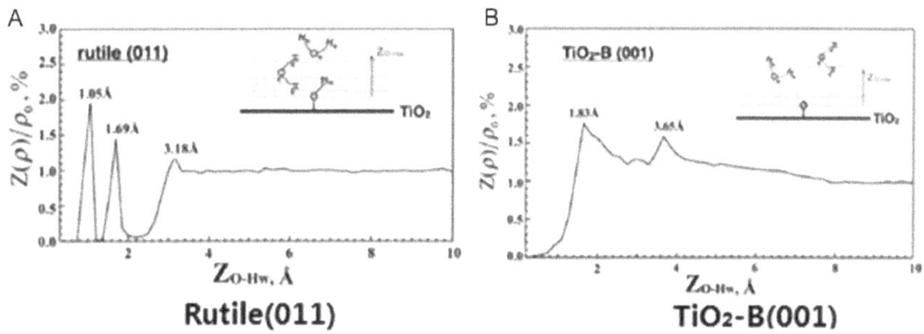

Figure 6 Density distribution of water molecules along the *z* direction: (A) rutile (011) with three distinct peaks and (B) TiO_2-B (001) with two distinct peaks. As indicated by the inset, the origin was set to be at the oxygen atoms of the TiO_2 surface. The peaks indicate an increased local density of water. *Figure taken from Huang et al (2014).*

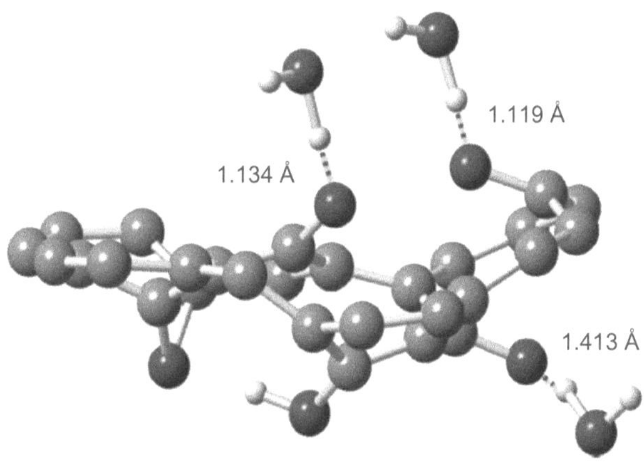

Figure 7 The local structure of the 30% (1:1) GO surface, where all three carbonyl groups are occupied by H_2O molecules. The mixture is made up of 100 H_2S and 100 H_2O molecules. A small portion of the GO structure is shown for clarity. *Figure taken from Huang et al (2013).*

We generated realistic atomistic GO structures using ReaxFF and temperature-programmed MD simulation. Epoxy and hydroxyl groups are grafted to the basal graphene surface. Using this surface structure model, the adsorptions of H_2S and H_2O/H_2S mixtures on GO were calculated (Huang et al, 2013). The results show that H_2S molecules dissociate on the carbonyl functional groups, and this reaction produces H_2O, CO_2, and CO molecules. For the H_2O/H_2S mixtures, H_2O molecules are preferentially adsorbed to the carbonyl sites and occupy the potential active sites for H_2S decomposition (Fig. 7). The results of simulation agree well with the experiments. The methodology and the procedure applied in this work can be extended to the research on other materials.

Based on the analysis mentioned above of ReaxFF simulation, we can extend the simulation to other catalyst systems. DFT simulation showed that H_2S molecules can strongly absorb on the surface of TiO_2-B (100) (0.475 eV). The absorbed H_2S molecules can be easily dissociated, and the energy barrier is only 0.155 eV. In the process of hydrodesulfurization, H_2S molecules can strongly absorb on the surface of TiO_2-B (100). The absorbed H_2S molecules can cover the reaction surface of catalyst for inhibiting the reaction. After the carbon surface modification over the surface of TiO_2, the absorption strength of H_2S on TiO_2 was significantly decreased, and H_2S molecules can be desorbed easily so that the hydrodesulfurization performance of TiO_2-supported catalyst can be improved. We found that the carbon surface modification can decline the absorption strength of H_2S on the surface of TiO_2-B by temperature-programmed reduction technique. The hydrodesulfurization conversion of surface-modified catalysts can reach up to 98%, which showed an excellent catalytic performance compared with unmodified catalysts (65%) (Li et al, 2012). It is feasible to research hydrodesulfurization catalytic reaction supported on TiO_2 or modified TiO_2 by ReaxFF simulation to see the hydrodesulfurization mechanism in mesoscale.

3.3 Simplify the Study: Pore Size and Surface Wettability

Gubbins et al using a simplified model treated a phase of nanoscale dimensions (the adsorbate) confined within a porous material (Gubbins et al, 2014). This model provides a useful framework for a qualitative understanding of many thermodynamic phenomena in confined nanophases. They proposed two main system variables: one is pore width expressed as the number of molecular diameters and another is microscopic wetting parameter, α_w, which emerges naturally from the corresponding state treatment. This model aids in interpreting experimental data, such as capillary condensation and freezing transitions in pores (Fig. 8).

However, this model does not include other factors, such as the roughness of the pore walls, their shape and connectivity, and nonspherical intermolecular interactions. The roughness and pore shape can be important for certain properties. Wall roughness leads to heterogeneity of the adsorption sites, and it has a strong effect on adsorption isotherms at low loading (Jagiello and Olivier, 2013a, 2013b) and the tangential pressures near the

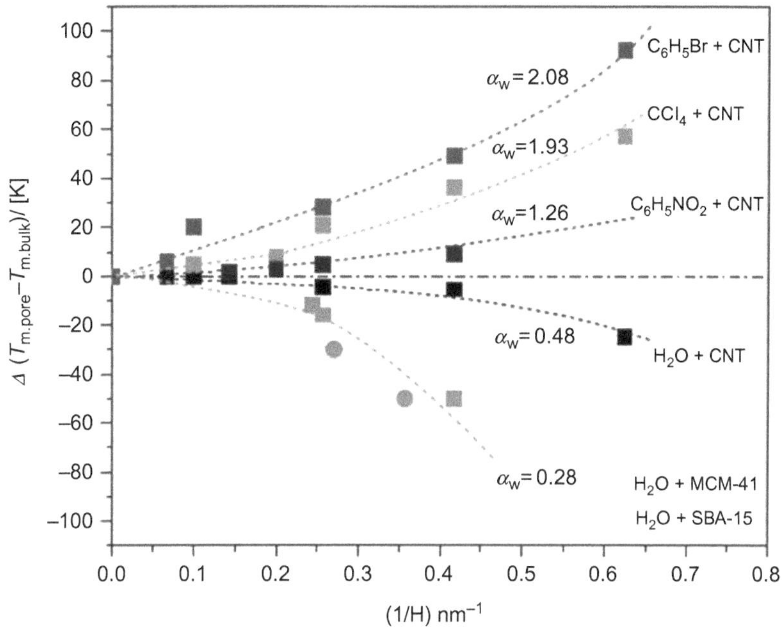

Figure 8 Experimentally measured shifts in the melting temperature for confined nanophases, showing the effect of pore width and wetting parameter. Here, CNT, multiwalled carbon nanotube. MCM-41 and SBA-15 are silica materials. *Figure taken from Gubbins et al (2014).*

pore walls (Long, 2012). The roughness disrupts the adsorbed contact layers next to the wall by lessening the compression of these layers. Water molecules tend to form molecular clusters when confined due to hydrogen bonding. This is particularly the case for adsorbed water in carbon materials (Long, 2012; Müller and Gubbins, 1998). Gubbins et al mainly considered slit-shaped pores. They also proposed that slit-shaped pores are easier to treat since the confinement is in only one direction. In curved pore walls, such as cylindrical and spherical pores, confinement will be in two or three dimensions, and it will have stronger steric hindrance by the walls. In small pores, phase changes are rounded and will not occur for very small pores. There are two distinct and independent pressures in the confined nanophase in simple slit pores or cylindrical geometry pores: one pressure is normal to the wall and another is parallel to it (the tangential pressure) (Long, 2012). There are also two such pressures for spherical pores; however, they are related through a first-order differential equation (Gray et al, 2011). The calculation results show that the normal and tangential pressures are very different in magnitude. Moreover, the influences of the main state and system variables on these two pressures are quite different.

From their work, Gubbins et al believed that the normal and tangential pressures play different roles in observed phenomena in confined nanophases. And it is not profitable to define a single scalar effective pressure for the confined nanophase.

4. EFFECT OF ROUGHNESS ON HETEROGENOUS SURFACE

Roughness in interface is an important factor to affect the catalytic reaction in nano- and microscale. The reason it affects the activity of the catalytic reaction is that the adhesive force and stability of the immobilized molecule and cluster are different with different roughness in interface. The way it affects the diffusion is that friction produced by roughness on the flowing fluids in the inner wall of the hole is different, thus affecting the transport of the liquids. When taking into account the dynamic behavior in the surface at the mesoscopic scale is needed, AFM will become a best selection. AFM has been widely used in micro- and nanoscale and even used to test the sample surface information accurately in atomic and molecular scale. The adhesive force was measured using the force-distance curve approach in the contact mode by AFM. In this technique, the AFM tip is brought into contact with the sample surface by extending the piezo vertically. The piezo is retracted next, and the force required to separate the tip from the sample is calculated. At this point, you can measure short-range repulsion between atoms. Lateral force microscopy can measure the lateral deflection caused by cutting forces that are parallel to the sample surface. It is mainly used to research the surface friction changes due to the physical structure and chemical nonuniformity of materials and thus get the coefficients of nanofriction of different sample surfaces. Therefore, the static adhesion force (the direction parallel to the sample surface) between the sample surface with different structures and the tip of the AFM or tip modified by the target (interpreted as clusters) and dynamic surface friction coefficient properties (the sample surface in the vertical direction) can be measured and thus reflects the change in the force on both directions, to obtain nanofrictional properties of different structured surfaces, and possible impact on its performance. The static behaviors of fluid or cluster molecules on TiO_2 surface can also be examined by static contact angle in macroscale.

By detecting the surface of TiO_2 film with different roughness using AFM, we found that friction coefficient of the surface of rough M-TiO_2 film is 26 times lower than the surface of smooth D-TiO_2 film. The construction of surface roughness changes the distribution of the surface liquid water molecules and further affects the adhesive force with the molecule of the tip. On the other hand, surface of TiO_2 film with different roughness has important effect on the wettability of the viscous ILs. The big difference of adhesive force can be detected from AFM.

4.1 Frictional and Adhesive Forces on a Heterogenous Surface

Ti material has been widely used for many applications such as medical implants, biosensor, catalysis, drug delivery, and heat exchangers because of its extraordinary mechanical strength, excellent anticorrosion properties, low density, and especially biocompatibility by forming the TiO_2 layer on its surface (Buettner and Valentine, 2012; Liu et al, 2014; Tijana et al, 2014; Wei et al, 2011a). However, the TiO_2 layer formed on the surface has high friction so that it is not beneficial for fluid flow for long time especially in certain extreme conditions (Liu et al, 2004). It can cause decreasing heat exchanger efficiency and the nonspecific adsorption of proteins. More seriously, it can cause thrombus and osteoarthritis and be even life-threatening in biomedical implants. Many methods were used to provide a heterogenous, stable TiO_2 surface with low friction (Souza et al, 2010). Surface modification with chemical functional groups cannot provide dependable and longtime protection in sliding or in fluidic systems (Sumant et al, 2005). However, by means of some physical strategies, increasing the roughness of the TiO_2 film sometimes will unavoidably lead to the retention of fluid flow, nonspecific protein adsorption, and low hemocompatibility of Ti surface (Jones et al, 1999; Liu et al, 2005a). It is critical for us to analyze the relationship between TiO_2 structure and its performance due to its excellent performance. The high performance for TiO_2 comes from not only a certain simple structure but also some characteristics of the structure interactions.

We present a simple physical strategy to reduce their frictional and adhesive forces on TiO_2 films by constructing TiO_2 films with heterogenously distributed nanopores on their surfaces, and TiO_2 films with densely packed nanoparticles were also prepared for comparison. The surface morphology of mesoporous and dense TiO_2 films examined by FESEM (field emission scanning electron microscopy) and AFM is shown in Fig. 9. The FESEM

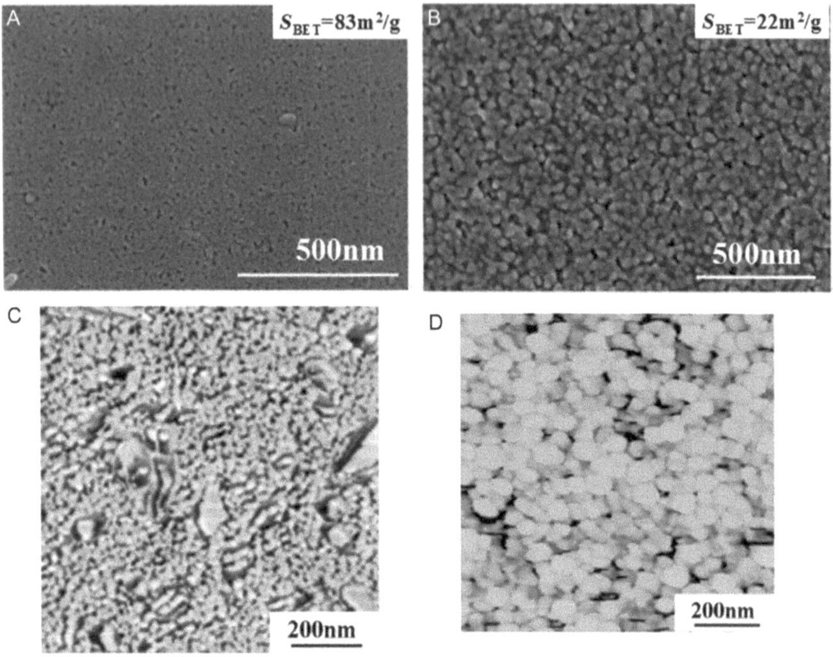

Figure 9 The FESEM images of (A) a mesoporous and (B) a dense TiO_2 film. The AFM phase images of (C) a mesoporous and (D) a dense TiO_2 film. *Figure taken from An et al (2012).*

image in Fig. 9A showed that nanostructured pores were created on the mesoporous TiO_2 film surface. Different surface morphologies for these two types of surfaces were demonstrated by the phase images by using AFM. In Fig. 9C, the heterogenously distributed nanopores were obviously observed on the mesoporous TiO_2 film surface. But in Fig. 9D, the homogeneously distributed nanoparticles were seen on the dense TiO_2 film surface (An et al, 2012).

The crystalline phases existing in the mesoporous and dense TiO_2 films were demonstrated using Raman spectroscopy. Figure 10 shows the Raman spectra of the mesoporous and dense TiO_2 films. It was found that the dense TiO_2 films had the strong peaks at 143.8, 195.4, 394.9, 513.1, and 638.2 cm^{-1} due to the anatase phase. The mesoporous TiO_2 were found two additional weak peaks at 119.4 and 242.2 cm^{-1}, corresponding to the TiO_2 (B) phase, which was different from the phases of anatase, rutile, and brookite TiO_2. The existence of the TiO_2 (B) phase is favorable to the thermal stability of the mesoporous TiO_2.

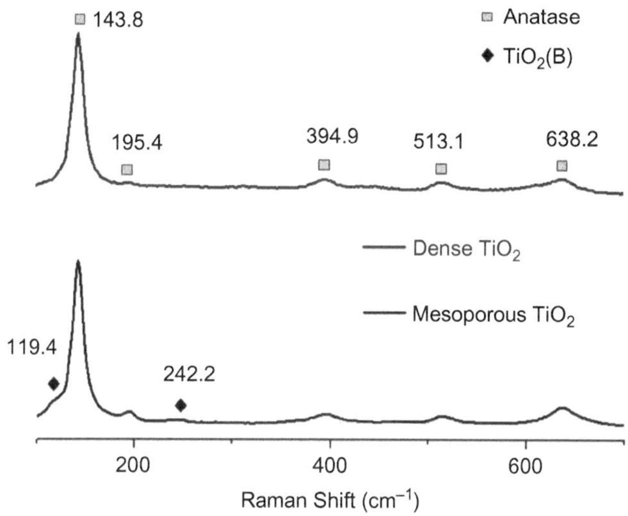

Figure 10 The Raman spectra of mesoporous and dense TiO_2 films. The Raman vibrational bands due to the anatase phase are marked in solid squares, while those due to the TiO_2 (B) phase are marked in solid diamonds. *Figure taken from An et al (2012).*

We can see the AFM topographic images of the mesoporous and the dense TiO_2 films in Fig. 11a and b, respectively. The line profiles (Fig. 11c and d) showed that the mesoporous TiO_2 film surface was much rougher than that of the dense TiO_2 film surface. (Average roughness (Ra) is 4.10 and 0.39 nm for the mesoporous and the dense TiO_2 film surfaces, respectively) The adhesive force measured at point A (top of the particle) on the mesoporous TiO_2 film surface was 12.6 nN, while at point B (deep void), it was only 3.2 nN. For point C on the dense TiO_2 film surface, the adhesive force was measured as 13.2 nN. The heterogeneity of the mesoporous TiO_2 film surface is demonstrated by the variation of the measured adhesive forces by AFM. The statistic histograms of the adhesive forces measured by AFM for the mesoporous and the dense TiO_2 film surfaces are shown in Fig. 11f and g. The distribution of adhesive forces measured on the mesoporous TiO_2 film surface showed a bimodal peak with one about 4 nN and another about 12 nN, on behalf of the adhesive forces measured in the deep voids and the particles, respectively. However, for the dense TiO_2 film surface, the histogram showed a single peak about 12 nN, indicating the dense TiO_2 film surface homogeneity.

The frictional forces on mesoporous TiO_2 film surface and dense TiO_2 surface were also investigated. Figure 12 shows the lateral forces on

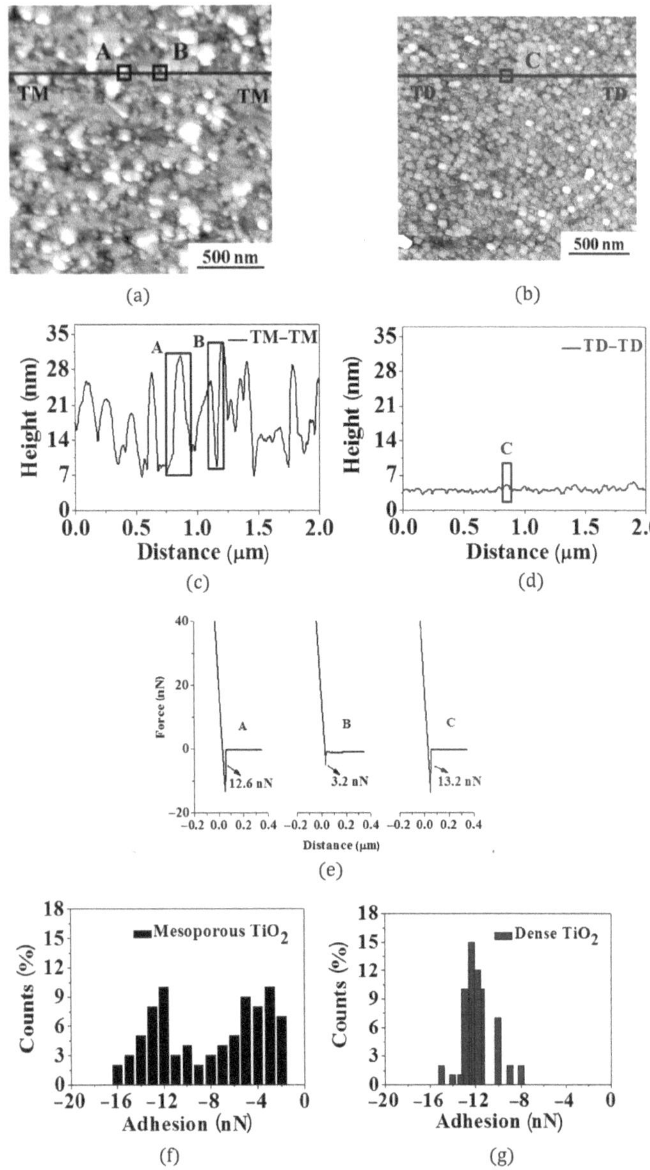

Figure 11 The AFM topographic images of the mesoporous (a) and the dense TiO_2 films (b). The line profiles along the TM–TM line in mesoporous TiO_2 film (c) and the TD–TD line in dense TiO_2 film (d). The adhesive forces measured at the points A and B on the mesoporous TiO_2 film surface and the point C on the dense TiO_2 film surface, where A and C are on the top of nanoparticles and B is right at the dip point (e). (f) and (g) The histogram of adhesive forces measured at ~80 different positions on the area of 2×2 μm of the mesoporous and the dense TiO_2 film surfaces. *Figure taken from An et al (2012).*

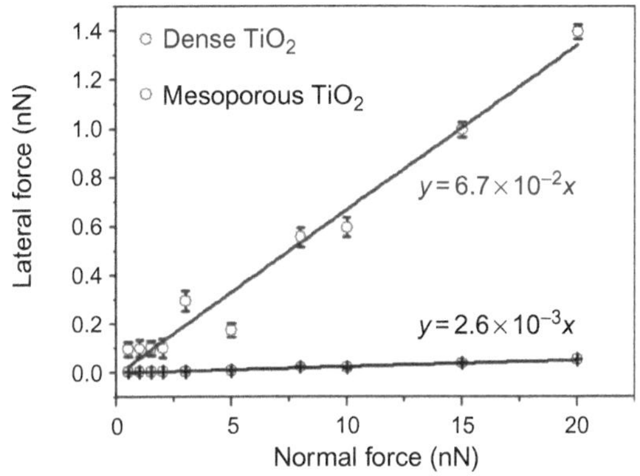

Figure 12 The lateral force measured at different normal loads using AFM on the mesoporous and the dense TiO_2 film surfaces. The slope of the line represents the friction coefficient. *Figure taken from An et al (2012).*

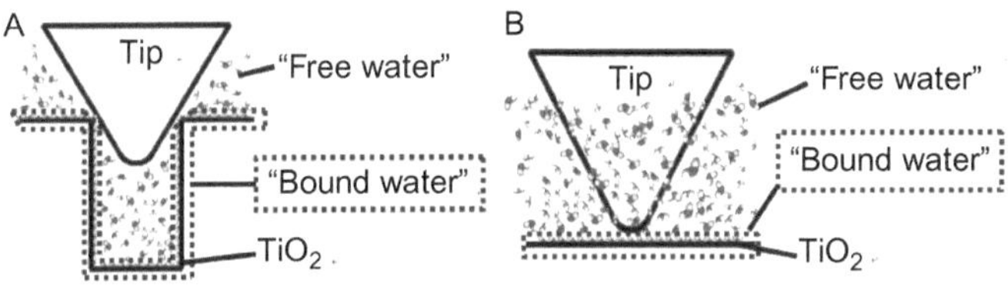

Figure 13 Illustrations of a Si_3N_4 tip on (A) the mesoporous and (B) the dense TiO_2 film surfaces in the presence of water molecules. The "bound water" means the first-layer water molecules on the surfaces, while the "free water" means those beyond the first layer. *Figure taken from An et al (2012).*

mesoporous TiO_2 film surface and dense TiO_2 film surface as a function of the normal load. The friction coefficient of these two surfaces is determined from the slope of each line. The friction coefficient for the mesoporous TiO_2 film surface is about 2.6×10^{-3} and for the dense TiO_2 film surface is about 6.7×10^{-2}, which means the friction coefficient for mesoporous TiO_2 is about 26 times lower than that of the dense TiO_2 film surface.

The simulation results from our group previous work (Liu et al, 2010; Wei et al, 2011a) and experimental results from Anpo's group (Takeuchi et al, 2005) together showed that the "bound water" molecules as part of TiO_2 surface have strong interaction with the TiO_2 surface. The desorption signal in the TPD spectra at 259 °C indicates the existence of the "bound water" on the TiO_2 surfaces, while signal at 127 °C corresponds to "free water." The larger difference in friction coefficient (nearly 26 times) is due to the nanostructured pores on mesoporous TiO_2 surface as shown in Fig. 13A. Nevertheless, the AFM tip prefers to walk on the stronger adsorbed "bound water" molecules on dense TiO_2 surface, which has been partially surrounded by "free water" as shown in Fig. 13B. As we know that water behavior is influenced by forming HBs on the chemically heterogenous surface, so it is necessary for the breakdown of HBs between water molecules in order to make the tip move on TiO_2 surface. While on mesoporous TiO_2 surface, due to the heterogenously nanostructured pores, the total HB numbers around the AFM tip are less than that on dense TiO_2 surface, which is shown in Fig. 13. Thus, the weaker interaction between the water molecules and the mesoporous TiO_2 surface makes less energy needed to break the HB network to separate the AFM tip from the TiO_2 surface, which results in a lower coefficient than dense TiO_2.

The mesoporous TiO_2 films show low frictional and adhesive forces due to the heterogenously distributed nanopores. The geometrically roughness-induced nanopores were obtained due to the TiO_2 (B) formed on the anatase surface. This work indicates a simple approach to addressing heterogenously distributed nanopores on TiO_2 surfaces for controlling surface friction properties. It opens up the way for the further examination on TiO_2 surfaces for applications such as heat exchangers, drug delivery, and medical implants.

4.2 CO_2 Absorption of ILs on Mesoporous Material

ILs are ionic compounds with large organic cations and various kinds of anions that exist in the liquid state over a wide temperature range (Luo et al, 2004). ILs have unique and desirable physicochemical properties, such as high thermal and chemical stabilities, low vapor pressure, and excellent solubilities for a very wide range of organic and inorganic compounds (Liu et al, 2005b; Stolte et al, 2012; Welton, 1999). Therefore, ILs have attracted considerable interest in various fields, such as CO_2 separation (Ramdin et al, 2012). To develop IL-based technology for CO_2 separation, lots of research work have been carried out (D'Alessandro et al, 2010; Hasib-ur-Rahman et al, 2010; Zhang et al, 2012). However, the use of ILs suffers two main drawbacks. One is the terribly low gas–liquid mass transport rate of CO_2 in ILs due to the high viscosities of ILs (Hasib-ur-Rahman et al, 2010). The other is the high price of ILs (Zhang et al, 2012). But usually a large amount of ILs is needed, due to the low absorption capacity (Xie et al, 2014). Thus, how to dramatically improve the CO_2 absorption mass transport rate in ILs and reduce the amount of ILs needed is very important in the application of ILs for CO_2 separation. Recently, IL immobilization into porous solid supports (Ren et al, 2012; Vicent-Luna et al, 2013; Wang et al, 2013; Zhang et al, 2006) has been found that they could significantly enhance the mass transport rate of CO_2 in ILs and could potentially reduce the amount of ILs needed for CO_2 capture (Samanta et al, 2012). For example, the equilibrium time was shortened to less than 15 min (Wang et al, 2013) for these reported IL-immobilized sorbents. The diffusion path was shortened, and the gas–liquid contact area was enlarged due to the much thinner film of ILs on support after immobilization (Ren et al, 2012; Vicent-Luna et al, 2013; Wang et al, 2013; Zhang et al, 2006). And this is the reason for the enhancement of the mass transport rate observed experimentally, as explained in all these studies (Ren et al, 2012; Vicent-Luna et al, 2013; Wang et al, 2013; Zhang et al, 2006). Therefore, the wettability of ILs on the interface of supporting material IL (i.e., solid–liquid) is essential in obtaining the effective contact area of CO_2 and IL (i.e., gas–liquid). While many other IL applications have been developed, fundamental aspects of the properties of ILs in solid–liquid interface have been relatively less investigated. Among all these applications, in essence, the performance strongly depends on the wetting behavior at interface between ILs and solid surfaces of the supporting materials (Bovio et al, 2009).

In our work (An et al, 2013), 1-butyl-3-methylimidazolium hexafluorophosphate ([Bmim][PF_6]) was used to wet the mesoporous and dense titanium dioxide (TiO_2) films, and their CO_2 absorption behavior was studied. It was found from the AFM analysis that [Bmim][PF_6] can form a wetting phase (with contact angle as 5°) on mesoporous TiO_2 (M-TiO_2) films, which has higher roughness and lower friction coefficient, but non-wetting and formed sphere-shaped droplets (with contact angle as 40°) on dense TiO_2 (D-TiO_2) films.

Figure 14 shows the AFM topographies of [Bmim][PF_6] coatings on M-TiO_2 and D-TiO_2 films and their corresponding statistic results. The multilayer stacking of [Bmim][PF_6] wetting on M-TiO_2 films is shown in Fig. 14A. It was different from the IL behavior shown on D-TiO_2 films, which exhibited nonwetting phase on rounded domains as in Fig. 14B. Figure 14C shows the height profile along the line T_1T_2 in Fig. 14A, and it could be found that the thickness of each [Bmim][PF_6] layer was in different multiples of 5–150 nm range on M-TiO_2 films. But [Bmim][PF_6] aggregated themselves as rounded domains on D-TiO_2 films, and the height could even reach 413.1 nm, as shown in the line profile T_3T_4 (Fig. 14D). Figure 14E provides the adhesion behavior of [Bmim][PF_6] on the M-TiO_2 films. For comparison, the adhesive force of

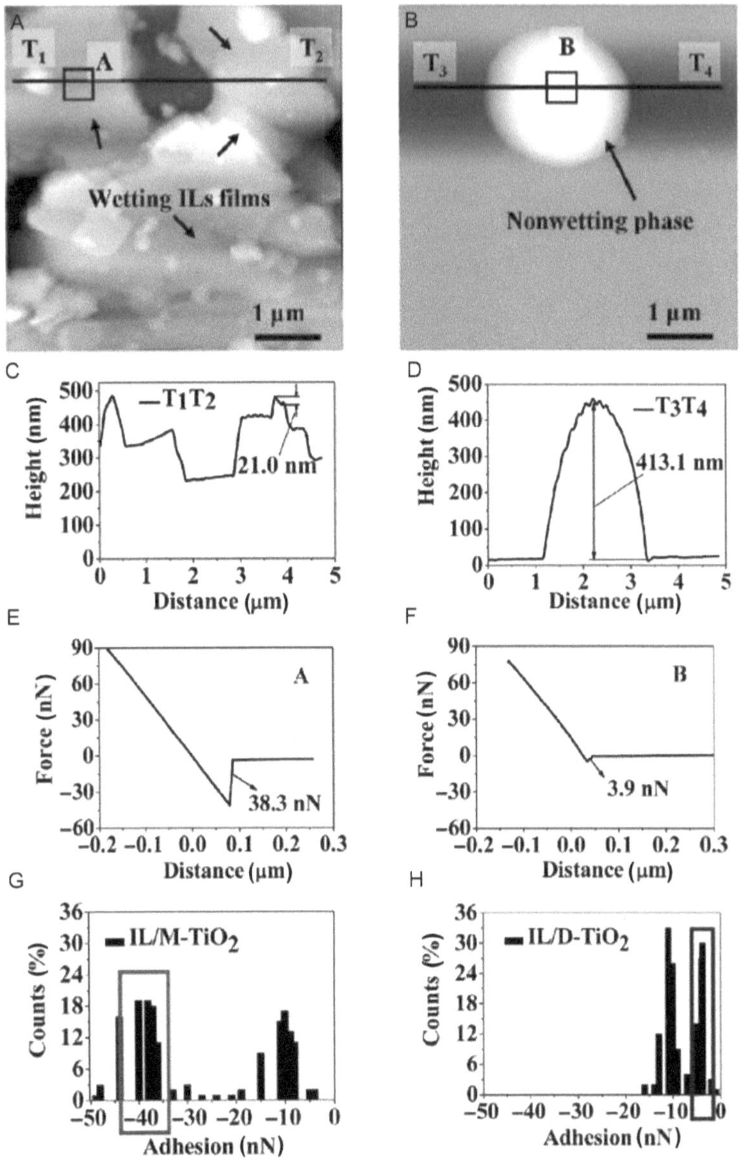

Figure 14 AFM topographic images of [Bmim][PF_6] on (A) M-TiO_2 and (B) D-TiO_2 films. The line profiles along the (C) T_1T_2 line and (D) T_3T_4 line. The adhesive forces measured at (E) the point A on M-TiO_2 films and (F) the point B on D-TiO_2 films, where A and B are on the top of ILs [Bmim][PF_6]. The histogram of adhesive forces measured at ~155 different positions on the area of 5 × 5 μm of (G) M-TiO_2 and (H) D-TiO_2 films. The vertical scales in panels (A) and (B) are 600 nm. *Figure taken from An et al (2013).*

[Bmim][PF_6] on D-TiO_2 films is also shown in Fig. 14F. It could be found that the adhesive force of [Bmim][PF_6] on M-$TiO_2$was stronger (38.3 nN) than on D-TiO_2 (3.9 nN). Therefore, the stronger adhesive force further leads to M-TiO_2 films being much more easily occupied by multilayer stack of [Bmim][PF_6] wetting phases, while [Bmim][PF_6] aggregated together on D-TiO_2 films and showed a more pronounced tendency to form nonwetting phases. Figure 14G shows the adhesion distribution of [Bmim][PF_6] on M-TiO_2 films. A bimodal distribution was observed, that is, heterogenous distribution (Zeller et al, 2010), centered around 40 and 10 nN, respectively, due to the [Bmim][PF_6] and the TiO_2 films (An et al, 2012). A heterogenously (Zeller et al, 2010) bimodal adhesion distribution of [Bmim][PF_6] on D-TiO_2 films is also shown in Fig. 14H, which centered around 10 and 4 nN due to the TiO_2 films (An et al, 2012) and ILs [Bmim][PF_6], respectively. Apparently, [Bmim][PF_6] can form wetting films on M-TiO_2 films because of stronger interactions, but a round nonwetting sphere on D-TiO_2 films formed for weaker interfacial interactions.

The friction coefficients of [Bmim][PF_6] on M-TiO_2 and D-TiO_2 films as a function of lateral force for Si tip at various normal loads are also measured and are shown in Fig. 15. It showed in Fig. 15 that the friction coefficients of [Bmim][PF_6] on both M-TiO_2 and D-TiO_2 films determined from the slopes of lateral force versus normal load plots were in the same scale (0.0025 for M-TiO_2 films and 0.0026 for D-TiO_2 films). The observation showed that tip can slip easily with the existence of [Bmim][PF_6] on

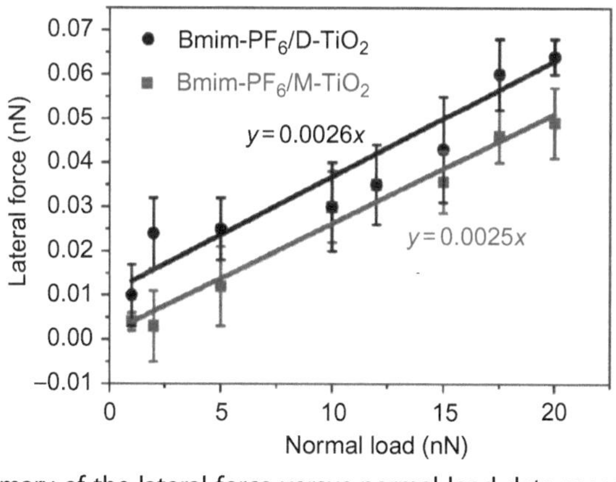

Figure 15 Summary of the lateral force versus normal load data recorded for [Bmim][PF_6] on the M-TiO_2 and D-TiO_2films and AFM tip. The slopes of the lines represent the corresponding friction coefficients. *Figure taken from An et al (2013).*

the films. On M-TiO_2 films, [Bmim][PF_6] had a more pronounced tendency to form wetting phases, acting like an excellent lubricant to make the friction coefficient lower. This would be beneficial to surface adsorption and the surface in equilibrium to update quickly to obtain a new adsorbent surface. While with tip scanning, the rounded and nonwetting [Bmim][PF_6] sphere on D-TiO_2 films tended to roll on the surface, leading to a low friction coefficient in same scale. The moving liquid-like IL sphere on D-TiO_2 films cannot interact with the TiO_2 surface. Therefore, it was not favorable for transmission and surface update. There were only some regions of D-TiO_2 surface that could be covered by the weakly interacted liquid-like IL. Therefore, the ILs on the film did not act as a stable and excellent lubricant for the whole system. The remaining bare D-TiO_2 surface still had high friction coefficient (An et al, 2012).

More importantly, the mass transport rate of CO_2 absorption in [Bmim][PF_6] immobilized on mesoporous TiO_2 was almost 2 and 10 times higher than that on dense TiO_2 and pure [Bmim][PF_6], respectively, as revealed in Fig. 16. This is because the stronger adhesive force on mesoporous TiO_2 films made [Bmim][PF_6] adhere to the mesoporous surface tightly. The stable spreading [Bmim][PF_6] films provided low friction coefficient (0.0025), large wetting areas, and short CO_2 diffusion distance on the whole surface of the mesoporous TiO_2 film, avoiding the significant decelerating effect through equilibrium limitations to increase CO_2 separation mass transport rate. Furthermore, ILs wetted on mesoporous TiO_2 decreased the time reaching to the maximum adsorption rate (2.8 min), much faster than that

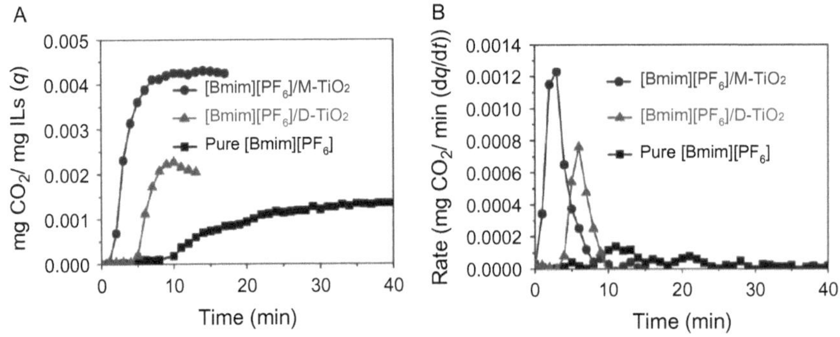

Figure 16 (A) Experimental CO_2 absorption behaviors of [Bmim][PF_6] immobilized on M-TiO_2 and D-TiO_2 at 35 °C, 1 bar. (B) Rate of CO_2 absorption of [Bmim][PF_6] and [Bmim][PF_6] immobilized on M-TiO_2 and D-TiO_2. Note the stoichiometry of ILs with M-TiO_2 and D-TiO_2 approaches 1:13. *Figure taken from An et al (2013).*

on dense TiO_2 films (6.1 min). This work was very important in guiding the improvement of the efficiency of the lubrication of micro-/nanoelectromechanical systems (M/NEMs), CO_2 capture, and gas separation.

On the other hand, although ILs have shown a promising prospect for CO_2 separation, some difficulties are still pending for their industrialization and commercialization due to their high viscosity and high cost (Zhao et al, 2010, 2011). To develop the process intensification IL-based technologies for CO_2 capture, it is necessary and urgent to study the kinetic process of CO_2 capture and to investigate the influence of different operation parameters on the mass transport rate of CO_2 in ILs.

Based on our previous studies of the dissolution and crystallization kinetics of potassium inorganic compounds based on linear nonequilibrium thermodynamics (Ji et al, 2010; Liu et al, 2009; Lu et al, 2011), we proposed to assume that the kinetic process of CO_2 absorption by ILs comprised two steps: surface reaction and diffusion, as shown in Fig. 17. Figure 17 demonstrates that when CO_2 in the vapor phase and the ILs were in contact, the chemical reaction of CO_2 with ILs occurred for the chemical absorption process of CO_2 by ILs in the first step, which was named as the surface reaction layer, while for the physical mass transport process of CO_2 by ILs in the first step, CO_2 in the vapor phase would be transported into the IL phase, which was also named as the assumed "surface reaction layer." As for the surface reaction layer, the driving force of the surface reaction was the chemical potential gradient of CO_2 between CO_2 at the vapor–liquid interface and gas CO_2. After that, in the second step, CO_2 in the IL phase would

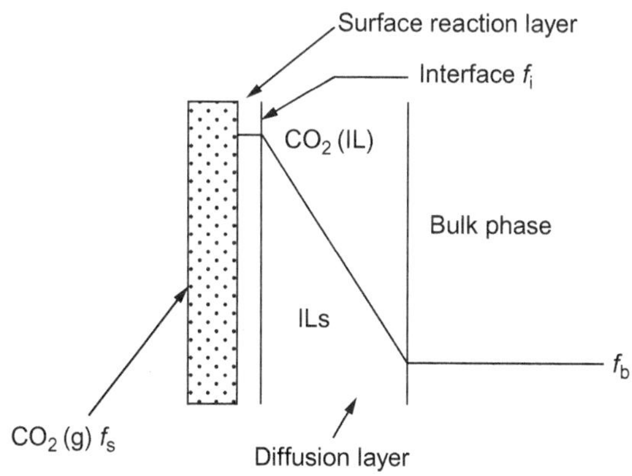

Figure 17 Schematic diagram of CO_2 absorption process by ILs. *Figure taken from Lu et al (2012).*

diffuse from the vapor–liquid interface into the bulk phase, which was named as the diffusion layer. In the diffusion layer, the driving force for the diffusion in solutions was the chemical potential gradient of CO_2 between the bulk phase and the vapor–liquid interface. The respective resistances for surface reaction and diffusion layers were the reciprocals of the surface reaction mass transport rate constant and the diffusion mass transport rate constant. Therefore, it is critical to study the mass transport rate of CO_2 at the vapor–liquid interface in order to describe the mass transport rate of the CO_2 capture process. Useful information for the analysis of the simple and controllable parameters for the process intensification for IL-based CO_2 capture could be provided by analyzing both the resistances of the surface reaction and diffusion layers in the kinetic process of CO_2 capture by ILs and by studying effects of different operation parameters on the respective resistances.

As mentioned before, nonequilibrium thermodynamics could be used to study the entropy generated by an irreversible process (Prigogine, 1945, 1947). The concept of linear nonequilibrium thermodynamics is that when the system is close to equilibrium, the linear relationship can be obtained between the flux and the driving force (Demirel and Sandler, 2004; Lu et al, 2011). Based on our previous linear nonequilibrium thermodynamic studies on the dissolution and crystallization kinetics of potassium inorganic compounds (Ji et al, 2010; Liu et al, 2009; Lu et al, 2011), the nonequilibrium thermodynamic model of CO_2 absorption and desorption kinetics by ILs could be studied. Figure 17 shows the schematic diagram of CO_2 absorption kinetic process by ILs. In our work, the surface reaction mass transport rate and diffusion mass transport rate were described using the linear nonequilibrium thermodynamic theory.

The mass transport rate of surface reaction was described as Eq. (1):

$$J = K_s(f_s - f_i) \quad (1)$$

The mass transport rate of diffusion was described as Eq. (2):

$$J = K_d(f_i - f_b) \quad (2)$$

In Eqs. (1) and (2), J was the CO_2 absorption or desorption rate (or flux) by ILs; K_s and K_d were the surface reaction mass transport rate constant and diffusion mass transport rate constant, respectively; and f_s, f_b, and f_i were the fugacities of CO_2 in ILs at equilibrium, at the vapor–liquid interface, and at the bulk phase of the ILs, respectively.

On the basis of the nonequilibrium thermodynamic-kinetic model of CO_2 absorption by ILs and our previous linear nonequilibrium thermodynamic studies on the dissolution and crystallization kinetics of potassium inorganic compounds (Ji et al, 2010; Liu et al, 2009; Lu et al, 2011), three critical scientific problems of nonequilibrium thermodynamic studies on CO_2 absorption kinetics by ILs were proposed. These problems are shown as follows:

(1) Reliable thermodynamic models and reliable solubility of CO_2 in ILs for describing the thermodynamic properties of systems
(2) Accurate experimental data of CO_2 absorption and desorption mass transport rate (or flux) by ILs
(3) Mass transport rate of CO_2 in ILs at the vapor–liquid interface in the CO_2 absorption and desorption process by ILs

Based on nonequilibrium thermodynamics and reaction–diffusion theory, in our preliminary work (Xie et al, Submitted; Xie et al, 2015), the kinetic studies of CO_2 absorption and desorption in several IL-immobilized sorbents were carried out. It was found that the apparent mass transport rate constants of CO_2 were in three regions with several differences in orders of magnitude for the IL film thicknesses in microscale, 100 nm scale, and 10 nm scale. By analyzing based on a diffusion–reaction theory, it was found that the process was altered from diffusion-controlled to reaction-controlled process by tailoring the IL film thickness from microscale to nanoscale. And this was the inherent mechanism for the dramatic rate enhancement. The extension to SILMs showed that for the membranes with nanoscale IL films, the significant improvement of CO_2 flux can be obtained theoretically. This made it feasible to implement CO_2 separation by ILs with low investment cost. Our work showed a promising future in reducing the investment cost on materials and equipment and shed light on solving the two pending critical problems in CO_2 capture with IL-based technologies.

Therefore, in order to solve the critical problems in IL application for CO_2 separation, a new methodology was established as mentioned in details, which not only explained the abnormal phenomenon in micro- and nanoscale interfaces but also gave guidance on how to enhance the processes.

4.3 Protein Adsorptive Behavior on Heterogenous Surface

As we know, protein adsorption is very important in engineer biosurfaces, such as nanobiointerface (Andre et al, 2009). Protein adsorption onto surfaces is fundamental for the development of nonfouling surfaces

(Vasudevan et al, 2014), protein purification and separation (Gu et al, 2006), and protein immobilization (Jens et al, 2015). Essentially, understanding the mechanism of protein adsorption onto surfaces is critical for the development of a lot of technologies such as biosensor, biomedicine, drug delivery, and implantation (Mou et al, 2014; Xu et al, 2014). Many surfaces have been designed to control protein adsorption in order to meet the various demands for applications. There are many methods for modification of surfaces to control protein adsorption, mainly the physical and chemical modification methods (Bhushan, 2011; Feng et al, 2002). Unfortunately, certain side effects are possible to have occurred when using these methods upon the biosurfaces, such as weakening protein responses and the adhesion strength of microorganisms (Brodbeck et al, 2002; Rosenhahn et al, 2010). In addition, it has been recognized that the functional groups modified on the surfaces are known to be affected by oxidation due to the presence of oxygen and transition metal ions in biochemically relevant solutions (Chen and Jiang, 2008). Surface chemical modifications can also induce conformational changes of proteins (Roach et al, 2005). The physical methods also have fewer side effects on the surrounding environment (Kirschner and Brennan, 2012). Nevertheless, the protein-surface interactions that control the behavior of adsorbed proteins have not been well established (Roach et al, 2006).

If the protein-surface interactions are not well quantified, it will result in different protein responses that restrict our ability to predict and control the behavior of proteins on surfaces accurately. We should focus on not only controlling the protein-surface interactions but also maintaining the biocompatibility of materials. Ti material has caused great interests recently due to its extraordinary mechanical strength, excellent anticorrosion properties, low density, and especially biocompatibility (Guo et al, 2015; Liu et al, 2004; Wei et al, 2011a).

Our group previous work chose the biocompatible TiO_2 as the substrate to study the protein adsorption. First, a series of geometric topographies with different pore sizes were obtained for mesoporous TiO_2. The precursor of $H_2O \cdot 2TiO_2$ was obtained with the hydrolytic and ion exchange (named TiO_2-0) processes. When the sintering temperature of $H_2O \cdot 2TiO_2$ is at 500 °C, about 5–7 nm pores (Fig. 18A and B) were generated. With the sintering temperature increasing, the pore size increased.

The Raman spectra (100–900 cm^{-1}) of hydrated titanate precursor $H_2O \cdot 2TiO_2$ sintered at different temperatures are shown in Fig. 18C. All of the mesoporous TiO_2 and P25 exhibit the characteristic bands at the

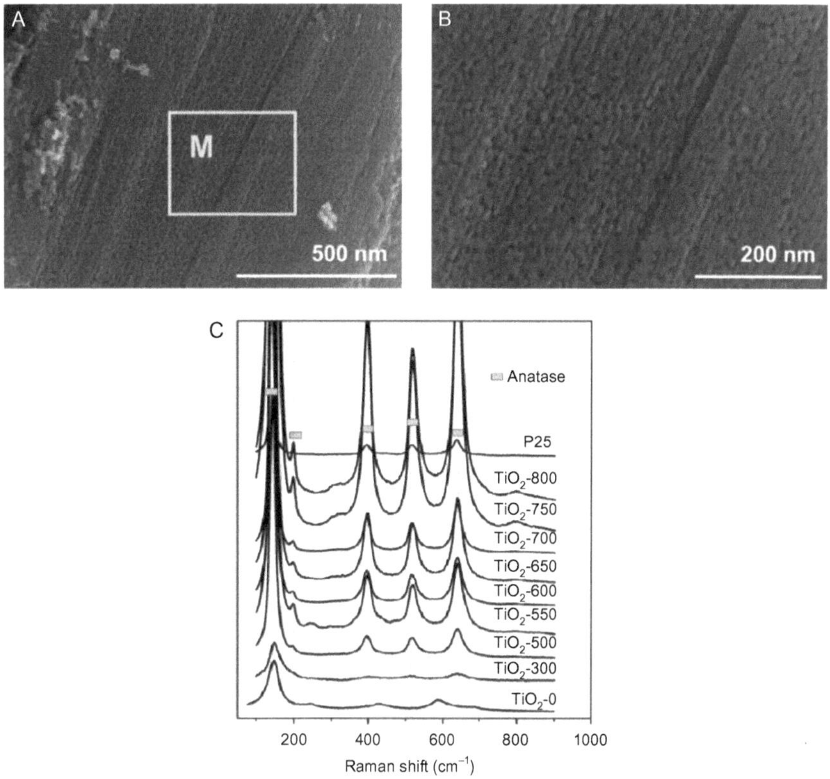

Figure 18 (A) FESEM images for mesoporous TiO_2 with 500 °C sintering procedure, (B) the enlarged image of the structure marked in part M, and (C) Raman spectra for TiO_2 with varying geometric topographies, that is, TiO_2-0, TiO_2-300, TiO_2-500, TiO_2-550, TiO_2-600, TiO_2-650, TiO_2-700, TiO_2-750, TiO_2-800, and P25. The Raman vibrational bands due to the anatase phase are marked with solid squares. *Figure taken from An et al (2014).*

frequency of 144.6, 397.7, 515.7, and 640.8 cm^{-1}, which represents the pure anatase TiO_2 nanocrystals without any other components.

The average pore size of mesoporous TiO_2 ranges from 3.51 to 24.5 nm. Figure 19 shows the equilibrium amount per unit area for BSA, myoglobin, and lysozyme adsorbed onto mesoporous TiO_2 with different pore sizes. In general, with increasing pore size, the equilibrium amount of protein adsorption on mesoporous TiO_2 per unit area (c_0) increases for all three proteins (BSA, myoglobin, and lysozyme). The mesoporous TiO_2 surface adsorbs higher equilibrium amounts of BSA than that of the other two proteins. However, there is no strong evidence that if the pore size of mesoporous TiO_2 is a dominant factor from this trend, no quantitative

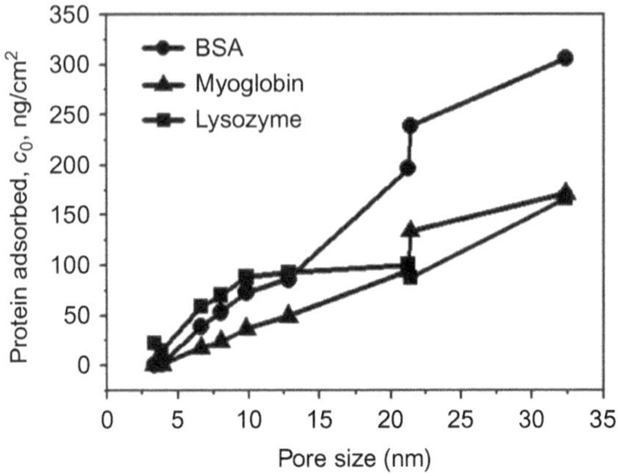

Figure 19 Equilibrium amount of proteins (BSA, myoglobin, and lysozyme) adsorbed on mesoporous TiO_2 with varying pore sizes per unit area. Each measurement of equilibrium amount of protein adsorption was carried out on five identical specimens, and the plotted data represent the average value. Mesoporous TiO_2 with varying pore size of 3.51 ± 0.12, 3.58 ± 0.20, 5.71 ± 1.24, 6.75 ± 1.45, 7.94 ± 1.38, 10.3 ± 1.55, 15.5 ± 1.29, 18.1 ± 1.30, and 24.5 ± 1.54 nm herein corresponding to the specimen TiO_2-0, TiO_2-300, TiO_2-500, TiO_2-550, TiO_2-600, TiO_2-650, TiO_2-700, TiO_2-750, and TiO_2-800, respectively. *Figure taken from An et al (2014).*

information can be regressed from the irregular trends for these three kinds of proteins.

AFM was used to estimate the interactions between proteins and mesoporous TiO_2 in order to further determine the affinity of BSA, myoglobin, and lysozyme onto mesoporous TiO_2 with varying geometric topographies. Figure 20A illustrates a schematic representation of covalently attached protein molecules on an AFM tip coated with gold. The AFM measurements were performed to quantify the adhesive forces between protein-modified tips and mesoporous TiO_2 surfaces with varying geometric topographies. Figure 20B shows representative curves of BSA-immobilized tip force during retraction versus the distance of separation for mesoporous TiO_2 surfaces with varying pore sizes. The adhesive forces increase with the increase in TiO_2 pore sizes.

Therefore, it is necessary to find an effective way to examine and quantify how the protein adsorption is influenced by geometric topographies on mesoporous TiO_2 surfaces with different pore sizes. As shown in Fig. 19, it seemed that the amount of adsorbed protein is related to the pore size of the mesoporous TiO_2. But we cannot get any regular information. The mechanism to govern the protein adsorption on the mesoporous

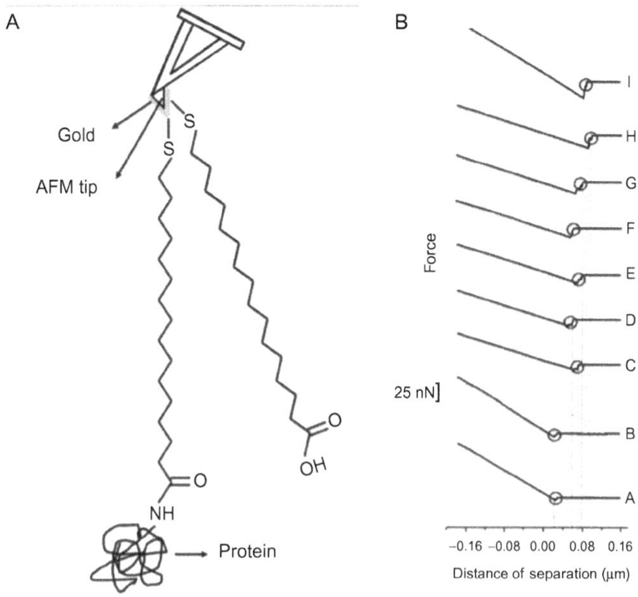

Figure 20 (A) Schematic representation of covalently immobilized protein molecules on the gold-coated AFM cantilever tip. (B) Representative curves of BSA covalently immobilized AFM tip force during retraction versus the distance of separation for mesoporous TiO_2 substrates with pore sizes, A, 3 nm; B, 4 nm; C, 6.5 nm; D, 8.8 nm; E, 13 nm; F, 15 nm; G, 20 nm; H, 21 nm; I, 30 nm (corresponding to the specimen TiO_2-0, TiO_2-300, TiO_2-500, TiO_2-550, TiO_2-600, TiO_2-650, TiO_2-700, TiO_2-750, TiO_2-800, respectively). *Figure taken from An et al (2014).*

TiO_2 surfaces has to be revealed. As exhibited in Fig. 21, the geometric factors such as the protein size α, pore size β, protein contact area S_t on TiO_2 available, and TiO_2 surface area S_T have been considered. It is surprising to find $\frac{b}{a}$ in Eq. (3) could be expressed by $\frac{S_t \cdot \alpha}{S_T \cdot \beta}$, like "volume effect" that could be further elucidated as possibility of proteins contacting with "bound water" layer on mesoporous TiO_2 surface. Here, α is protein maximum size, β is TiO_2 pore size, and $a = \alpha/\beta$ represents the critical value of protein matched with the pore on mesoporous TiO_2. S_t is protein contact area on TiO_2 available ($S_t = (c/M_W) \times N_A \times$ protein cross-sectional area), S_T is TiO_2 BET surface area, and $b = S_t/S_T$ represents the effective utilization coefficient:

$$\frac{b}{a} = \frac{S_t}{S_T} \cdot \frac{\beta}{\alpha} = \frac{S_t \cdot \alpha}{S_T \cdot \beta} \cdot \left(\frac{\beta}{\alpha}\right)^2 \tag{3}$$

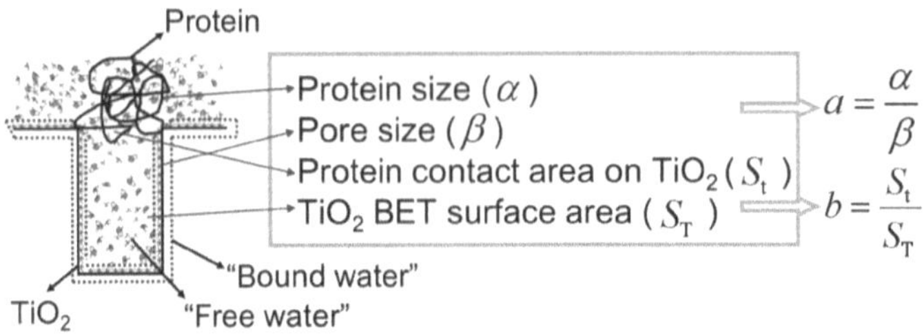

Protein size α = intermediate axis length for protein

c as measured equilibrium amount of protein adsorption

$$S_t = \frac{c}{M_W} \times N_A \times \text{protein cross-sectional area}$$

Protein cross-sectional area = π × short axis radius × long axis radius

S_T as desorption BET surface area

Figure 21 Scheme of the linear treatment process of the equilibrium amount of protein adsorption on mesoporous TiO_2. α as protein maximum size, β as TiO_2 pore size, S_t as protein contact area on TiO_2, S_T as TiO_2 BET surface area, c as measured equilibrium amount of protein adsorption (mg protein/g TiO_2), M_W as protein molecular weight, N_A as Avogadro's constant. As one part of TiO_2 surface, the "bound water" layer has considerably strong interactions with the TiO_2 surface. In our previous study, the existence and strong interactions of the "bound water" layer have been indicated on the TiO_2 surfaces, while the weak interaction with TiO_2 surface corresponds to "free water" layer. *Figure taken from An et al (2014).*

Inspired by the analysis on the adhesive forces (Fig. 20), the experimental data of adsorbed protein on 1 cm^2 TiO_2 surface (the value of c_0) were transformed as one-dimensional interaction by square root processing. Also, the value of $\frac{b}{a}$ was transformed as the possibility of protein contacting with "bound water" layer on TiO_2 surface by cubic root processing. After the transformations, we can see the X-axis suggests the geometrically topographic effects, and the Y-axis suggests the one-dimensional interaction effects in Fig. 22. The line is divided into three sections, which contain the nonfouling, protein separation, and immobilization by distinguishing the amount of adsorption on surfaces. The straight line segments in the part of the nonfouling levels (just the green (gray in the print version) part of the graph) of the mesoporous TiO_2 surfaces are less than 4 ng/cm^2, which demonstrates the surface is resistant to nonspecific protein adsorption. Therefore, according to the gray part, based on the differing characteristics for different proteins, mesoporous TiO_2 surfaces could be extensively used as candidates for an effective separation of protein mixtures. Of course, the mesoporous

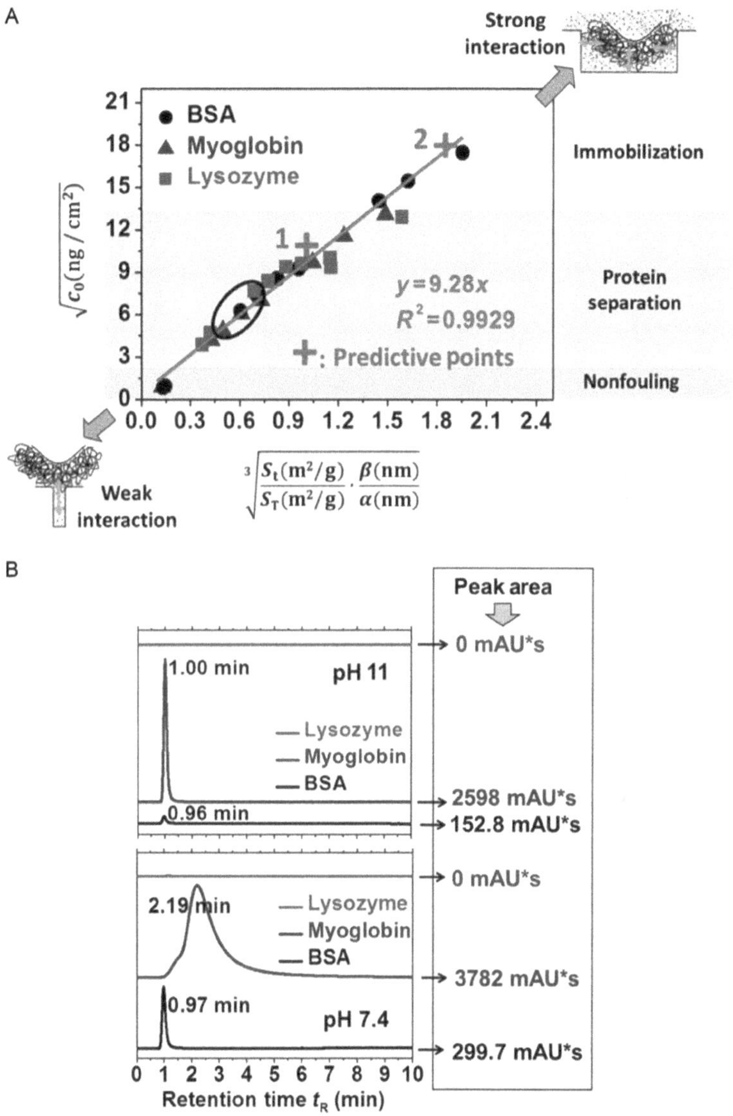

Figure 22 (A) Comparison of one-dimensional interaction translated from the experimental data for protein adsorption on mesoporous TiO_2 with varying geometric topographies. (B) HPLC adsorption profiles of individual protein samples after passing through the imprinted TiO_2-500 (5.71 $\pm$ 1.24 nm pore size) column. *Figure taken from An et al (2014).*

TiO_2 surfaces could also be designed to provide higher binding affinity with proteins, which can mediate cell attachment and enzymatic activity (just the yellow (white in the print version) part in this graph), so it will be convenient to create the surfaces readily immobilizing proteins. Combining all of the proteins (BSA, myoglobin, and lysozyme) adsorbed on the mesoporous TiO_2, the adsorption lines focus on a linear slope with the coefficient of 9.28. The question about the exact effect of geometric topography on the protein adsorption can be resolved by combining the protein size α, TiO_2 pore size β, protein contact area S_t on TiO_2 available, and TiO_2 surface area S_T to obtain the following Eq. (4). The adsorption of protein chicken ovalbumin has confirmed the linear curve (just the pink (gray in the print version) point in the graph):

$$\left[\sqrt{c_0\left(\mathrm{ng/cm^2}\right)} = k \cdot \sqrt[3]{\frac{S_t(\mathrm{m^2/g})}{S_T(\mathrm{m^2/g})} \cdot \frac{\beta(\mathrm{nm})}{\alpha(\mathrm{nm})}}\right] \tag{4}$$

To verify the predictive model for protein adsorption in Fig. 22A, we choose TiO_2-500 with pore size of 5.71 ± 1.24 nm as stationary phase in HPLC (high-performance liquid chromatography) system. The mobile-phase protein BSA after passing through the column has no significant retention time increase at pH 7.4 and pH 11. However, compared with myoglobin and lysozyme, the peak areas for myoglobin are greatest at pH 7.4 and pH 11, whereas no peak has been observed in lysozyme adsorption profiles (Fig. 22B). The HPLC results show lowest adsorption amount and weakest interaction for myoglobin onto TiO_2-500. On the contrary, lysozyme adsorbed on TiO_2-500 with highest amount has the strongest interactions. These findings verify the data in the parts marked as ellipse in Fig. 22A (adsorption amount on TiO_2-500: lysozyme > BSA > myoglobin), also in agreement with the adhesive force results in Fig. 20B.

According to this model, TiO_2 surfaces with well-designed geometric topographies can be constructed efficiently and expected for various applications such as nonfouling, protein separation, and enzyme immobilization. It is critical for us to realize the relationship between mesoporous material surface structures and the transmission performance, and finally, we can achieve the regulation between the structure and performance. This general protein-surface topography interaction behavior could generate a significant impact on the fields of implantation, biosensor, drug delivery, and biomedicine.

5. SURFACE/INTERFACE COARSE-GRAINED SIMULATIONS

All-atom simulation combined with coarse-grained simulation method can be used to study mesoscale structures of surface/interface of catalytic material in different scales. In the all-atom simulation model, the most basic element is a single atom or a group of united atoms; the interactions between them are parameterized or obtained ab initio (Baschnagel et al, 2000). In these simulations, there is integration of the Newton equation based on a femtosecond scale step, in which, combined with coarse-grained simulation, the system can be in nanometer to micron level. United-atom simulation is succeeded in predicting local kinetic properties, such as interfacial phase behavior of nanoparticles and aggregate of nanoparticles (Dong and Zhou, 2013).

In current coarse-grained simulation, the construction of force fields is in many ways. The most widely used method is to map a group of atoms or function groups of united atoms into a joint coarse-grained bead (Baschnagel et al, 2000; Fukunaga et al, 2002). The interactions between the beads are obtained from all-atom or united-atom simulation results or by fitting the experiment results. This specific coarse-grained model by this coarse-grained strategy is based on the key chemical structure information and can be used in prediction quantitatively (or semiquantitatively) for a specific system. In principle, coarse-grained force field and the interaction potential function should be able to reproduce the structure property observed in united-atom simulation and sometimes also require to reproduce the dynamic and thermodynamic properties (Muller-Plathe, 2002).

The dynamic physical and chemical interaction between nanomaterials and biologically active molecules in the interface forms nanobiological interface; it plays an important role in the performance of active biological molecules (Nel et al, 2009). We have researched the TiO_2 surface and its catalytic performance by structure regulation and surface modification on the surface. For nanomaterials with high specific surface area, the effects of nanoscale (similar to biological molecular scale) structure (pore size, roughness, and specific surface area) and surface properties (charge, groups, and polar), nonuniformity on the conformation, and activity of biomolecules in nanobiointerface are very important. There are many factors that affect the amount of immobilized biomolecules and stability. Studies have shown that the geometry of pore size and other parameters have a significant effect on the amount of immobilized biomolecules and stability (Bayne et al, 2013). Other researchers believe the surface charge and chemical modification have greater impact (Vashist et al, 2014). The key is to establish interaction mechanisms of TiO_2 surface nonuniformity and biologically active molecules in nanobiointerface. We have clarified the relationship between adsorption properties of proteins and TiO_2 structures by atomic force microscope characterize (An et al, 2014). We also studied a variety of proteins immobilized on the surface of TiO_2 and found adjusting the pore size and the surface area can significantly increase the amount of immobilization. However, for the presence of affect of surface charge induced by pH, ionic strength, and hydrogen bonding, covalent binding, and dipole interaction, mechanism of interaction between biological molecules and the nonuniform surface of TiO_2 is not yet clear; there is no reliable theoretical basis for designing the structural properties of the medium for immobilization.

Nanoparticles' phase transfer behaviors at the oil–water interface have many in common with lipid bilayer crossing behavior and the Pickering emulsion formation. The phase transfer behavior and interfacial behavior are intuitive indicators for the application potential of nanoparticle materials. Polymer brush modification enables nanoparticles to behave differently in hydrophilic solvent, hydrophobic solvent, and their interface region.

With the rise of bionics, people get inspiration from nature of superhydrophobic surface of the lotus leaf and prepared a large amount of hydrophobic material and apply it. Because the apparent contact angle of droplets on a hydrophobic surface exceeds 90° and the contact angle hysteresis is very small, artificial hydrophobic surface materials have good prospects in industrial catalysis, self-cleaning surface areas, fluid drag, protein adsorption, and micro-/nanoelectromechanical system (NEMS), it has been widespread concerned (Mi and Jiang, 2014; Yao et al, 2011).

In this section, we present the coarse-grained simulation work of protein adsorption, interfacial behavior, and surface wettability.

5.1 Mesoscopic Simulations of Protein Adsorption at Different Surfaces

Understanding the interaction of proteins with material surfaces is of great significance in a broad range of applications, such as biotechnology, medicine, biocatalysis, protein separation and purification, and other fields (Gary, 2004; Jaeger and Eggert, 2004; Rabe et al, 2011). Despite its significance, the mechanism of interfacial protein adsorption is not fully understood because several complex chemical and physical factors affect this complicated process, especially for the surface structure and properties (Vertegel et al, 2004; Shrivastava et al, 2012).

A series of experimental techniques (Chen et al, 2003; Wang et al, 2004; Zhang et al, 2009a, 2009b) and theoretical models (Liu et al, 2013; Peng et al, 2014; Xie et al, 2010; Yu et al, 2014; Yu et al, 2015; Zhao et al, 2015; Zhou et al, 2003, 2004a) have been widely used to explore the interactions of proteins with surfaces; however, it is difficult to directly obtain the relevant atomistic information from these explorations, while molecular simulation (He et al, 2008; Kubiak and Mulheran, 2010; Wei et al, 2011b; Zheng et al, 2005; Zhou et al, 2004b) is an effective approach to complement this.

Because of its limitations in computational power, QM is hard to explore the systems of interfacial protein adsorption because, when used for MD simulations, QM methods are limited to no more than a few tens of atoms in spatial scale and a few picoseconds in timescale. However, it can be widely used to develop parameterization for the all-atom empirical force field methods due to their high level of accuracy (Latour, 2008). Similarly, all-atom molecular dynamics (AAMD) simulations can provide the detailed molecular information at atomistic level but still be restricted to systems containing tens to hundreds of thousands of atoms or undergoing a sub-microsecond timescale. Thus, the sampling of AAMD simulations always is insufficient. Therefore, many interesting phenomena of interfacial protein adsorption (i.e., the whole process of protein adsorption and protein unfolding and the formation of higher-order protein complexes) are beyond the capabilities of AAMD simulations. That is to say, much longer simulations are needed to gain a complete understanding of the adsorption process. Lately, as a complement of AAMD simulations, the mesoscopic methods are becoming more and more popular, which can be employed for the complex adsorption phenomenon in much larger spatial scale and longer timescale.

It should be pointed out that various large-scale simulation methods can be considered as mesoscopic methods, such as Brownian dynamics (BD), dissipative particle dynamics (DPD), mesoscopic coarse-grained Monte Carlo (CGMC), parallel tempering Monte Carlo (PTMC), and mesoscopic coarse-grained molecular dynamics (CGMD). Recently, many efforts have been conducted on the protein adsorption behaviors at different surfaces via mesoscopic coarse-grained methods. In the following, we will pick out some typical examples to introduce these mesoscopic methods in brief.

5.1.1 Brownian Dynamics

The BD simulation is a mesoscopic method in which explicit solvent molecules are replaced instead by a stochastic force (Dunweg and Paul, 1991; Plimpton, 1995; Schneider and Stoll, 1978). Ravichandran and Talbot (2000) simulated the adsorption of lysozyme on a solid surface using a mesoscopic method of BD. Lysozyme was represented by a uniformly charged sphere and interacted with other lysozymes through screened coulombic and double-layer forces. The different steps involved in the simulation are illustrated in Fig. 23. They have investigated the effect of lateral diffusion on the adsorption kinetics, the structure of the adsorbed layer, the bulk protein concentration, and the ionic strength (IS). Their simulation results indicated that the coulombic interaction is effectively shielded at high ionic strengths, leading to increased surface coverage. This effect is quantified with an effective particle radius. They also figured out that the high ionic strength and low bulk concentrations can promote the clustering of the adsorbed molecules. Additionally, they found that lateral protein mobility decreases with increasing surface coverage. Meanwhile, these observed trends are consistent with previous theoretical and experimental studies. Subsequently, Ravichandran et al (2001) investigated the initial stages of lysozyme adsorption at a charged solid interface at neutral

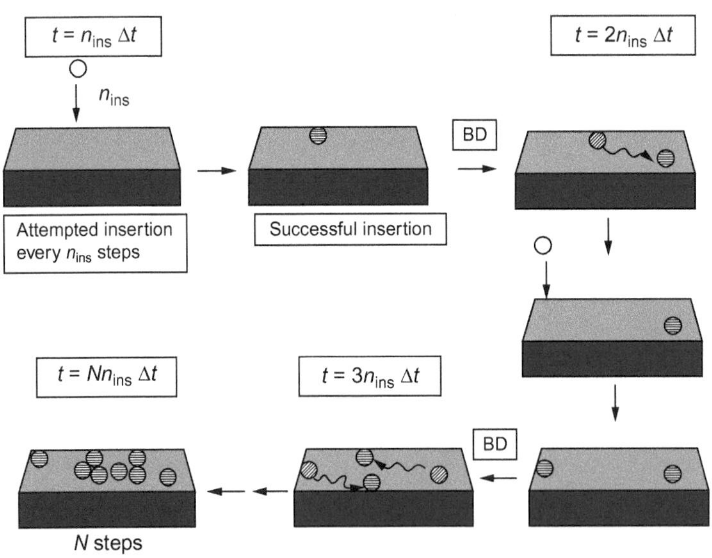

Figure 23 Schematic diagram of the different steps involved in the BD simulation. *Figure taken from Ravichandran and Talbot (2000).*

pH using BD simulations. The lysozyme was modeled at the atomistic level, and the adsorption surface was represented by a planar array of positively charged sites (see Fig. 24). They found that the lysozyme, which has a net positive charge (+8 e) at the neutral pH, could adsorb onto the positively charged surface both at low slat concentration (I=0.1 M) and at high slat concentration (I=0.3 M). The orientational distribution of the adsorbed protein was more nonuniform at I=0.3 M, while there is considerable preference for ASP119 and ARG128. Thus, they concluded that even a single residue still can drive the protein adsorption. The electric field distribution around the protein is the best predictor of the adsorption behavior.

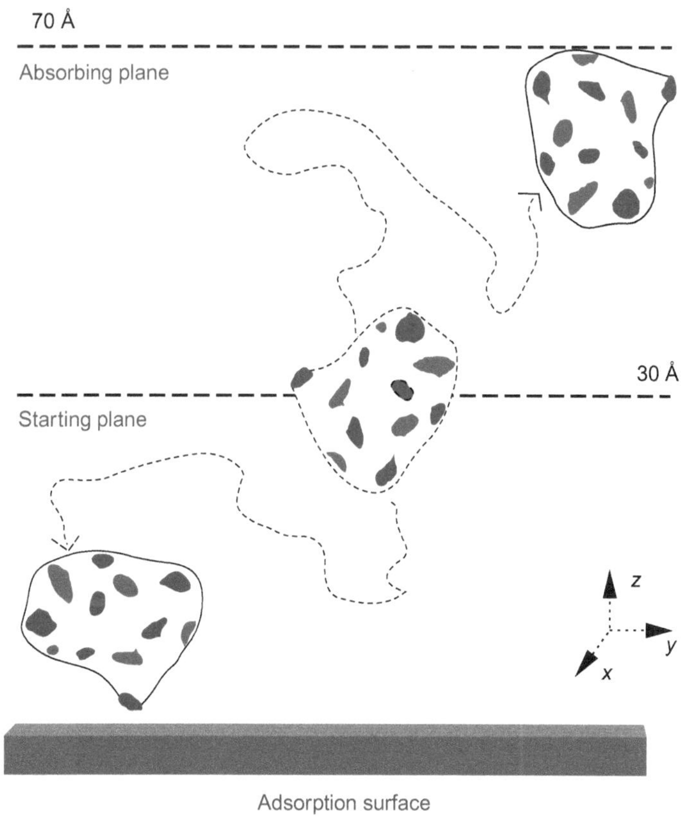

Figure 24 Schematic diagram showing the details of the BD simulation method. *Figure taken from Ravichandran et al (2001).*

5.1.2 Dissipative Particle Dynamics

Dissipative particle dynamics (DPD), first introduced by Hoogerbrugge and Koelman and then revised by Espanol and Warren (Espanol and Warren, 1995; Groot and Warren, 1997; Hoogerbrugge and Koelman, 1992), is one of the most common coarse-grained methods. Patterson et al (2011) employed the mesoscopic simulation method of DPD to investigate the adsorption behaviors of model proteins with different shapes and sizes on different hydrophobic surfaces. As shown in Fig. 25, the model proteins are modeled as semiflexible rodlike objects consisting of a bundle of linear chains formed by coarse-grained beads that represent lumps of proteins. The distinct differences in the adsorption (and desorption) kinetics of the two types of model proteins were presented. In systems containing small proteins, they observed rapid diffusion and adsorption kinetics, as well as frequent desorption events. However, the protein diffusion is rather slow in systems containing large proteins. Meanwhile, once these large proteins adsorb on the hydrophobic surfaces, desorption events become rather rare.

5.1.3 Mesoscopic Coarse-Grained Monte Carlo

Carlsson et al (2004) investigated several key factors of lysozyme adsorption onto negatively charged surfaces using Monte Carlo simulations. As displayed in Fig. 26, lysozyme was represented as a hard sphere with embedded positive and negative surface charges depending on the solution pH. The mica surface was selected as the negatively charged surface that was described by a hard wall with embedded charges. Most of these simulation findings are consistent with previous experimental results. The lysozyme adsorption on

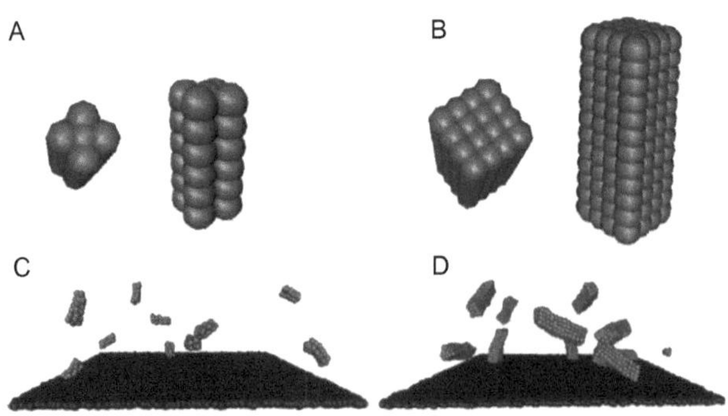

Figure 25 Schematic of model proteins used in this study: (A) initial structure of small elongated and (B) large elongated proteins and examples of typical structures of proteins before adsorption (C) and after adsorption (D). *Figure taken from Patterson et al (2011).*

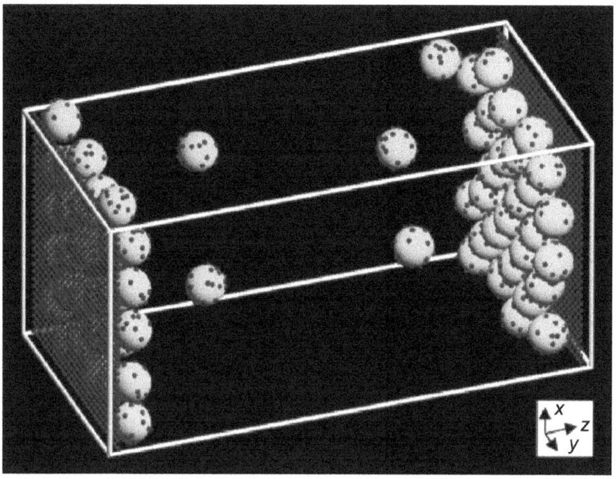

Figure 26 Illustration of the simulation box including the external coordinate frame *xyz* and a typical configuration taken from the end of the simulation. *Figure taken from Carlsson et al (2004).*

the mica surface was favored by high protein concentration, high protein net charge, low ionic strength, and high surface charge density. In addition, as the amount of protein at the surface increases, the lateral structure within the adsorbed layer becomes more ordered. Furthermore, the adsorbed lysozyme molecules showed a preferential orientation at the mica surface. However, it was not sensitive on the adsorbed amount. Meanwhile, a strong preferential orientation may prevail even at weak adsorption. That is to say, their simplified model for interfacial lysozyme adsorption can be used to study many important features of lysozyme adsorption at a hydrophilic and negatively charged surface; however, it cannot clearly figure out the reorientation and the conformational changes of lysozyme.

Antibodies have found many applications in clinical medicine and biotechnology. Moreover, antibodies are always required to immobilize on a solid interface. Meanwhile, the natural activity of adsorbed antibodies depends on its orientation on the solid interface. Therefore, Zhou et al (2003, 2004a) focused on the orientation of adsorbed antibodies on charged surfaces by performing multiscale simulations based on colloidal, united-residue, and all-atom models.

Zhou et al (2003) developed a new residue-based protein-surface interaction potential model to explore the adsorption and orientation of two antibodies, IgG1 and IgG2a. The Monte Carlo simulation results showed that when electrostatic interactions dominate, there are preferred orientations for these two antibodies on both positively and negatively charged surfaces. In addition, IgG2a has more possible orientations than IgG1 because of a smaller dipole moment. Furthermore, IgG1 and IgG2a also exhibit multiple orientations when vdW interactions dominate. As a result, vdW and electrostatic interactions codetermine the orientation of adsorbed antibodies. Interestingly, the "end-on" orientation, which is the preferred orientation of IgG1 on a positively charged surface, is well suited for biosensor applications.

Also, Zhou et al (2004a) focused on the orientation of adsorbed antibodies on charged surfaces by performing Monte Carlo simulations based on both colloidal and all-atom models. In the colloidal model (see Fig. 27), the 12 domains of an antibody molecule were represented by 12 connected beads. The all-atom structure model of antibody was taken from the protein data bank. Simulation results indicated that at high surface

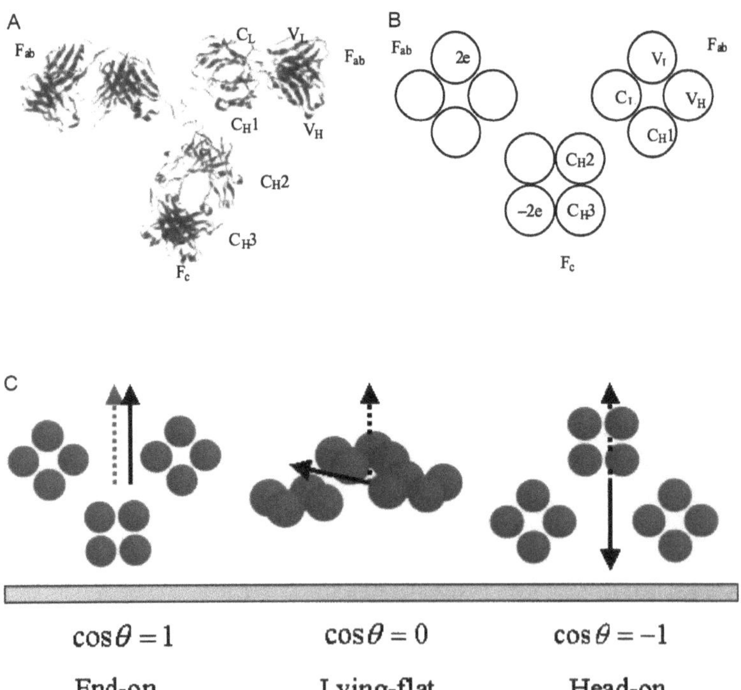

Figure 27 Schematic diagram of IgG: (A) ribbon-representation antibody, (B) 12-bead model antibody, (C) definition of the orientation of adsorbed antibody. *Figure taken from Zhou et al (2004a).*

charge density and low solution ionic strength where electrostatic interactions dominate, the preferred orientation of antibody is the desired "end-on" orientation on positively charged surfaces and undesired "head-on" orientation on negatively charged surfaces, while the antibody tends to have "lying-flat" orientation on surfaces at low surface charge density and high solution ionic strength where vdW interactions dominate. As a result, the antibody orientation on surfaces depends on the compromise between electrostatic and vdW interactions. Especially, it should be pointed out that when electrostatic interactions dominate, the dipole moment of an antibody is a key factor for the orientation of the adsorbed antibody. Based on both the 12-bead and the all-atom models, the charge-driven protein orientation hypothesis was verified. Meanwhile, it was further confirmed by experiments (Chen et al, 2003; Wang et al, 2004) and multiscale simulations (Liu et al, 2013).

5.1.4 Parallel Tempering Monte Carlo

Xie et al (2010) developed the PTMC algorithm to probe the primary orientation of adsorbed proteins on charged surfaces, based on the coarse-grained united-residue model (Zhou et al, 2003). As in PTMC, both lowest-energy orientation and whole orientation distribution could be figured out just in a single simulation. Meanwhile, it can break the timescale limit and more efficiently acquire the desired orientation. PTMC simulation results also indicated that the orientation of adsorbed lysozyme is codetermined by electrostatic interactions and vdW interactions. When driven by dominant electrostatic interactions, lysozyme tends to be adsorbed on negatively charged surfaces with the side-on orientation for which the active site of lysozyme faces sideways. As the contribution from vdW interactions gradually dominates, the back-on orientation becomes the preferred one. It was also suggested that despite its net positive charge, lysozyme could be adsorbed on positively charged surfaces with both "back-on" and "end-on" orientations due to the screening effect from ions in solution and the nonuniform charge distribution over lysozyme surface. However, owing to the too strong repulsive electrostatic interactions, lysozyme could not be adsorbed on positively charged surfaces at low IS. These simulation results agree well with experimental data. The PTMC simulation method could provide an efficient way to point out the preferred orientation of adsorbed proteins on charged surfaces for biosensor and biomaterial applications.

5.1.5 Mesoscopic CGMD Based on MARTINI Force Field

Hung et al (2011) investigated the interaction of the simple, globular protein cytochrome C (Cyt C) with monolayer-protected metal nanoparticle (MPMN) surfaces using experimental protein assays and mesoscopic CGMD simulations (see Fig. 28) based on the MARTINI force field (Marrink et al, 2007; Monticelli et al, 2008), which allows simulations to be run on longer timescales and larger spatial scales and still can preserve much more basic structural detail of proteins than BD, DPD, and CGMD based on the united-residue model. Protein–surface adsorption enthalpies indicated a monotonic increase in adsorption enthalpy with respect to MPMN surface polarity, which was calculated according to the results of mesoscopic simulations. Meanwhile, reveled by experimental assays, the adsorption of Cyt C generally increased with increasing surface polar ligand content. Thus, simulation results are in qualitative agreement with experimental results. In addition, the mesoscopic simulations showed that the lysine is especially important for Cyt C interactions with MPMN surfaces owing to the amphipathic character of the lysine side chain. Furthermore, they also considered that arginine may display a similar capacity for "dual binding." However, arginine is not present on the wild-type Cyt C surface; thus, there is no direct observation that could be found from this work. Therefore, Hung et al (2013) performed another AAMD simulation of lysozyme on nanostructured surfaces to further confirm that amphiphilic amino acids play a special role in the adsorption of proteins on surfaces. In

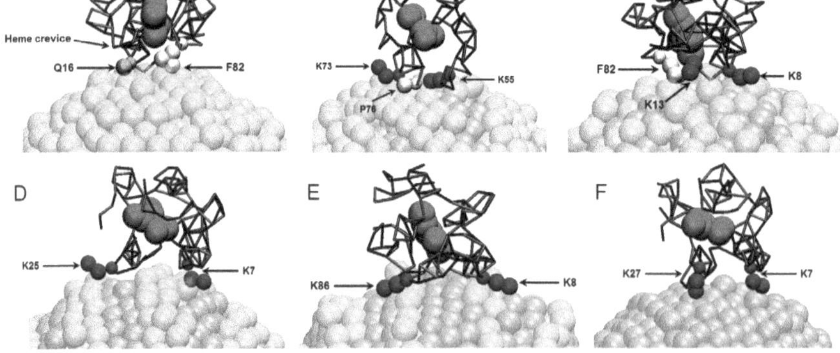

Figure 28 Lowest-energy-binding conformation for the Cyt C-MPMN systems. (A) OT Homo, (B) 1:2 MH:OT, (C) 1:1 MH:OT "thin", (D) 1:1 MH:OT "thick", (E) 2:1 MH:OT, and (F) MH Homo MPMNs. *Figure taken from Hung et al (2011).*

this work (see Fig. 29), another amphipathic amino acid, arginine, enabled the lysozyme to form close contacts with both polar and nonpolar surface ligands simultaneously, and arginine is also capable of forming close contacts with homogeneous hydrophobic and hydrophilic ligand surfaces, while

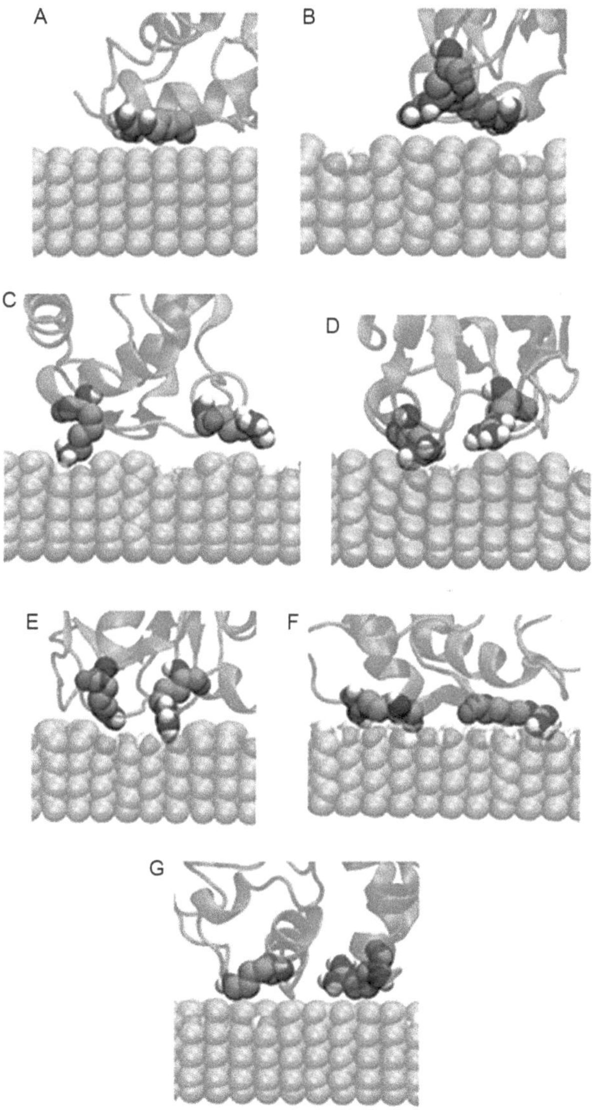

Figure 29 Arginine and lysine side-chain orientations at (A) pure OT, (B) 2:1, (C) 1:1 "thin," (D) 1:1 "thick," (E) 1:2 OT:MH surfaces, (F) pure MH, and (G) "random" surfaces. *Figure taken from Hung et al (2013).*

Hung et al (2011) had previously showed that the selective adsorption behavior of Cyt C on nanostructured surfaces is due to the amphiphilic amino acid lysine. The simulation results also figured out other amphiphilic amino acids (i.e., tyrosine and tryptophan) that interacted with surfaces via water-mediated contacts. As we know, nanomaterials can be designed to selectively interact with different proteins. Now, with these findings of importance of amphiphilic residues, proteins can also be engineered to specifically interact with nanopatterned surfaces by targeted incorporation of synthetic amino acids possessing multiple affinities just like amphiphilic amino acids.

5.1.6 Mesoscopic CGMD Based on BMW-MARTINI Force Field

As we know, MARTINI water beads, just as many other CG water models, are blind to electrostatic fields and polarization effects (Yesylevskyy et al, 2010). Therefore, in order to deal with electrostatic interactions more realistically, a modified MARTINI force field, called BMW-MARTINI, was developed (Wu et al, 2010, 2011), which could be applied to predict the properties at the interfaces and two orders of magnitude more efficient than atomistic models (Wu et al, 2011). For interfacial protein adsorption, the oriented immobilization and the antifouling surfaces are two major concerns. However, the mechanism of protein adsorption is still far from being completely understood because various factors may affect this complicated process. Yu et al (2014) adopted coarse-gained simulation method based on BMW-MARTINI force field to investigate the interfacial behaviors of protein adsorption on different surfaces. The whole adsorption process and adsorption mechanisms of lysozyme on different (hydrophobic, neutral hydrophilic, zwitterionic, negatively charged, and positively charged) surfaces were studied (see Fig. 30). Simulation results indicate that (i) the conformation change of lysozyme on the hydrophobic surface is bigger than any other studied surfaces; (ii) lysozyme adsorbs on the hydrophobic surface and hydrophilic surface with different orientations; (iii) neutral hydrophilic surface can induce the adsorption of lysozyme, while the nonspecific protein adsorption can be resisted by zwitterionic surface; (iv) when the solution ionic strength is low, lysozyme can anchor on the negatively charged surface easily, but cannot adsorb on the positively charged surface; (v) when the solution ionic strength is high, the positively charged lysozyme can also adsorb on the like-charged surface; (vi) the major positive potential center of lysozyme, especially the residue ARG128, plays a vital role in leading the adsorption of lysozyme on charged surfaces; and (vii) when ionic strength is

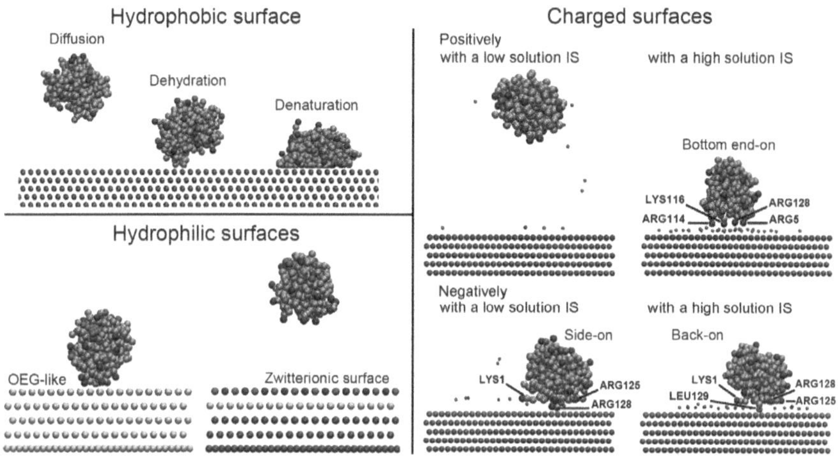

Figure 30 Adsorption states of lysozyme on different surfaces. *Figure taken from Yu et al (2014).*

high, a counterion layer is formed above the positively charged surface, which is the key factor why lysozyme can adsorb on like-charged surface.

Hydrophobic charge induction chromatography (HCIC) is a new type of mixed-mode chromatography, in which the hydrophobic attraction controls protein adsorption, whereas the electrostatic repulsions regulate protein desorption by adjusting pH. To deeply understand the interfacial mechanisms of HCIC at the molecular level, Yu et al (2015) performed mesoscopic coarse-grained simulations to investigate the adsorption/desorption behaviors of lysozyme on 4-mercaptoethylpyridine. Simulation results indicate that (i) lysozyme can be adsorbed mainly with "top end-on" and "bottom end-on" orientation on hydrophobic surfaces, dominated by the two hydrophobic regions located at both ends of lysozyme's long axis. Elution from the "top end-on" orientation is more difficult than that from the "bottom end-on" orientation; (ii) a higher ligand density can get a faster adsorption rate and stronger adsorption. Interestingly, the effect of ligand density on the desorption is mainly determined by the distribution probability of the positively charged groups of ligands; (iii) a higher ionic strength can lead to a wider orientation distribution, a stronger adsorption, and a lower elution rate.

Besides, Liu et al (2015) combined three different simulation methods, including PTMC, AAMD, and mesoscopic CGMD simulations, to study the orientations and conformation of ribonuclease A (RNase A) adsorbed on oppositely charged self-assembled monolayers (SAMs). Simulation results indicated that RNase A is adsorbed on oppositely charged surfaces with opposite orientations. The active site of RNase A is oriented toward the surface when it adsorbs on a negatively charged surface, while for RNase A adsorbed on a positively charged surface, the active site is oriented toward the solution. Negatively charged surfaces could be used for RNase A removal since the catalytic active site is blocked. To bring the enzymatic catalysis of RNase A into play, positively charged surfaces can be used to control the orientation of RNase A with the active site accessible. The dipole moment and side chains of RNase A on both surfaces are slightly changed, whereas the backbone structure of RNase A is well preserved. That is to say, RNase A preserves its native conformation during the adsorption process (Fig. 31).

5.2 Mesoscopic Simulations on the Interfacial Behaviors

Understanding the interfacial behaviors of polymer nanomaterials at liquid interface has received much attention due to their excellent properties and potential applications in areas such as controlled drug delivery, biocompatible materials, Pickering emulsion, biosensing, and other fields (Shan and Tenhu, 2007). We mainly focus on the interfacial behaviors of three nanomaterials at liquid interface, which consists of polymer, nanoparticles, and polymer brush-modified nanoparticles at liquid interface. Numerous relevant experiments and simulation works have been published; however, it is still difficult to interpret the experimental observation at the atomistic level; therefore, molecular simulation is an effective complemental approach. Due to the limitation at longer timescale and bigger space scale for all-atom molecular simulations, a lot of works based on coarse-grained molecular simulations have been reported.

5.2.1 Polymers at Liquid Interface

In the past decades, the interfacial behaviors of polymers at liquid–liquid interface have been widely explored due to its increasingly technological importance in recent years. For example, many self-assembly processes of macromolecules and some chemical reactions usually occur at such interfaces, polymer-based drug delivery vehicles, and other industrial applications (Hu et al, 2012).

Dendrimers, a class of macromolecules featured with uniform structure and easy surface functionality, have been widely used as anticancer drugs and gene delivery vectors. To investigate the stability of amphiphilic dendrimers at the liquid–liquid interface, the free energy of adsorption of model

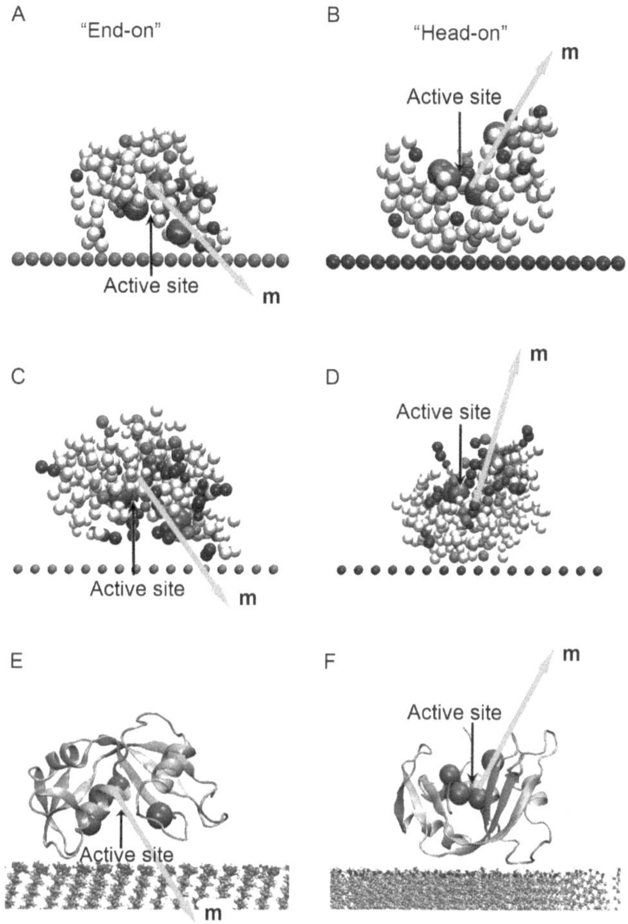

Figure 31 Preferred configurations of protein orientations on COOH-SAM (A), (C), (E) and NH_2-SAM (B), (D), (F) from PTMC (above), mesoscopic CGMD (middle), and AAMD (below) simulations. *Figure taken from Liu et al (2015).*

dendrimers with monomers of different chemical affinities and polymers with linear and star architecture at fluid interface was calculated by MD simulations (Cheung and Carbone, 2013; Taddese et al, 2015). Simulation results showed that amphiphilic dendrimers can stably locate at the interface, and Janus dendrimer does not show a higher interfacial stability (Fig. 32). Besides, it was found that changing the topology of polymers had a minor effect on the desorption-free energy but can be seen for polymers with relatively high molecular weight at fluid between two good solvents.

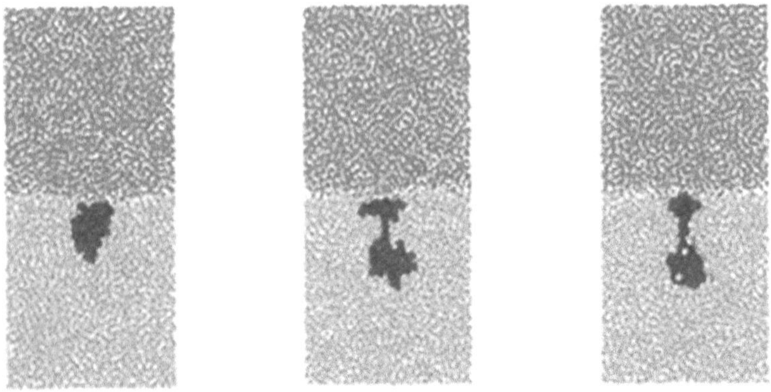

Figure 32 Simulation snapshots of different dendrimers near liquid–liquid interface. *Figure taken from Cheung and Carbone (2013).*

Although most polymers tend to accumulate at the fluid interface, reports involving the transfer of polymeric micelles (micellar shuttle) between two immiscible phases have been published. Poly(*N*-isopropylacrylamide) (PNIPAM), a thermally responsive polymer, is insoluble and can undergo a conformation change above its lower critical solution temperature of 32 °C. The thermoreversible micellization–demicellization process and micellar shuttle of PNIPAM–PEO diblock copolymer at a water-IL interface were investigated by dissipative particle dynamics (DPD) simulations (Soto-Figueroa et al, 2012). Simulation results confirm that the phase transfer behavior of polymeric micelles is controlled by the temperature effect that changes the diblock copolymer from hydrophilic to hydrophobic (as shown in Fig. 33).

It has been reported that a series of polymer blends in aqueous solution could self-assemble into phase-separated structures such as core–shell or Janus-type polymer nanoparticles (Motoyoshi et al, 2010). Moreover, Guo et al (2013) performed DPD simulations to systematically investigate the effects of hydrophobicity and compatibility, and both play important roles in controlling the self-assembled structures of polymer blends in aqueous solution. Most importantly, the temperature-dependent core–shell to Janus structure transition of thermosensitive polymer blends was observed by DPD simulations for the first time.

Besides, it is of great importance to understand the interfacial behaviors of polymers at liquid interface for industrial applications, such as emulsion formation and extraction process. It is known that polymers can be used

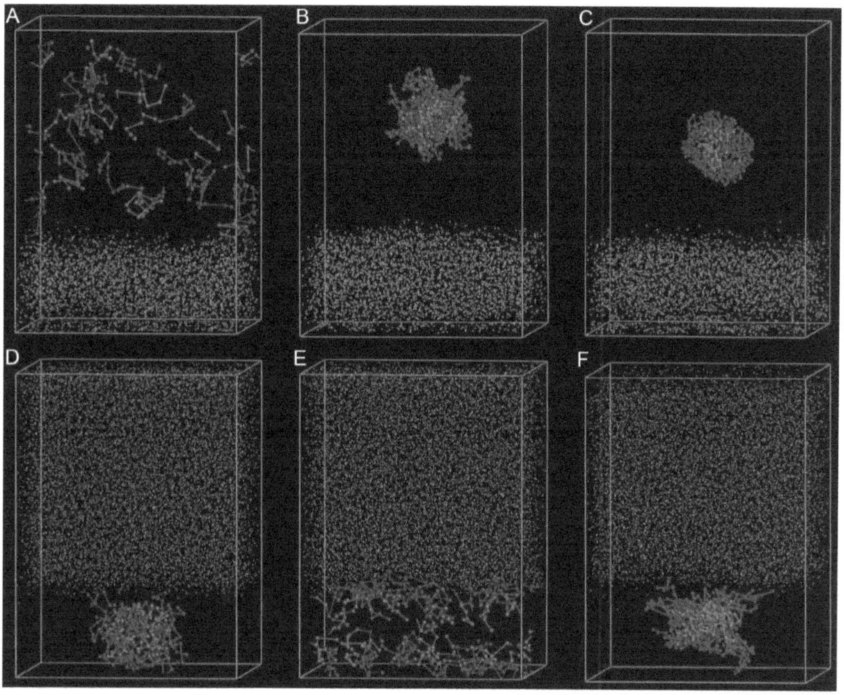

Figure 33 Simulation snapshots of the phase transfer process of the PNIPAM–PEO micellar shuttle in a water–ionic liquid system at different temperatures. *Figure taken from Cesar et al (2012).*

as emulsifier for emulsion formation and stability. Combining experimental methods with DPD simulations, the microstructures of an emulsion with alternating styrene–maleic acid copolymers as emulsifier were studied (Fig. 34); it is found that the concentration and the stiffness of polymer emulsifier can affect the mesostructures of the emulsified oil droplets to a large extent, and the simulation results can be well used to interpret the experiment results (Lin et al, 2012). The properties of water/benzene/caprolactam (CPL) ternary system in the absence or presence of nonionic surfactants were studied by DPD simulation (Shi et al, 2015); the results demonstrated that the benzene/water interfacial tension with CPL can be predicted quantitatively by molecular simulations, and the addition of surfactants can drive CPL from water/benzene interface to water phase; this work may provide a guideline for the design of more efficient industrial extraction process for recycling CPL.

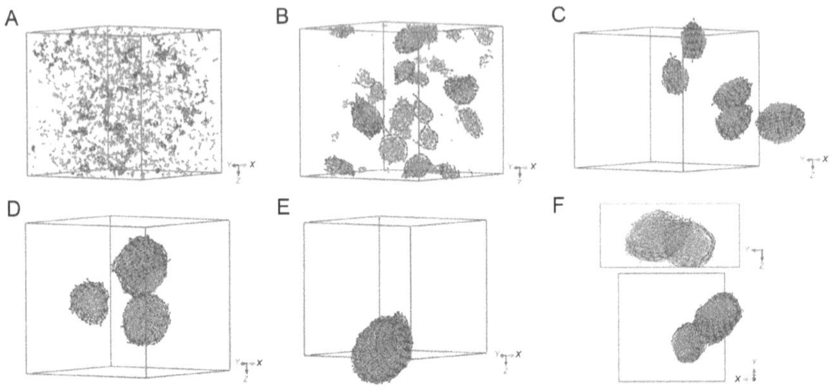

Figure 34 Simulation snapshots of the phase transfer process of the PNIPAM–PEO micellar shuttle in a water–ionicliquid system. *Figure taken from Lin et al (2012).*

5.2.2 Nanoparticles at Liquid Interface

Recently, nanoparticles with different surface chemistries have been widely used in Pickering emulsion field. A series of works to study the interfacial behaviors of nanoparticles adsorbed at the water–oil interface have been conducted by using DPD simulations (Luu and Striolo, 2014; Luu et al, 2013a, 2013b). The structural and dynamic properties of spherical homogenous and Janus nanoparticles with different surface compositions accumulated at the water–oil interface were investigated (Fig. 35). It is found that the surface density plays an important role in the aggregation state of nanoparticles.

However, the shape of nanoparticles also plays an important role in the interfacial properties. It was found that the orientation of ellipsoidal Janus nanoparticles at oil–water interface was influenced by two nanoparticle aspect ratios, namely, the amount of polar with respect to nonpolar groups and the interactions between the nanoparticle surface groups and aqueous and nonaqueous solvents. Most importantly, when the nanoparticles lay with their longer axis parallel to the water–oil interface, the interface tension will be reduced with nanoparticles with sufficiently high surface coverage. Besides, the equilibrium behavior of ellipsoidal Janus nanoparticles adsorbed at spherical water–oil interface was also investigated by DPD simulations. It is found that several phenomena that happened on planar water–oil interface were not observed, demonstrating that the curvature of interface can also influence the state of nanoparticles adsorbed at water–oil interface (Fig. 36).

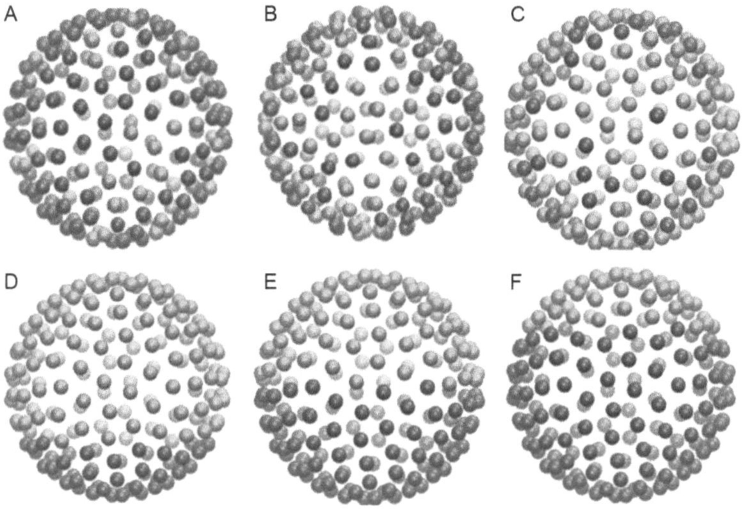

Figure 35 Schematic representation of the nanoparticles (NPs) studied in this work. Top panel represents homogeneous NPs, while bottom panel represents Janus NPs. Purple (gray in the print version) and green (light gray in the print version) spheres are non-polar and polar beads, respectively. *Figure taken from Luu et al (2013a, 2013b).*

5.2.3 Polymer Brush-Modified Nanoparticles at Liquid Interface

Polymer brush-modified inorganic nanoparticles, featured with the stability of inorganic nanoparticles and the versatility of polymer brushes, have received much attention due to its unique properties. It has been indicated that many factors could cause differences in physical or chemical properties of brushes, such as grafting density, composition, chain length, asymmetry, and solvent quality (Wang and Mueller, 2009); therefore, they could also influence the properties of polymer brush-modified nanoparticles. To further investigate the interfacial behaviors of polymer brush-modified nanoparticles at liquid interface, coarse-grained MD simulations were performed to study the effects of polymer block sequence and ratio on the interfacial and phase transfer behaviors of ABA-type triblock polymer brush-modified gold nanoparticles (AuNPs) at oil–water interface (Dong et al, 2014). Simulation results showed that most of AuNPs modified by ABA polymer brushes made up with hydrophobic and weak hydrophilic blocks can break the energy barrier of the interface region and forced the particle transferred from the water phase to the oil phase (as shown in Fig. 37). When using hydrophobic blocks as head and tail ends, in most cases, the particle can wholly move into the oil phase, while with weak

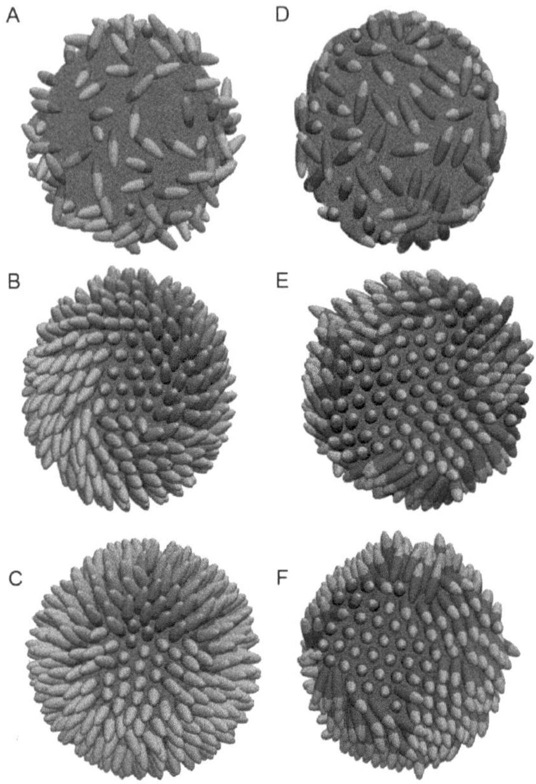

Figure 36 Representative simulation snapshots of ellipsoidal Janus NPs adsorbed at water–oil interface with different curvatures. Panels (A) and (D) are for flat interface; (B) and (E) for spherical interface with a large diameter, while (C) and (F) for spherical interface with a small diameter. *Figure taken from Luu and Striolo (2014).*

Figure 37 The transfer process of ABA-type polymer brush-modified AuNP at oil–water interface. *Figure taken from Dong et al (2014).*

hydrophilic blocks as head and tail ends, the particle would be trapped by the vicinity of interface region because of the affinity between weak hydrophilic and water phase. However, for AuNPs modified by strong hydrophilic blocks, due to the affinity between block and their own good solvent, the particle also was trapped at the interface region.

Polymer brush-modified nanoparticles in solutions are also worth exploring. Serials of PS–PEO block copolymers and mixed PS/PEO polymer brush-grafted gold nanoparticles were designed to investigate the solvent-responsive behaviors of amphiphilic gold nanoparticles (Dong and Zhou, 2013). The simulation results show that typical core–shell, Janus-type, buckle-like, ring-like, jellyfish-like, and octopus-like morphologies were formed in five solvents with different polarities (Fig. 38).

By using coarse-grained molecular simulations, AB diblock copolymer-grafted particles (DBCGPs) as compatibilizers in an immiscible blend of A and B homopolymers were studied (Estridge and Jayaraman, 2015). The results show that the fraction of the A block in the graft can tune the location of the particles within the blend. The desorption energy to leave the interface and the drop in interfacial tension are larger for the particles than ungrafted diblock copolymers and can be used as compatibilizers (Fig. 39). The conformations and effective interactions of polymer-coated nanoparticles adsorbed at a model liquid–liquid interface were also investigated. The polymer shells strongly deform at the interface. The effective interaction of nanoparticles at the liquid-liquid interface differs quantitatively from the bulk and can be significantly affected by the length of the polymer chains and by the solvent quality.

5.3 Mesoscopic Simulations on the Wetting Behaviors

The hydrophobicity of material surfaces is a common interfacial property, which is of great importance in many related fields, such as self-cleaning surface, biocompatible medical materials, industrial catalysis, and protein adsorption (Mi and Jiang, 2014; Yao et al, 2011). Numerous studies have shown that the wettability of solid surface is influenced by two factors, namely, chemical property and topological structure. Quan et al (2014) adopted CGMD simulations to study the effect of surface topology of hydrophobic surfaces on their wetting states. Simulation results show that the increase in surface roughness has little effect on the hydrophobicity of a hydrophobic surface but does have effect on its wetting states. The Wenzel–Cassie wetting transition is related to the ratio of pillar spacing to pillar height (d/h), which is mainly dependent on the vdW interaction (Fig. 40).

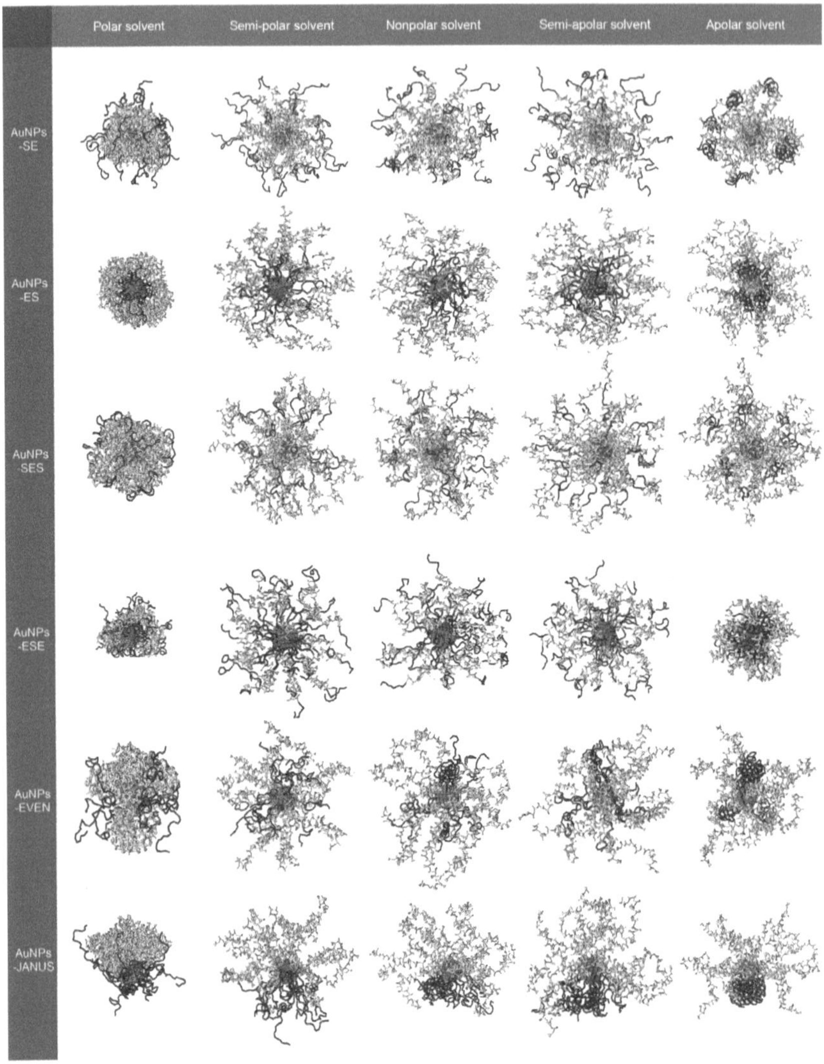

Figure 38 Equilibrium morphologies of six polymer brush-grafted AuNP systems in five solvents. *Figure taken from Dong and Zhou (2013).*

However, understanding the wetting behaviors of oil nanodroplets on surfaces with different patternings was also meaningful. Jabbarzadeh (2013) performed MD simulations to investigate the effect of nanopatterning on the wettability of a surface by oil droplets. The simulation results show that the surface becomes more oleophobic as the lattice constant of surface is increased. Then the contact angle of oil droplets on roughened surface

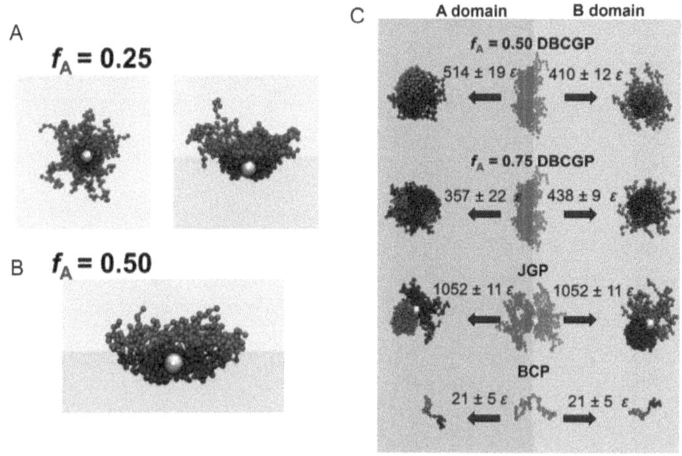

Figure 39 Representative simulation snapshots of (A) $f_A = 0.25$ DBCGP in B domain and at interface; (B) $f_A = 0.50$ DBCGP at the interface. (C) Energetic penalty for leaving the interface. *Figure taken from Estridge and Jayaraman (2015).*

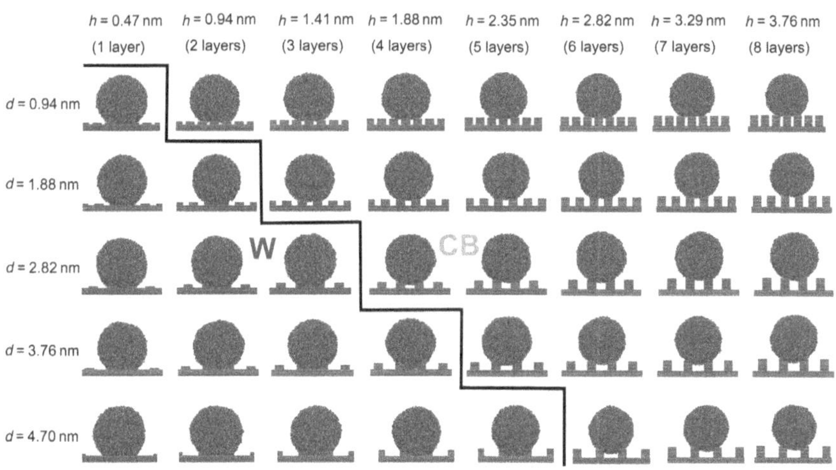

Figure 40 Snapshots of the final conformation of water bead on pillar surface. *Figure taken from Quan et al (2014).*

was calculated as a function of surface roughness, and the results were compared with theoretical predictions based on Wenzel and Cassie-Baxter models (Fig. 41).

The wetting behaviors of surface can also be tuned by chemical modifications that generally involve the introduction of active functional groups

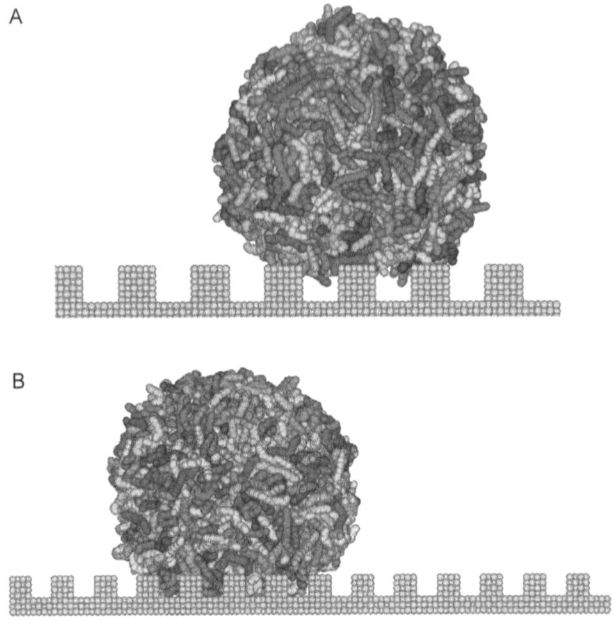

Figure 41 Snapshots of the final conformation of oil bead on pillar surface; (A) Cassie-Baxter and (B) Wenzel states. *Figure taken from Jabbarzadeh (2013).*

to the surface. Experiment and dissipative particle dynamics simulations to explore the wetting property changes of water on the surface of polydimethylsiloxane (PDMS) modified with different amounts of acrylic acid (AA) have been performed (Ramirez et al, 2015). The results showed that the surface transferred from hydrophobic to hydrophilic with the increase in AA percentage (Fig. 42) and a good agreement between experimental contact angles and those from molecular simulations.

In addition, a lot of experiments and theoretical works indicate that surfactant molecules dissolved in the droplets can reduce the contact angles and enhance the spreading of aqueous drops on nonpolar surfaces; this intriguing ability of certain surfactant molecules can be used in many applications such as enhanced oil recovery, coating technologies, drug delivery, pesticide science, and other fields. In order to study the wetting properties of graphitic surfaces by aqueous solutions that contain different concentrations of commercially available nonionic surfactants with long hydrophilic chains, linear or T-shaped, systematically molecular simulations were performed (Sergi et al, 2012). It was found that the wetting behaviors of graphitic surface are mainly influenced by many factors, such as the length and apolarity of

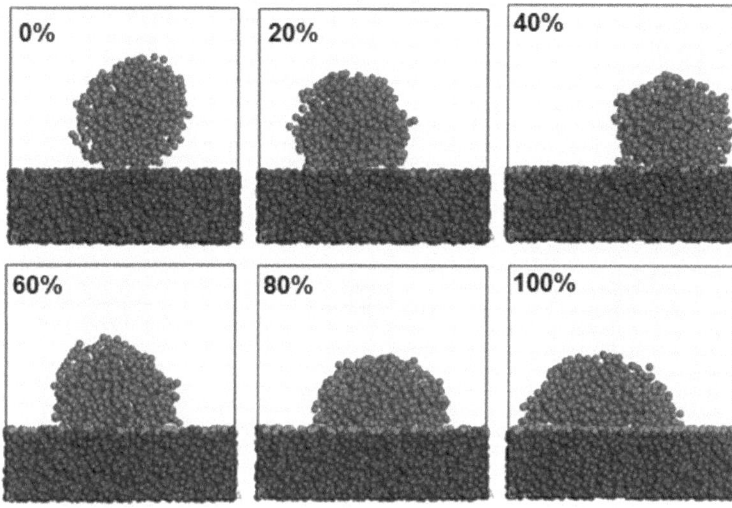

Figure 42 The final snapshots of water droplets on PDMS surface modified with different amounts of AA. *Figure taken from Ramirez et al (2015).*

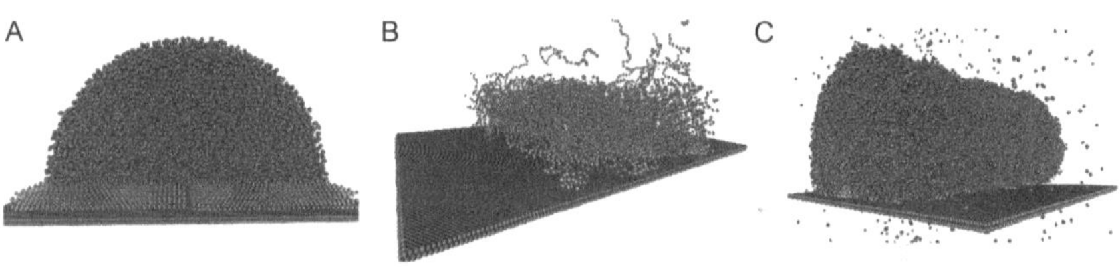

Figure 43 From left to right: the initial configuration and final configuration without/with water beads of droplets with surfactant. *Figure taken from Sergi et al (2012).*

the hydrophobic tail for linear surfactants and the length of the hydrophilic head group for T-shaped surfactants. Moreover, the T-shaped topology appears to facilitate the adsorption of surfactants onto the graphitic surface and faster spreading (Fig. 43).

Despite numerous experimental efforts have shown that certain surfactant molecules possess the ability to drive the superspreading of liquids to complete wetting on hydrophobic substrates, the precise mechanisms underlying superspreading remain unknown. Theodorakis et al (2015) conducted CGMD simulations to study the mechanism of superspreading of liquids with the addition of surfactant molecules with varying molecular architecture and substrate affinity (Fig. 44). They concluded that two key conditions must be satisfied simultaneously if the superspreading phenomenon occurs, the adsorption of surfactants from the liquid–vapor surface

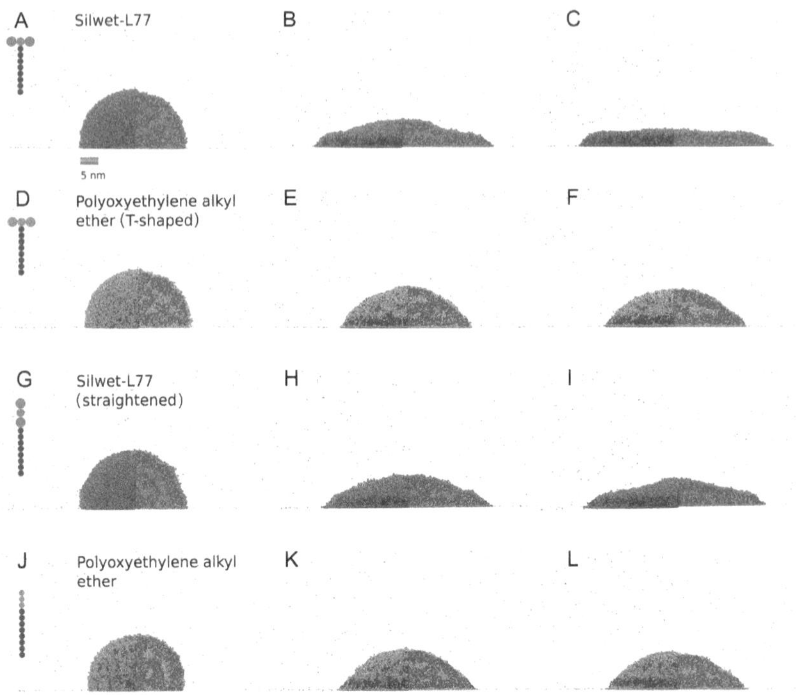

Figure 44 The state of the superspreading process at increasing time for different surfactants. *Figure taken from Theodorakis et al (2015).*

onto the three-phase contact line and the rapid replenishment of liquid–vapor and solid–liquid interfaces with surfactants from the interior of the droplet.

6. CONCLUSIONS

Mesoporous materials, such as meso-TiO_2, r-Al_2O_3, and MCM-41, as a catalyst support have good prospects in important heterogenous reactions. But the transport of fluid in mesoporous materials is different from that of macroscopic; there will be enrichment, condensation, and other phase transfers; the transport is of the same importance as reaction. In order to reveal the regulatory mechanism, we must clarify the complex structure and function at the interface and analyze the main factors affecting reaction and transport.

TiO_2 as the "active support" can significantly improve the activity of the catalyst, but there are many crystal forms and planes of TiO_2; people know its surface far less than r-Al_2O_3 and SiO_2. We prepared mesoporous TiO_2

and modified it with carbon to change the surface hydrophilic and hydrophobic properties and found that its hydrodesulfurization performance is greatly improved. Current experimental characterization still cannot fully describe the surface/interface with nano-/microstructure and also cannot accurately detect the complex interaction in the mesoscale. By molecular modeling, we found that the polar fluid mobility was improved significantly on strongly hydrophilic surface of titanium oxide. It indicates that the interfacial properties of titanium material are probably an important factor that improves its catalytic performance.

Wetting or nonwetting of ILs on the mesoporous or dense TiO_2 surfaces can be investigated quantitatively by AFM measurement; ILs could be supported in the form of stable films on the mesoporous TiO_2 surface due to the interfacial properties of ILs. The ILs could wet the mesoporous TiO_2 surface when there is strong adhesive force between ILs and mesoporous TiO_2. In contrast, the adhesive force is weak between ILs and dense TiO_2 surface; the nonwetting IL spheres on dense TiO_2 surface are apt to move along the surface, leaving the bare dense TiO_2 surface still highly frictional. While on the whole mesoporous TiO_2 surface, the ILs are employed to form stable and lowly frictional films. These results are expected to provide a new insight for wetting and nanofriction phenomena and extend the usage of AFM.

Mesoporous TiO_2 films exhibit low frictional and adhesive forces as a result of heterogenously distributed nanopores. The geometrically roughness-induced nanopores were achieved by TiO_2 (B) formed on the anatase surface. It is noteworthy that there are strong interactions of the TiO_2 surface with bound water and weak interactions with free water on the mesoporous and dense TiO_2 films. By AFM measurement, we knew TiO_2 surface provides a hydrophobic-like force to push the tip of the AFM up and make the tip slip easily. That could result in low surface frictional and adhesive forces on the mesoporous TiO_2 films. This work shows a simple approach to addressing heterogenously distributed nanopores on TiO_2 surfaces to control surface frictional properties. It opens the way to further examinations of TiO_2 surfaces for applications.

By molecular modeling technique, we can obtain the microscopic information that could not be detected by current experimental methods. But conventional molecular simulation method still has inherent limitations for various complex forces and complex mesoscopic structures under special mesoscale. For example, quantum chemistry cannot calculate the systems containing a large number of atoms. We have to ignore the role of chemical forces for large-scale simulation. This makes the results of simulation incomparable with that of experiment.

Some coarse-grained simulation methods have been established to understand the structure–property relationships of material interfaces, which include BD, dissipative particle dynamics (DPD), and CGMD based on MARTINI force field, and many efforts have focused on the protein adsorption, interfacial behavior and surface wettability, and so on.

The new mesoscale method for large systems takes account of chemical reaction, and transport should be established to examine the reaction mechanism quantitatively and the relationship with the changes of concentration of reactants and products. AFM study shows that the interfacial roughness of catalyst support has a significant impact for clusters on its chemical and physical adsorption stability; this problem is in the nanometer and micrometer mesoscale; traditional simulation also met the problem that studied system is too big to handle. A new medium-scale method is needed to be established to study effects of interfacial roughness on chemical and physical phenomena.

The experimental phenomenon is the result of complex structures and forces interaction of different scales, to design catalyst with nano-/micro-/mesoscale channels; the transport and reaction mechanism should be concerned at the same time. Experimental characterization alone cannot accurately analyze microscopic phenomena; simulation is a powerful tool for analyzing in microscopic scale, but for the behavior of the fluid at the rough interface with the reaction coupling current, simulation method is demanded; using the new mesoscale simulation approach, the transport model for heterogenous catalysis in mesoscale can be established eventually. ReaxFF simulation method can be used to investigate the transport and reaction properties in the pores of TiO_2, C, and TiO_2/C composites. Coarse-grained simulation could be used to study the effects of interface roughness under nano-/micro-/mesoscale structures on the reactivity, stability, and transport properties. AFM is an effective method for the study of solid surface structure and performance. By testing the deformation quantity of the AFM microcantilever, it can gain the interaction between tip and sample. It also can acquire the adhesion between the clusters and roughness surfaces by modification of the AFM tips. The adhesion and friction tested by AFM will be changed when the surface structure and surface chemical properties have been changed. These changes contain the surface information that can be used for modifying the surface coarse-grained force field.

ACKNOWLEDGMENTS

The work presented includes studies of former and current PhD students in our group at Nanjing Tech University; we would like to mention the PhD theses by Rong An, Mingjie Wei, and Xiaojing Guo. We would like to thank the financial support by the National Basic Research Program of China (Grant Nos. 2013CB733500 and 2015CB655300), the National Natural Science Foundation of China Grants (Grant Nos. 21176113, 21206070, 91334240, 21376089, and 21490584), and the Natural Science Foundation of the Jiangsu Higher Education Institutions of China (Grant No. 14KJB530008).

REFERENCES

Alexiadis A, Kassinos S: Molecular simulation of water in carbon nanotubes, *Chem Rev* 108:5014–5034, 2008.

An R, Yu Q, Zhang L, et al: Simple physical approach to reducing frictional and adhesive forces on a TiO_2 surface via creating heterogeneous nanopores, *Langmuir* 28:15270–15277, 2012.

An R, Zhu Y, Wu N, et al: Wetting behavior of ionic liquid on mesoporous titanium dioxide surface by atomic force microscopy, *ACS Appl Mater Interfaces* 5:2692–2698, 2013.

An R, Zhuang W, Yang ZH, et al: Protein adsorption behavior on mesoporous titanium dioxide determined by geometrical topography, *Chem Eng Sci* 117:146–155, 2014.

Andre E Nel, Lutz M, et al: Understanding biophysicochemical interactions at the nano-bio interface, *Nat Mater* 8:543–557, 2009.

Baschnagel J, Binder K, Doruker P, et al: Advances in polymer science: viscoelasticity, atomistic models, *Stat Chem* 152:41–156, 2000.

Bayne L, Ulijn RV, Halling PJ: Effect of pore size on the performance of immobilised enzymes, *Chem Soc Rev* 42:9000–9010, 2013.

Bhushan B: Biomimetics inspired surfaces for drag reduction and oleophobicity/philicity, *Beilstein J Nanotechnol* 2:66–84, 2011.

Bonn D, Eggers J, Indekeu J, et al: Wetting and spreading, *Rev Mod Phys* 81:740–805, 2009.

Bovio S, Podesta A, Lenardi C, et al: Evidence of extended solidlike layering in [BMIM] [NTf_2] ionic liquid thin films at room-temperature, *J Phys Chem B* 113:6600–6603, 2009.

Brodbeck WG, Nakayama Y, Matsuda T, et al: Biomaterial surface chemistry dictates adherent monocyte/macrophage cytokine expression in vitro, *Cytokine* 18(6):311–319, 2002.

Buettner KM, Valentine AM: Bioinorganic chemistry of titanium, *Chem Rev* 112:1863–1881, 2012.

Carlsson F, Hyltner E, Arnebrant T, et al: Lysozyme adsorption to charged surfaces. A Monte Carlo study, *J Phys Chem B* 108:9871–9881, 2004.

Chen SF, Jiang SY: A new avenue to nonfouling materials, *Adv Mater* 20(2):335–338, 2008.

Chen SF, Liu L, Zhou J, et al: Controlling antibody orientation on charged self-assembled monolayers, *Langmuir* 19:2859–2864, 2003.

Cheung DL, Carbone P: How stable are amphiphilic dendrimers at the liquid-liquid interface? *Soft Matter* 9:6841–6850, 2013.

Costa D, Arrouvel C, Breysse M, Toulhoat H, Raybaud P: Edge wetting effects of γ-Al_2O_3 and anatase-TiO_2 supports by MoS_2 and CoMoS active phases: a DFT study, *J Catal* 246:325–343, 2007.

D'Alessandro DM, Smit B, Long JR: Carbon dioxide capture: prospects for new materials, *Angew Chem Int Edit* 49:6058–6082, 2010.

Demirel YA, Sandler SI: Nonequilibrium thermodynamics in engineering and science, *J Phys Chem B* 108:31–43, 2004.

Dong J, Zhou J: Solvent-responsive behavior of polymer-brush-modified amphiphilic gold nanoparticles, *Macromol Theory Simul* 22:174–186, 2013.

Dong J, Li J, Zhou J: Interfacial and phase transfer behaviors of polymer brush grafted amphiphilic nanoparticles: a computer simulation study, *Langmuir* 30:5599–5608, 2014.

Dunweg B, Paul W: Brownian dynamics simulations without Gaussian random numbers, *Int J Mod Phys C* 2:817–827, 1991.

Dupont C, Lemeur R, Daudin A, Raybaud P: Hydrodeoxygenation pathways catalyzed by MoS_2 and NiMoS active phases: a DFT study, *J Catal* 279:276–286, 2011.

Eijsbouts S, Mayo SW, Fujita K: Unsupported transition metal sulfide catalysts: from fundamentals to industrial application, *Appl Catal A Gen* 322:58–66, 2007.

Espanol P, Warren P: Statistical mechanics of dissipative particle dynamics, *Europhys Lett* 30:191–196, 1995.

Estridge CE, Jayaraman A: Diblock copolymer grafted particles as compatibilizers for immiscible binary homopolymer blends, *ACS Macro Lett* 4:155–159, 2015.

Fang W, Liu WJ, Guo XJ, et al: Theoretical investigation of CO adsorption on clean and hydroxylated TiO2-B (100) surfaces, *J Phys Chem C* 115:8622–8629, 2011.

Feng L, Jiang L, et al: Super-hydrophobic surfaces: from natural to artificial, *Adv Mater* 14(24):1857–1860, 2002.

Fukunaga H, Takimoto J, Doi M: A coarse-graining procedure for flexible polymer chains with bonded and nonbonded interactions, *J Chem Phys* 116:8183–8190, 2002.

Gary JJ: The interaction of proteins with solid surfaces, *Curr Opin Struct Biol* 14:110–115, 2004.

Gray CG, Gubbins KE, Joslin CG: Theory of molecular fluids. In *Applications*, vol. 2, Oxford, 2011, Oxford University Press, p 940.

Groot RD, Warren PB: Dissipative particle dynamics: bridging the gap between atomistic and mesoscopic simulation, *J Chem Phys* 107:4423–4435, 1997.

Gu HW, Xu KM, et al: Biofunctional magnetic nanoparticles for protein separation and pathogen detection, *Chem Commun* 12:941–949, 2006.

Gubbins KE, Long Y, Małgorzata Bartkowiak S: Thermodynamics of confined nano-phases, *J Chem Thermodyn* 74:169–183, 2014.

Guo XJ, Liu WJ, Fang W, et al: DFT study of coverage-depended adsorption of NH3 on TiO2-B (100) surface, *Phys Chem Phys* 14:16618–16625, 2012.

Guo HY, Qiu XQ, Zhou J: Self-assembled core-shell and Janus microphase separated structures of polymer blends in aqueous solution, *J Chem Phys* 139:084907, 2013.

Guo SG, Zhang JH, et al: Selective adsorption of bovine hemoglobin on functional TiO_2 nano-adsorbents: surface physic-chemical properties determined adsorption activity, *Mater Res Express* 2:045101, 2015.

Hasib-ur-Rahman M, Siaj M, Larachi F: Ionic liquids for CO_2 capture—development and progress, *Chem Eng Process* 49:313–322, 2010.

Hayashi T, Sano KI, Shiba K, et al: Mechanism underlying specificity of proteins targeting inorganic materials, *Nano Lett* 6:515–519, 2006.

He Y, Hower J, Chen SF, et al: Molecular simulation studies of protein interactions with zwitterionic phosphorylcholine self-assembled monolayers in the presence of water, *Langmuir* 24:10358–10364, 2008.

Hoogerbrugge PJ, Koelman JMVA: Simulating microscopic hydrodynamic phenomena with dissipative particle dynamics, *Europhys Lett* 19:155–160, 1992.

Hu L, Chen M, Fang X, Wu L: Oil-water interfacial self-assembly: a novel strategy for nanofilm and nanodevice fabrication, *Chem Soc Rev* 41:1350–1362, 2012.

Huang LL, Seredych M, Bandosz Teresa J, Van Duin Adri CT, Lu XH, Gubbins KE: Controllable atomistic graphene oxide model and its application in hydrogen sulfide removal, *J Chem Phys* 139:194707, 2013.

Huang LL, Gubbins KE, Li LC, Lu XH: Water on titanium dioxide surface: a revisiting by reactive molecular dynamics simulations, *Langmuir* 30:14832–14840, 2014.
Hung A, Mwenifumbo S, Mager M, et al: Ordering surfaces on the nanoscale: implications for protein adsorption, *J Am Chem Soc* 133:1438–1450, 2011.
Hung A, Mager M, Hembury M, et al: Amphiphilic amino acids: a key to adsorbing proteins to nanopatterned surfaces, *Chem Sci* 4:9928–9937, 2013.
Jabbarzadeh A: Effect of nano-patterning on oleophobic properties of a surface, *Soft Matter* 9:11598–11608, 2013.
Jaeger KE, Eggert T: Lipases for biotechnology, *Curr Opin Biotechnol* 14:110–115, 2004.
Jagiello J, Olivier JP: 2D-NLDFT adsorption models for carbon slit-shaped pores with surface energetical heterogeneity and geometrical corrugation, *Carbon* 55:70–80, 2013a.
Jagiello J, Olivier JP: Carbon slit pore model incorporating surface energetical heterogeneity and geometrical corrugation, *Adsorption* 19:777–783, 2013b.
Jens Meissner, Findenegg Gerhard H, et al: Protein immobilization in surface-functionalized SBA-15: predicting the uptake capacity from the pore structure, *J Phys Chem C* 119:2438–2446, 2015.
Ji Y, Ji X, Liu C, et al: Modelling of mass transfer coupling with crystallization kinetics in microscale, *Chem Eng Sci* 65:2649–2655, 2010.
Jones MI, McColl IR, Grant DM, Parker KG, Parker TL: Haemocompatibility of DLC and TiC-TiN interlayers on titanium, *Diam Relat Mater* 8:457–462, 1999.
Karan S, Samitsu S, Peng X, Kurashima K, Ichinose I: Ultrafast viscous permeation of organic solvents through diamond-like carbon nanosheets, *Science* 335:444–447, 2012.
Kim SY, Van Duin ACT, Kubicki JD: Simulations of the Interactions between TiO_2 nanoparticles and water with Na^+ and Cl^-, methanol and formic acid using a reactive force field, *J Mater Res* 28:513–520, 2013.
Kirschner CM, Brennan AB: Bio-inspired antifouling strategies. In Clarke DR, editor: *Annual review of materials research*, 2012, pp 211–229.
Koktysh DS, Liang X, Yun BG, et al: Biomaterials by design: layer-by-layer assembled ion-selective and biocompatible films of TiO2 nanoshells for neurochemical monitoring, *Adv Funct Mater* 12:255–265, 2002.
Kresge C, Leonowicz M, Roth WJ, Vartull JC, Beck JS: Ordered mesoporous molecular sieves synthesized by a liquid-crystal template mechanism, *Nature* 359:710–712, 1992.
Kubiak K, Mulheran PA: Mechanism of hen egg white lysozyme adsorption on a charged solid surface, *Langmuir* 26:15954–15965, 2010.
Latour RA: Molecular simulation of protein-surface interactions: benefits, problems, solutions, and future directions, *Biointerphases* 3:FC2–FC12, 2008 (Review).
Li JH: Perspectives on chemical engineering in the 21st century, *J Chem Ind Eng (China)* 59:1879–1883, 2008.
Li Q, Yang H, Qiu F, Zhang X: Promotional effects of carbon nanotubes on V2O5/TiO2 for NOx removal, *J Hazard Mater* 192:915–921, 2011.
Li LC, Zhu YD, Lu XH, et al: Carbon heterogeneous surface modification on a mesoporous TiO_2-supported catalyst and its enhanced hydrodesulfurization performance, *Chem Commun* 48:11525–11527, 2012.
Lin S-L, Xu M-Y, Yang Z-R: Dissipative particle dynamics study on the mesostructures of n-octadecane/water emulsion with alternating styrene-maleic acid copolymers as emulsifier, *Soft Matter* 8:375–384, 2012.
Liu XY, Chu PK, Ding CX: Surface modification of titanium, titanium alloys, and related materials for biomedical applications, *Mater Sci Eng R Rep* 47:49–121, 2004.
Liu CL, Yang DZ, et al: Corrosion resistance and hemocompatibility of multilayered Ti/TiN-coated surgical AISI 316L stainless steel, *Mater Lett* 59:3813–3819, 2005a.
Liu Y, Wang M, Li Z, et al: Preparation of porous aminopropylsilsesquioxane by a non-hydrolytic sol-gel method in ionic liquid solvent, *Langmuir* 21:1618–1622, 2005b.

Liu C, Ji Y, Shao Q, et al: Thermodynamic analysis for synthesis of advanced materials, *Struct Bond* 131:193–270, 2009.

Liu WJ, Wang JG, Li W, et al: A shortcut for evaluating activities of TiO_2 facets: water dissociative chemisorption on TiO_2-B (100) and (001), *Phys Chem Phys* 12: 8721–8727, 2010.

Liu WJ, Wang JG, Guo XJ, et al: Dissociation of methanol on hydroxylated TiO2-B (1 0 0) surface: insights from first principle DFT calculation, *Catal Today* 165:32–40, 2011.

Liu J, Liao CY, Zhou J: Multiscale simulations of protein G B1 adsorbed on charged self-assembled monolayers, *Langmuir* 29:11366–11374, 2013.

Liu K, Cao MY, Fujishima Akira, Jiang L: Bio-inspired titanium dioxide materials with special wettability and their applications, *Chem Rev* 114:10044–10094, 2014.

Liu J, Yu GB, Zhou J: Ribonuclease A adsorption onto charged self-assembled monolayers: a multiscale simulation study, *Chem Eng Sci* 121:331–339, 2015.

Long Y: *Pressure tensor of adsorbate in nanoporous materials: molecular simulation studies,* Ph.D. Thesis, 2012, North Carolina State University.

Lu L, Zhu Y, Li F, Zhuang W, Chan KY, Lu X: Carbon titania mesoporous composite whisker as stable supercapacitor electrode material, *J Mater Chem* 20:7645–7651, 2010.

Lu X, Ji Y, Liu H: Non-equilibrium thermodynamics analysis and its application in interfacial mass transfer, *Sci China: Chem* 54:1659–1666, 2011.

Lu X, Ji Y, Feng X, et al: Methodology of non-equilibrium thermodynamics for kinetics research of CO2 capture by ionic liquids, *Sci China: Chem* 55:1079–1091, 2012.

Luo H, Dai S, Bonnesen PV, et al: Extraction of cesium ions from aqueous solutions using calix[4]arene-bis(tert-octylbenzo-crown-6) in ionic liquids, *Anal Chem* 76:3078–3083, 2004.

Luu X-C, Striolo A: Ellipsoidal Janus nanoparticles assembled at spherical oil/water interfaces, *J Phys Chem B* 118:13737–13743, 2014.

Luu X-C, Yu J, Striolo A: Nanoparticles adsorbed at the water/oil interface: coverage and composition effects on structure and diffusion, *Langmuir* 29:7221–7228, 2013a.

Luu X-C, Yu J, Striolo A: Ellipsoidal Janus nanoparticles adsorbed at the water-oil interface: some evidence of emergent behavior, *J Phys Chem B* 117:13922–13929, 2013b.

Marrink SJ, Risselada HJ, Yefimov S, et al: The MARTINI force field: coarse grained model for biomolecular simulations, *J Phys Chem B* 111:7812–7824, 2007.

Mi L, Jiang S: Integrated antimicrobial and nonfouling zwitterionic polymers, *Angew Chem Int Edit* 53:1746–1754, 2014.

Monticelli L, Kandasamy SK, Periole X, et al: The MARTINI coarse-grained force field: extension to proteins, *J Chem Theory Comput* 4:819–834, 2008.

Motoyoshi K, Tajima A, Higuchi T, et al: Static and dynamic control of phase separation structures in nanoparticles of polymer blends, *Soft Matter* 6:1253–1257, 2010.

Mou CY, et al: Enhanced activity and stability of lysozyme by immobilization in the matching nanochannels of mesoporous silica nanoparticles, *J Phys Chem* C118:6734–6743, 2014.

Müller EA, Gubbins KE: Molecular simulation study of hydrophilic and hydrophobic behavior of activated carbon surfaces, *Carbon* 36:1433–1438, 1998.

Muller-Plathe F: Coarse-graining in polymer simulation: from the atomistic to the mesoscopic scale and back, *ChemPhysChem* 3:754–769, 2002.

Mutin PH, Guerrero G, Vioux A: Hybrid materials from organophosphorus coupling molecules, *J Mater Chem* 15:3761–3768, 2005.

Nair RR, Wu HA, Jayaram PN, Grigorieva IV, Geim AK: Unimpeded permeation of water through helium-leak-tight graphene-based membranes, *Science* 335:442–444, 2012.

Nel AE, Madler L, Velegol D, et al: Understanding biophysicochemical interactions at the nano-bio interface, *Nat Mater* 8:543–557, 2009.

Ng YH, Lightcap IV, Goodwin K, Matsumura M, Kamat PV: To what extent do graphene scaffolds improve the photovoltaic and photocatalytic response of TiO2 nanostructured films? *J Phys Chem Lett* 1:2222–2227, 2010.

Patterson K, Lisal M, Colina CM: Adsorption behavior of model proteins on surfaces, *Fluid Phase Equilib* 302:48–54, 2011.

Peng CW, Liu J, Zhao DH: Adsorption hydrophobin on different self-assembled monolayers: the role of the hydrophobic dipole and the electric dipole, *Langmuir* 30:11401–11411, 2014.

Plimpton S: Fast parallel algorithms for short-range molecular dynamics, *J Comput Phys* 117:1–19, 1995.

Popat KC, Leoni L, Grimes CA, Desai TA: Influence of engineered titania nanotubular surfaces on bone cells, *Biomaterials* 28:3188–3197, 2007.

Prigogine I: Modération et transformations irreversibles des systemes ouverts, *Bull Cl Sci Acad R Belg* 31:600–606, 1945.

Prigogine I: *Etude thermodynamique des processus irréversibles,* Liege, 1947, Desoer.

Quan X, Dong J, Zhou J: Effect of topology of hydrophobic surfaces on their wetting states by coarse-grained simulations, *Acta Chim Sin* 72:1075–1078, 2014.

Rabe M, Verdes D, Seeger S: Understanding protein adsorption phenomena at solid surfaces, *Adv Colloid Interf Sci* 162:87–106, 2011.

Ramdin M, de Loos TW, Vlugt TJ: State-of-the-art of CO_2 capture with ionic liquids, *Ind Eng Chem Res* 51:8149–8177, 2012.

Ramirez GD, Nieto DC, Pannacci N, et al: Surface photografting of acrylic acid on poly(dimethylsiloxane). Experimental and dissipative particle dynamics studies, *Langmuir* 31:1400–1409, 2015.

Ravichandran S, Talbot J: Mobility of adsorbed proteins: a Brownian dynamics study, *Biophys J* 78:110–120, 2000.

Ravichandran S, Madura JD, Talbot J: A Brownian dynamics study of the initial stages of hen egg-white lysozyme adsorption at a solid interface, *J Phys Chem B* 105: 3610–3613, 2001.

Ren J, Wu L, Li B: Preparation and CO_2 sorption/desorption of N-(3-aminopropyl) aminoethyl tributylphosphonium amino acid salt ionic liquids supported into porous silica particles, *Ind Eng Chem Res* 51:7901–7909, 2012.

Roach P, Farrar D, Perry CC: Interpretation of protein adsorption: surface-induced conformational changes, *J Am Chem Soc* 127:8168–8173, 2005.

Roach P, Farrar D, Perry CC: Surface tailoring for controlled protein adsorption: effect of topography at the nanometer scale and chemistry, *J Am Chem Soc* 128(12):3939–3945, 2006.

Rosenhahn A, Schilp S, et al: The role of "inert" surface chemistry in marine biofouling prevention, *Phys Chem Phys* 12(17):4275–4286, 2010.

Samanta A, Zhao A, Shimizu GKH, et al: Post-combustion CO_2 capture using solid sorbents: a review, *Ind Eng Chem Res* 51:1438–1463, 2012.

Schneider T, Stoll E: Molecular dynamics study of a three-dimensional one-component model for distortive phase transitions, *Phys Rev B* 17:1302–1322, 1978.

Sergi D, Scocchi G, Ortona A: Coarse-graining MARTINI model for molecular-dynamics simulations of the wetting properties of graphitic surfaces with non-ionic, long-chain, and T-shaped surfactants, *J Chem Phys* 137:094904, 2012.

Shan J, Tenhu H: Recent advances in polymer protected gold nanoparticles: synthesis, properties and applications, *Chem Commun* 44:4580–4598, 2007.

Shi K, Lian C, Bai Z, et al: Dissipative particle dynamics study of the water/benzene/caprolactam system in the absence or presence of non-ionic surfactants, *Chem Eng Sci* 122:185–196, 2015.

Shrivastava S, Nuffer JH, Siegel RW, et al: Position-specific chemical modification and quantitative proteomics disclose protein orientation adsorbed on silica nanoparticles, *Nano Lett* 12:1583–1587, 2012.

Skelton AA, Liang T, Walsh TR: Interplay of sequence, conformation, and binding at the peptide–titania interface as mediated by water, *ACS Appl Mater Interfaces* 1:1482–1491, 2009.

Soto-Figueroa C, del Rosario Rodriguez-Hidalgo M, Vicente L: Dissipative particle dynamics simulation of the micellization-demicellization process and micellar shuttle of a diblock copolymer in a biphasic system (water/ionic-liquid), *Soft Matter* 8:1871–1877, 2012.

Souza JCM, Henriques M, Oliveira R, et al: Biofilms inducing ultra-low friction on titanium, *J Dent Res* 89:1470–1475, 2010.

Stolte S, Steudte S, Areitioaurtena O, et al: Ionic liquids as lubricants or lubrication additives: an ecotoxicity and biodegradability assessment, *Chemosphere* 89:1135–1141, 2012.

Sumant A, Grierson DS, Gerbi JE, et al: Toward the ultimate tribological interface: surface chemistry and nanotribology of ultrananocrystalline diamond, *Adv Mater* 17:1039–1045, 2005.

Taddese T, Carbone P, Cheung DL: Thermodynamics of linear and star polymers at fluid interfaces, *Soft Matter* 11:81–93, 2015.

Takeuchi M, Martra G, Coluccia S, Anpo M: Investigations of the structure of H2O clusters adsorbed on TiO2 surfaces by near-infrared absorption spectroscopy, *J Phys Chem B* 109:7387–7391, 2005.

Tauster SJ, Fung SC, Garten RL: Strong metal-support interactions. Group 8 noble metals supported on titanium dioxide, *J Am Chem Soc* 100:170–175, 1978.

Theodorakis PE, Mueller EA, Craster RV, et al: Superspreading: mechanisms and molecular design, *Langmuir* 31:2304–2309, 2015.

Thorpe IF, Zhou J, Voth GA: Peptide folding using multiscale coarse-grained models, *J Phys Chem B* 112:13079–13090, 2008.

Tijana Rajh, Dimitrijevic Nada M, et al: Titanium dioxide in the service of the biomedical revolution, *Chem Rev* 114:10177–10216, 2014.

Van Duin ACT, Dasgupta S, Lorant F, et al: ReaxFF: a reactive force field for hydrocarbons, *J Phys Chem A* 105:9396–9409, 2001.

Vashist SK, Lam E, Hrapovic S, Male KB, Luong JHT: Immobilization of antibodies and enzymes on 3-aminopropyltriethoxysilane-functionalized bioanalytical platforms for biosensors and diagnostics, *Chem Rev* 114:11083–11130, 2014.

Vasudevan R, Kennedy AJ, et al: Microscale patterned surfaces reduce bacterial fouling-microscopic and theoretical analysis, *Collids Surf B: Biointerfaces* 117:225–232, 2014.

Vertegel AA, Siegel RW, Dordick JS: Silica nanoparticle size influences the structure and enzymatic activity of adsorbed lysozyme, *Langmuir* 20:6800–6807, 2004.

Vicent-Luna JM, Gutierrez-Sevillano JJ, Anta JA, et al: Effect of room-temperature ionic liquids on CO_2 separation by a Cu-BTC metal-organic framework, *J Phys Chem C* 117:20762–20768, 2013.

Wang J, Mueller M: Microphase separation of mixed polymer brushes: dependence of the morphology on grafting density, composition, chain-length asymmetry, solvent quality, and selectivity, *J Phys Chem B* 113:11384–11402, 2009.

Wang H, Castner DG, Ratner BD, et al: Probing the orientation of surface-immobilized immunoglobulin G by time-of-flight secondary ion mass spectrometry, *Langmuir* 20:1877–1887, 2004.

Wang X, Akhmedov NG, Duan Y, et al: Immobilization of amino acid ionic liquids into nanoporous microspheres as robust sorbents for CO_2 capture, *J Mater Chem* 1:2978–2982, 2013.

Wei MJ, Zhou J, Lu XH, et al: Diffusion of water molecules confined in slits of rutile TiO_2(110) and graphite(0001), *Fluid Phase Equilib* 302:316–320, 2011a.

Wei T, Carignano MA, Szleifer I: Lysozyme adsorption on polyethylene surfaces: why are long simulations needed, *Langmuir* 27:12074–12081, 2011b.
Wei MJ, Zhang LZ, Lu LH, Zhu YD, Gubbins KE, Lu XH: Molecular behavior of water in TiO2 nano-slits with varying coverages of carbon: a molecular dynamics simulation study, *Phys Chem Chem Phys* 14:16536–16543, 2012.
Welton T: Room-temperature ionic liquids. Solvents for synthesis and catalysis, *Chem Rev* 99:2071–2084, 1999.
Williams G, Seger B, Kamat PV: TiO2-graphene nanocomposites. UV-assisted photocatalytic reduction of graphene oxide, *ACS Nano* 2:1487–1491, 2008.
Wu Z, Cui Q, Yethiraj A: A new coarse-grained model for water: the importance of electrostatic interactions, *J Phys Chem B* 114:10524–10529, 2010.
Wu Z, Cui Q, Yethiraj A: A new coarse-grained force field for membrane-peptide simulations, *J Chem Theory Comput* 7:3793–3802, 2011.
Xie Y, Zhou J, Jiang SY: Parallel tempering Monte Carlo simulations of lysozyme orientation on charged surfaces, *J Chem Phys* 132:065101, 2010.
Xie Y, Zhang Y, Lu X, et al: Energy consumption analysis for CO_2 separation using imidazolium-based ionic liquids, *Appl Energy* 136:325–335, 2014.
Xie W, Ji X, Feng X, et al: Mass transfer rate enhancement for CO_2 separation using ionic liquids: theoretical study on the mechanism of rate enhancement, *AIChE J*, 2015. Accepted.
Xie W, Ji X, Feng X, et al: Mass transfer rate enhancement for CO_2 separation by ionic liquids: experimental study on the effect of film thickness. Ind Eng Chem Res, Accepted.
Xu YJ, Zhuang Y, Fu X: New insight for enhanced photocatalytic activity of TiO2 by doping carbon nanotubes: a case study on degradation of benzene and methyl orange, *J Phys Chem C* 114:2669–2676, 2010.
Xu LJ, et al: Synthesis of mono-dispersed mesoporous SBA-1 nanoparticles with tunable pore size and their application in lysozyme immobilization, *RSC Adv* 4:37470, 2014.
Yao X, Song Y, Jiang L: Applications of bio-inspired special wettable surfaces, *Adv Mater* 23:719–734, 2011.
Yesylevskyy SO, Schäfer LV, Sengupta D, et al: Polarizable water model for the coarse-grained MARTINI force field, *PLoS Comput Biol* 6:e1000810, 2010.
Yu GB, Liu J, Zhou J: Mesoscopic coarse-grained simulations of lysozyme adsorption, *J Phys Chem B* 118:4451–4460, 2014.
Yu GB, Liu J, Zhou J: Mesoscopic coarse-grained simulations of hydrophobic charge induction chromatography (HCIC) for protein purification, *AIChE J* 61:2035–2047, 2015.
Zeller A, Musyanovych A, Kappl M, et al: Nanostructured coatings by adhesion of phosphonated polystyrene particles onto titanium surface for implant material applications, *ACS Appl Mater Interfaces* 2:2421–2428, 2010.
Zhang J, Zhang S, Dong K, et al: Supported absorption of CO_2 by tetrabutylphosphonium amino acid ionic liquids, *Chem Eur J* 12:4021–4026, 2006.
Zhang AJ, Xie Y, Zhou J: Experimental control and characterization of protein orientation on surfaces, *Prog Chem* 21:1408–1417, 2009a.
Zhang L, Zhao GF, Sun Y: Molecular insight into protein conformational transition in hydrophobic charge induction chromatography: a molecular dynamics simulation, *J Phys Chem B* 113:6873–6880, 2009b.
Zhang XP, Zhang XC, Dong HF, et al: Carbon capture with ionic liquids: overview and progress, *Energy Environ Sci* 5:6668–6681, 2012.
Zhao DY, Feng JL, Huo QS, et al: Triblock copolymer syntheses of mesoporous silica with periodic 50 to 300 Angstrom pores, *Science* 279:548–552, 1998.
Zhao Y, Zhang X, Zeng S, et al: Density, viscosity, and performances of carbon dioxide capture in 16 absorbents of amine plus ionic liquid + H_2O, ionic liquid + H_2O, and amine + H_2O systems, *J Chem Eng Data* 55:3513–3519, 2010.

Zhao Y, Zhang X, Dong H, et al: Solubilities of gases in novel alcamines ionic liquid 2,2-hydroxyethyl (methyl) amino ethanol chloride, *Fluid Phase Equilib* 302:60–64, 2011.
Zhao DH, Peng CW, Zhou J: Lipase adsorption on different nanomaterials: a multi-scale simulation study, *Phys Chem Chem Phys* 17:840–850, 2015.
Zheng J, Li LY, Tsao HK, et al: Strong repulsive forces between protein and oligo (ethylene glycol) self-assembled monolayers: a molecular simulation study, *Biophys J* 89:158–166, 2005.
Zhou J, Chen SF, Jiang SY: Orientation of adsorbed antibodies on charged surfaces by computer simulation based on a united-residue model, *Langmuir* 19:3472–3478, 2003.
Zhou J, Tsao HK, Sheng YJ, et al: Monte Carlo simulations of antibody adsorption and orientation on charged surfaces, *J Chem Phys* 121:1050–1057, 2004a.
Zhou J, Zheng J, Jiang SY: Molecular simulation studies of the orientation and conformation of cytochrome c adsorbed on self-assembled monolayers, *J Phys Chem B* 108:17418–17424, 2004b.
Zhou J, Thorpe IF, Izvekov S, Voth GA: Coarse-grained peptide modeling using a systematic multiscale approach, *Biophys J* 92:4289–4303, 2007.

CHAPTER THREE

Role of Interfacial Force on Multiphase Microflow—An Important Meso-Scientific Issue

Kai Wang, Jianhong Xu, Guotao Liu, Guangsheng Luo[1]

Department of Chemical Engineering, The State Key Laboratory of Chemical Engineering, Tsinghua University, Beijing, PR China

[1]Corresponding author: e-mail address: gsluo@tsinghua.edu.cn

Contents

Abstract

An understanding of multiphase microflows is critical for the development and application of microstructured chemical systems in the chemical industry. As one of the most important meso-scientific issues, interfacial science could be a bridge connecting microscopic molecular components and macroscopic fluid behaviors in these systems. Working together with viscous and inertial forces, the interfacial force also dominates complicated multiphase flow patterns and well-controlled droplets and bubbles. In this review, the generation mechanisms of different flow patterns and the break-up rules for droplets and bubbles in microchannels are introduced first. The effects of the adjustable fluid/solid interfaces, or so-called wetting properties, of microchannels on multiphase flow patterns, as well as microchannel surface modification methods, are then discussed. The dynamic fluid/fluid interfaces in multiphase microflows with variable

Advances in Chemical Engineering, Volume 47
ISSN 0065-2377
http://dx.doi.org/10.1016/bs.ache.2015.10.003

interfacial tensions are also presented, including the scientific reasons, measuring methods, their effect on multiphase microflows, and further works.

1. INTRODUCTION

During the last 20 years, microstructured chemical systems have consistently been a rapidly developing technology within chemical engineering (Hessel and Lowe, 2003a, 2003b, 2003c), which have exhibited strong competencies in implementing chemical reactions and substance separation processes. In the early study of microstructured chemical systems, researchers found that microreactors containing microchannels or other precise fabricated structures have the advantages of enhancing reactant mixing and heat exchange (Ehrfeld et al, 2000; Worz et al, 2001). These characteristics are important for many homogeneous-phase chemical processes, such as the selectivity promotion of organic chemical reactions and the explosive nucleation of nanoparticle precipitation (Luo et al, 2011; Yoshida et al, 2005). In most multiphase chemical processes, microreactors have also exhibited advantages in mass transfer enhancement, since those microstructures are convenient for dispersing droplets, bubbles, or other composite fluid particles into micrometer scales (Xu et al, 2014), which have huge specific areas on the order of 10^3 to 10^6 m^2/m^3. The excellent mass and heat transfer performances also promoted the development of microstructured devices for separation processes, such as distillation, adsorption, and extraction (Choi et al, 2010; Hartman et al, 2009; Tan et al, 2013). It has been proven that the volumetric mass and heat transfer coefficients in liquid/liquid or gas/liquid microdispersed systems can be raised 10 to 10^3 times those of common extractors, adsorption columns, and heat exchangers (Tan et al, 2013; Wang et al, 2008; Zhao et al, 2007).

As a new type of equipment, a microstructured chemical system cannot change the chemical process by using bubbles or droplets at the micrometer scale, but the apparent transport and reaction performances at that scale are significantly improved from traditional chemical processes. According to recent studies of Hessel et al (2013) and Tonhauser et al (2012), some obviously increased reaction performances, such as narrow residence time distributions, regular flow patterns, and easily designed reactant adding sequences, are from the improved flow characteristics of multiphase flows in microchannels or small tubes (Haase and Brujic, 2014; Han et al, 2009; Nightingale et al, 2014). The studies of microstructured chemical systems also show that microfluidic devices, which have been used in many other areas, such as biochips, microanalysis, and material fabrication (Dittrich and Manz, 2006; Serra and Chang, 2008; Song et al, 2006), are good tools for understanding the basic rules of microflow, micromixing, and mass transfer in microchannels. Microfluidic devices work as single dispersion, mixing, or reaction units in complicated microstructured devices. With the help of microfluidic devices that facilitate observation of conditions, the most significant characteristics of multiphase microflows have been concluded to be variable flow patterns, small dispersed sizes, and large specific areas (Ciceri et al, 2014), as the fluid dynamic basements of rapid mass and heat transfer, and well-controlled reaction time and yields.

In traditional chemical engineering processes, the effects of surfaces or interfaces are easily ignored for the low interfacial surface areas in huge chemical equipment, but in microstructured chemical devices, it is decisive, since the influence of interfacial forces becomes larger than the effect of gravity with the reduction of channel diameter (Gunther and Jensen, 2006). Not only is the interfacial force between fluids important but also the interfacial force between the fluid and the solid wall of the device, which is usually expressed as the contact angle of the fluid on the solid surface, is important for the control of flow patterns, residence time distribution, and dispersed size (Fidalgo et al, 2007; Kuhn et al, 2011; Tostado et al, 2011). The dynamic adsorption of the surfactant, the mass transfer of the reagent, and chemical reactions can further affect the composition of interfaces in time and space, which creates variable interfacial tensions and causes more complicated flow phenomena, such as the Marangoni effect and droplet tip-streaming (Babatunde et al, 2013; Ward et al, 2010). All of these flow characteristics point to an important meso-scientific issue between the microscopic molecular component and the macroscopic flow behavior of droplets and bubbles in microstructured chemical systems—multiphase microflow interface science (Li and Huang, 2014; Li et al, 2013), which is critical for the design and operation of microdevices. In this review, the effects of interfacial force on multiphase microflow patterns are introduced first. Then the generation mechanisms and the scaling laws of droplets and bubbles in different microchannel devices are discussed. In order to show the variable interfacial force in multiphase microflows, the important effect of the wetting property of a solid–fluid interface is exhibited. The typical dynamic interface phenomena caused by dynamic surfactant adsorption and mass transfer of the reagent, which are famous meso-science issues in the studies of bubble and droplet dispersion in microchannel devices, are introduced last.

2. THE ROLE OF INTERFACIAL FORCE ON MULTIPHASE MICROFLOWS

2.1 Multiphase Microflows in Microchannels

The hydrodynamic diameters of microchannels usually range from 10 to 1000 μm. In such confined flowing spaces, the multiphase flow patterns of Newtonian fluids are more variable compared with the common bubbly or droplet flows in larger vessels and columns. The confined flowing channel first affects the shape of droplets; therefore, the flow patterns of liquid/liquid dispersed systems are usually categorized as plug flow and droplet flow, as shown in Fig. 1A and B. Usually, the droplet flow has larger specific surface area, which is fit for the mass transfer enhancement process (Mary et al, 2008). The plug flow has narrow residence time distributions and it is benefit for the reaction time control (Song and Ismagilov, 2003). Similar flow patterns in gas/liquid systems are named slug flow and bubbly flow (Xu et al, 2006a, 2006b). Although both plug flow and droplet flow contain dispersed liquids, the flow fields in them are very different. There are strong circular flows in liquid plugs (Malsch et al, 2008), but the circular flows are not obvious in round droplets, owing to the low velocity gradients in them (Wang et al, 2013a). The average velocity of moving droplets and liquid plugs are also different. Wong et al (1995a, 1995b) reported that there were gutter fluxes around gas slugs in a rectangular cross-section microchannel, and that this phenomenon showed that the slug-shaped fluid particles moved more slowly than the bulk continuous phase. However, the smaller droplets or bubbles moved faster than the continuous phase due to their center channel position and the higher velocities at that position (Castro-Hernandez et al, 2009). Besides the plug and droplets flows, the liquid/liquid parallel flow is also common in microchannels, especially for highly viscous or low interfacial tension fluids (Pohar et al, 2012). They are good template to preparing fiber materials (Lan et al, 2009). Figure 1C is an example of a parallel flow generated from a T-junction microchannel. Parallel flow hardly exists in large vessels because of the strong turbulence in them, while it is stable under the laminar flow conditions in microchannels. For gas/liquid systems, which have larger interfacial tensions (usually $\gamma > 20$ mN/m), parallel flows are normally generated at gas velocities greater than 1 m/s (Gunther and Jensen, 2006), and usually have annular cross-sections in capillaries. Furthermore, at both high gas and high liquid velocities, such as when both of them

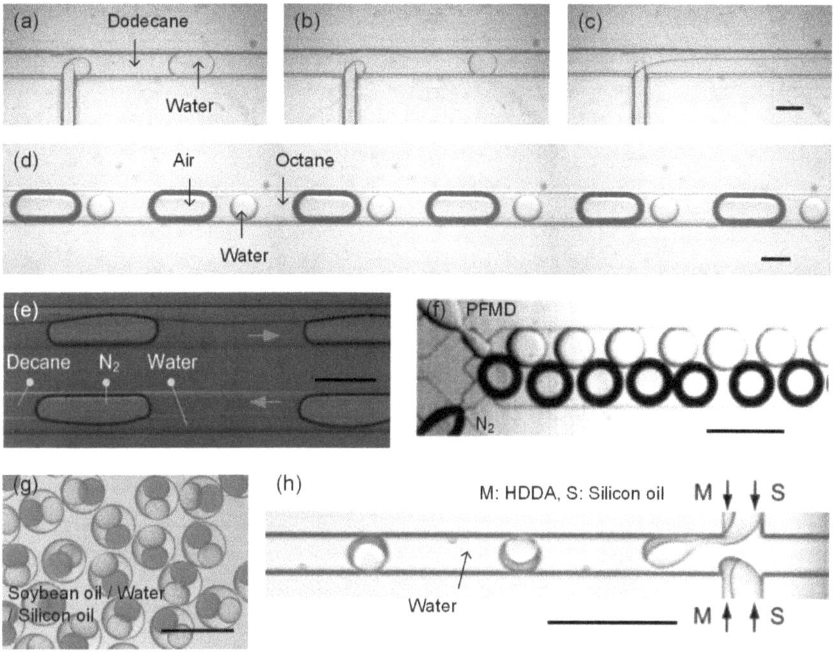

Figure 1 Varied flow patterns in microchannels. (A) Liquid plug flow. (B) Droplet flow. (C) Liquid/liquid parallel flow. (D) Bubble droplet alternate flow. (E) Parallel water/oil flow containing gas slugs. (F) Bubbles embedded in liquid/liquid annular flow. (G) Multiemulsions. (H) Janus fluid particles generated in a cross-junction microchannel. All scale bars are 0.5 mm. *Panel (D): Adapted from Wang et al (2015a) with permission of Wiley, panel (E): adapted from Yue et al (2014) with permission of The Royal Society of Chemistry, panel (G): adapted from Deng et al (2013a) with permission of The Royal Society of Chemistry, and panel (H): adapted from Nisisako and Hatsuzawa (2010) with kind permission from Springer Science and Business Media.*

are greater than 1 m/s, the parallel gas/liquid flow is unstable and the interface is easy to break in the flowing period.

The flow patterns of gas/liquid and liquid/liquid flows are important for control of chemical processes in microchannels. For instance, the liquid plugs in plug flow, which can provide narrow residence time distribution, can be used to droplet reactors in the study of flow chemistry (He and Jamison, 2014). More complicated systems, such as gas/liquid/liquid or liquid/liquid/liquid systems, accounting for a large proportion of fine chemical processes in pharmaceuticals and emulsions, are also a popular issue in the research of microstructured chemical processes. For gas/liquid/liquid systems in microchannels, some flow patterns are combinations of gas/liquid and liquid/liquid two-phase flows, such as the bubble droplet alternate flow given in Fig. 1D (Wang et al, 2015a), while other flow patterns are difficult to describe. For example, Fig. 1E shows a parallel water/oil flow that contains gas slugs in the oil phase (Yue et al, 2014). A similar flow pattern in a liquid/liquid/liquid system was observed in a study by Feng et al (2010). Figure 1F shows regularly organized droplets and bubbles flowing together in a microchannel (Hashimoto et al, 2007). Besides these special flow patterns, gas/liquid/liquid and liquid/liquid/liquid systems are generally more often used to generate multiemulsions, as shown in Fig. 1G, which are good templates for the preparation of hollow materials (Chu et al, 2007; Deng et al, 2013a). Xu et al (2012a) exhibited that the dual-coaxial capillary is an efficient microfluidic device for preparing monodispersed gas/liquid/liquid double emulsion with controllable structure, which provided the new idea of preparing porous microspheres with microbubbles as the direct core/pores templates (Chen et al, 2012). For generating a stable double emulsion, the three-phase interfacial tensions obey the rule of $\gamma_{io} > \gamma_{im} + \gamma_{om}$, where i, o, and m refer to the inner, outer, and middle phases, respectively. If the three-phase interfacial tensions do not obey this rule, Janus fluid particles will form as shown in Fig. 1H (Nisisako and Hatsuzawa, 2010; Thiele and Seiffert, 2011), which are theoretical models for asymmetric materials (Yu et al, 2014).

2.2 Control Mechanisms of Versatile Flow Patterns

Other complicated flow patterns, besides multiemulsions, have been generated as the number of phases in microchannels has increased (Wang et al, 2011b). Normally, we should know which kind of flow pattern can be formed in a microchannel at a certain operating condition. In order to predict these flow patterns, the control mechanisms of multiphase microflows are a key issue of concern for many researchers. First of all, flow patterns are strongly dominated by channel structure, with T-junction, Y-junction, cross-junction, and coflowing microchannels being most commonly used (Fu et al, 2012; Humphry et al, 2009; Ushikubo et al, 2014). The channel shape determines the direction of viscous ($\sim \mu u L$) or inertial ($\sim \rho u^2 L^2$) forces in or between phases (Tan et al, 2009), which withstand the total interfacial force ($\sim \gamma L$) on droplets and bubbles. These forces act to break up interfaces and the following flow of dispersed fluid. The force competition can be used to characterize flow pattern by a number of dimensionless parameters, such as the capillary number ($\mathrm{Ca} = \mu u / \gamma$) and Webber number ($\mathrm{We} = L \rho u^2 / \gamma$), as well as some other none dimensionless parameter, such as phase ratio (Q_D/Q_C). Different droplet and bubble break-up methods, such as dripping, jetting, and squeezing (Christopher et al, 2008; Fu et al, 2010), in the joined microchannels affect the size and shape of the dispersed phases that determine the flow patterns in the following microchannel. An early report on the generation mechanism of liquid plug flow was given by Garstecki et al (2006), who found that the squeezing pressure in the T-junction microchannel dominated the liquid plug pinch-off, and that the plug flow could only generate when $Ca < 0.01$. Although this criterion is somewhat arbitrary, according to succeeding research, it is a creative work owing to the drawing of a flow map using dimensionless numbers. In T-junction microchannels, except for the viscous shearing force of the continuous phase ($\sim\mu_C u_C L$) and the interfacial force ($\sim\gamma L$) between phases, the viscous force of dispersed phase ($\sim \mu_D u_D L$) is next most important, showing the degree of difficulty of the fluid break-up (Guillot et al, 2007; Tice et al, 2004). Therefore, the capillary numbers of both the continuous phase and the dispersed phase are effective in showing the flow patterns, as the example in Fig. 2A depicts. Figure 2B is a flow map of gas/liquid flow, which shows that the ratio of inertial force to interfacial force ($We_G = d_G \rho_G u_G^2/\gamma$, where d_G is the hydraulic diameter of gas phase feeding channel) is also important for determining the squeezing or dripping flow of gas phase in microchannels (Wang et al, 2013b). We believed this rule is important, since almost all the other flow patterns of gas/liquid flows are drawn with flow rates (Q) or fluid viscosities (u) (Shui et al, 2007; Zhao and Middelberg, 2011), which did not exhibit the relation between forces. In coflowing microchannels, the inertial force of the dispersed phase works similarly. Utada et al (2007) suggested a flow map using the capillary ($C_{out} = \mu_C u_C/\gamma$) and Webber ($W_{in} = d_{in} \rho_D u_D^2/\gamma$, d_{in} is the diameter of inner tube of the microdevice) numbers to distinguish the dripping and jetting of dispersed liquids.

For gas/liquid/liquid and liquid/liquid/liquid working systems, the flow patterns in the downstream microchannel of channel junctions are affected more by the feeding sequence of fluids, which determine their package structures (Hashimoto et al, 2007; Rajesh and Buwa, 2012). The adjustment of the feeding sequence should be coordinated with the control of channel wetting properties (Chen et al, 2009), which will be discussed further in the next section. When three-phase fluids are mixed together in a microchannel junction, the flow patterns are primarily determined by the force balance between fluids, if the effect of the channel wall is neglected. For example, Nie et al (2005) provided a triangle flow map to show the effect of flow rate

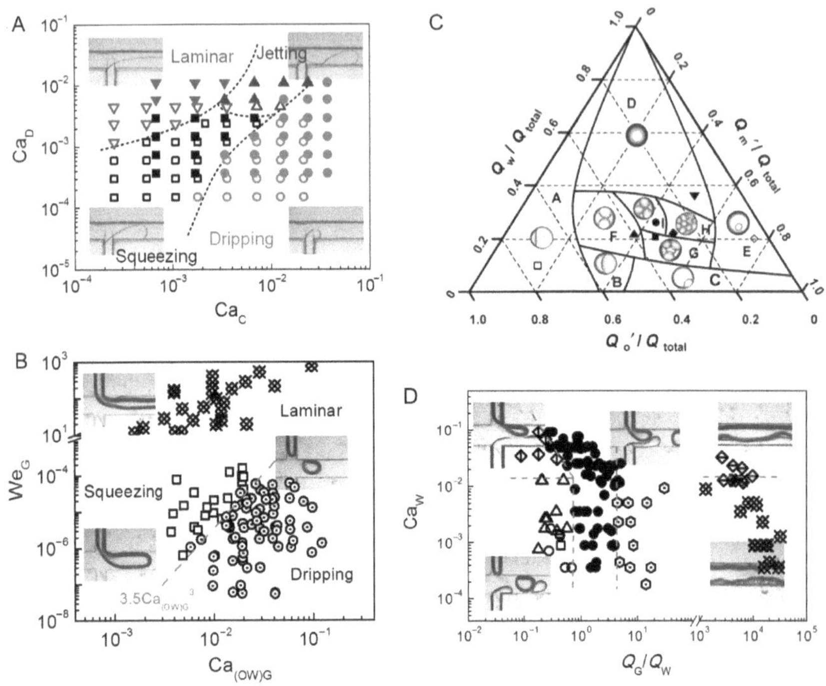

Figure 2 Flow maps of T-junction microchannels. (A) Liquid/liquid two-phase flow in a T-junction microchannel, whose cross-section is 0.52×0.2 mm^2 for the main channel and 0.27×0.2 mm^2 in for the side channel. The solid dots are from the experiment with water/2 wt% span80-dodecane and the hollow dots are from the experiment with octane/3 wt% SDS (sodium dodecyl sulfonate)—water. (B and D) Gas/liquid two-phase and gas/liquid/liquid three-phase flows in a cross-junction microchannel. (C) Liquid/liquid/liquid three-phase flows in a cross-junction microchannel in a flow-focusing microfluidic device. *Panels (B and D): These figures are adapted from Wang et al (2013b) with permission of Wiley. Panel (C): Reprinted from Nie et al (2005) with permission of American Chemical Society.*

ratio on the shape and structure of double emulsions, as shown in Fig. 2C. In a study of a cross-junction microchannel, Wang et al (2013b) gave the gas/liquid/liquid three-phase flow patterns using the capillary number of water phase ($Ca_W = \mu_W u_W / \gamma$) in the side channel and phase ratio of gas to water phase (Q_G / Q_W) from both side channels, as shown in Fig. 2D. It should be noted that some dimensionless numbers need to be modified to give a clearer force relation in the three-phase system. An example is the modified capillary number defined by Wang et al (2013b), which shows that the total shearing force comes from two phases, as shown in Fig. 2B and Eq. (1).

$$\mathrm{Ca}_{(\mathrm{OW})\mathrm{G}} = \frac{\mu_{\mathrm{OW}} u_{\mathrm{OW}}}{\gamma_{\mathrm{OG}}} = \frac{\mu_{\mathrm{O}} Q_{\mathrm{O}} + \mu_{\mathrm{W}} Q_{\mathrm{W}}}{\gamma_{\mathrm{OG}} \cdot wh} \quad (1)$$

where W, O, and G represent water, oil, and gas, respectively. For the gas–liquid, liquid–liquid and three-phase microflows, the force working mechanism should be similar, but the range of forces caused by the different viscosities and interfacial tensions of gas and liquid is the most difference of those systems.

2.3 The Action of Interfacial Force on Fluid Break-Up

In the operation of multiphase microflows, smaller dispersed droplets and bubbles are sometimes preferable to increase the surface area between fluids. The interfacial force ($\sim\gamma d$, where d is the droplet or bubble diameter in analysis of the break-up process) is decisive for the generation of micrometer-scaled droplets and bubbles, where a large amount of interfacial surface area can be produced. The interfacial force usually resists the break-up of a fluid, which mostly conflicts with the viscous shearing force ($\sim\mu_{\mathrm{C}} u_{\mathrm{C}} d_{\mathrm{i}}$, where d_{i} is the characterized size of microchannel) of the continuous phase before a force balance is reached (Christopher and Anna, 2007). The fluid break-up process in a dripping flow at the T-junction microchannel obeys this mechanism, as shown by Fig. 3A (Wang et al, 2011a). With the growth of the droplet or bubble, the shearing force continues to increase from the narrowing of the passageway in the continuous phase; droplet break-up occurs the moment after the balance between the shearing force and the interfacial force is reached. For this reason, the droplet size is inversely proportion to the strength of the shearing force, but is proportional to the interfacial force (Thorsen et al, 2001) at the break-up moment. Therefore, the droplet or bubble sizes (d) are a function of the capillary number of the continuous phase ($\mathrm{Ca} = \mu_{\mathrm{C}} u_{\mathrm{C}} / \gamma$), as shown by Eq. (2) (Xu et al, 2006a, 2006b),

$$\frac{d}{d_{\mathrm{i}}} = k\mathrm{Ca}^{\mathrm{n}} \quad (2)$$

Usually the channel width (w) is used as the characterized channel size d_{i} (Xu et al, 2006a, 2006b). Equation (2) is more suitable for the situation in which droplet size is much smaller than the channel width. For larger droplets, whose sizes are close to the channel width (some of the literature calls this flow a transition flow between dripping and squeezing, Fu et al, 2010; Xu et al, 2008a), Eq. (3) is more commonly used, where the effect of the

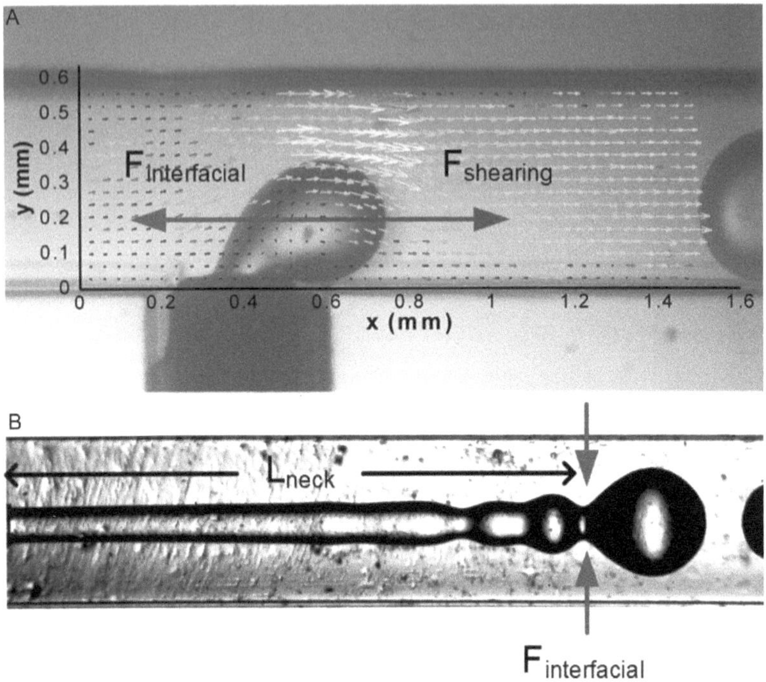

Figure 3 Forces acting on dispersed fluids. (A) The balance of shearing force and interfacial force on a dripping ruptured droplet. The small arrows show the flow field around the droplet. (B) A jetting liquid in a coflowing microchannel. The interfacial force dominates the break-up of droplets under the Rayleigh–Plateau effect. *Panel (A): This figure is adapted from Wang et al (2011a) with permission of Wiley.*

relaxation time between the force balance moment and droplet detaching moment from the side channel is considered by the flow rate ratio (Q_D/Q_C). Wang et al (2013c) showed that Eq. (3) is also workable in predicting the bubble size in the dripping flow of a coflowing microchannel.

$$\frac{d}{d_i} = k\left(\frac{Q_D}{Q_C}\right)^m \mathrm{Ca}^n \tag{3}$$

Besides its resistance effect in the droplet/bubble growing processes, the interfacial force is also important in finishing the break-up process. In jetting flow, droplet break-up is mainly caused by Rayleigh–Plateau instability, which is a spontaneous pinch-off of the interface (Guillot et al, 2007). The neck length of the dispersed phase in jetting flow is an important parameter of droplet generation, as shown in Fig. 3B. Higher interfacial tension causes faster interface break-up, which is the opposite of the effect of the shearing rate of the continuous phase that extends the neck; therefore, neck length increases with the rising of the capillary number of continuous phase (Utada et al 2008). A similar effect has been observed in other kinds of microchannel devices. For example, at the end of the generation of a gas slug in the squeezing flow at the T-junction of a microchannel, the higher interfacial tension causes faster interface contraction, which promotes the interface break-up (van Steijn et al, 2009). Wang et al (2014) showed that the irreversible surface break-up of a liquid plug only occurred at a constrictive ratio larger than 3.78 in the constrictive microchannels, which is the result of the interfacial tension caused by interface constriction. When the interfacial tension is very low, such as in the two aqueous system of polyethylene glycol (PEG)/K_3PO_4, whose interfacial tension is less than 10 μN/m, enhanced instability by artificial fluctuation is useless for causing the break-up of the fluid, but it produces beautiful patterns of liquid jets (Sauret et al, 2012).

3. THE ADJUSTABLE SOLID–FLUID INTERFACE IN MULTIPHASE MICROFLOWS

3.1 Effect of Wetting Properties on Multiphase Flow Patterns

In most flow maps reported in the literature, only the force ratios acting on fluids are exhibited using dimensionless numbers. However, these flow maps are all based on the assumption that the effect of the solid–fluid interface does not need to be considered. However, in such a confined space, the wetting property of the channel wall is crucial because of its determination of phase continuity. In traditional chemical vessels, the continuous-phase fluid is usually determined by the volume ratios, but in microchannels only the channel wetting fluid can act as the continuous phase, no matter what kinds of break-up methods occur. The flow of foams with extreme phase ratios of dispersed phase to continuous phase is a good example, as shown in Fig. 4A (Marmottant and Raven, 2009; Shui et al, 2008). The wetting property of the channel wall is usually characterized by the fluid contact angle against it. This contact angle is a comparison of the wettability of fluids on a solid surface (Ben Said et al, 2014). In an early study, Li et al (2007) found that stable multiphase flow only existed at a contact angle of the dispersed phase greater than 105°; otherwise, the interfacial force from the solid–liquid interface significantly affected the generation of droplets, and could change the droplet flow to liquid/liquid parallel flow. The attached solid–liquid interfacial force strongly affects the shape of dispersed fluid particles. When

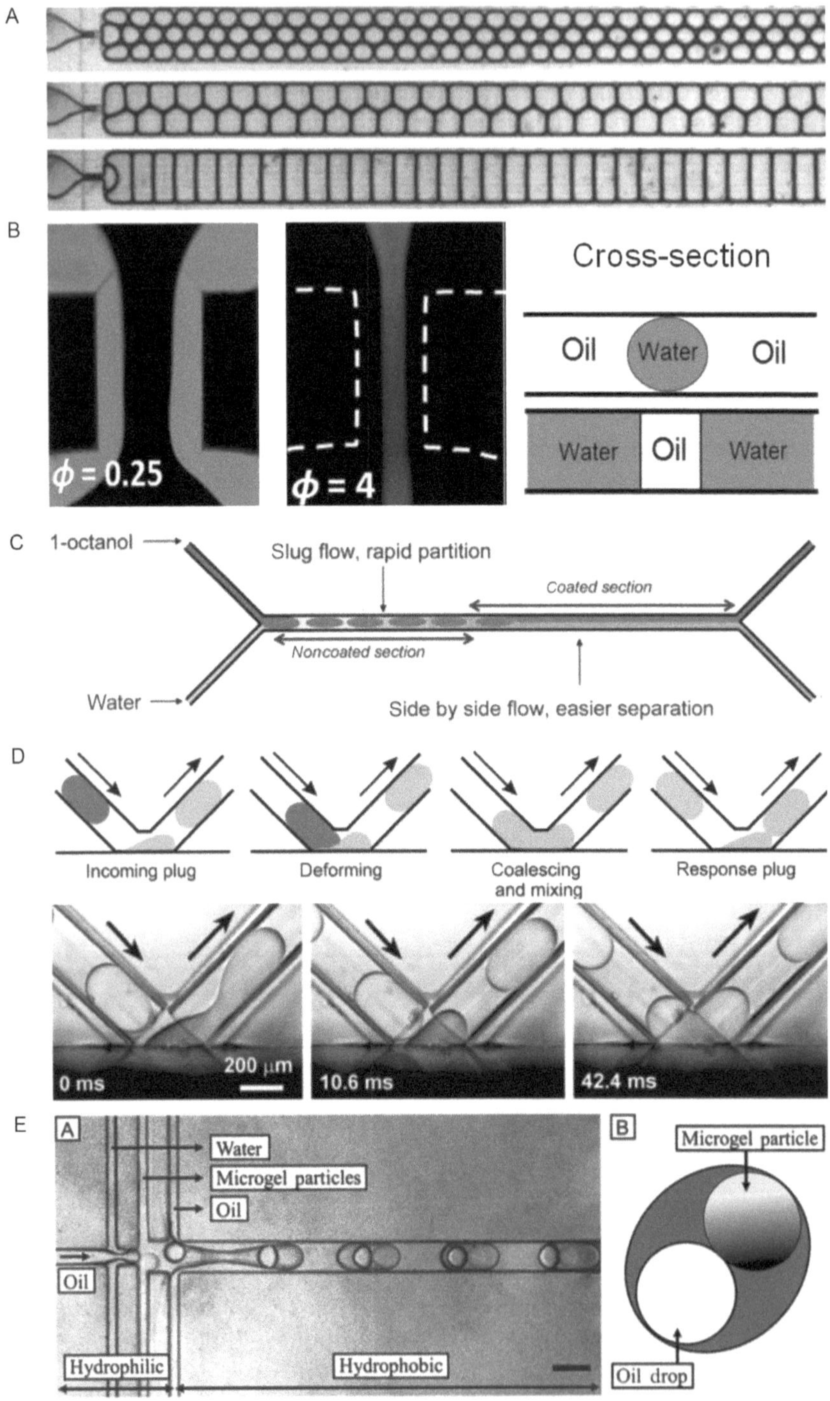

Figure 4 See legend on opposite page.

the contact angle is close to 90°, such as in the octane/water system on a polymethyl methacrylate (PMMA) plate (Xu et al, 2006c), it is hard to distinguish which phase is the continuous phase. In a similar example, the water phase of a gas/water system in a hydrophobic microchannel is easy to disperse into tiny droplets that adhere to the channel wall and are hard to flush out (Cubaud et al, 2006).

The attached solid–liquid interfacial force on the flowing fluid in a microchannel is not always an instability factor for multiphase microflows (Roberts et al, 2012). It is also a good tool for controlling flow patterns. For example, parallel flow is very hard to obtain in a microchannel at low capillary numbers (Cubaud and Mason, 2008), but it is very easy to realize in a local modified microchannel device (Zhao et al, 2001), as shown in Fig. 4B (Roberts et al, 2012). The stable parallel flow can be used in liquid/liquid extraction without worrying about phase separation (Ju et al, 2014), which is convenient for chemical analysis (Lu et al, 2011). Figure 4C shows that the local modified microchannel can not only stabilize liquid/liquid parallel flow, it is also a good tool for promoting rapid phase separation (Logtenberg et al, 2011). Using local dispersed phase wetting channels, the flowing droplet can be trapped and even experience phase inversion (Deng et al, 2013b; Liu and Ismagilov, 2009), as shown in Fig. 4D. Using a fluid-permeable membrane as the wetting medium, the membrane wetting fluid can be drained out directly (Kralj et al, 2007). For preparing multiemulsions, the wetting property of microchannel is also important for its determination of the emulsion structure. Usually, the dispersed system of the middle and inner phases is first generated and then encapsulated by the outer phase fluid (Chu et al, 2007). Since the middle

Figure 4 Effect of wetting properties on microflow. (A) Foam flows in polydimethylsiloxane (PDMS) microchannel. The air bubbles contact each other in the microchannel. (B) Having both hydrophilic and hydrophobic surfaces microchannel that allows for creation of oil-in-water or water-in-oil emulsions. (C) Fast phase separation under the effect of different local wetting properties. (D) Trapping droplet with the microchannel wall, with wetting of the dispersed phase. (E) Generation of multiemulsions with the help of microchannel modification. *Panel (A): This figure is adapted from Marmottant and Raven (2009) with permission of The Royal Society of Chemistry, panel (B): this figure is adapted from Roberts et al (2012) with permission of The Royal Society of Chemistry, panel (C): this figure is adapted from Logtenberg et al (2011) with permission of The Royal Society of Chemistry, panel (D): this figure is adapted from Liu and Ismagilov (2009) with permission of the American Chemical Society, and panel (E): this figure is adapted from Thiele and Seiffert (2011) with permission of The Royal Society of Chemistry.*

and outer phases have opposite properties, the channels are usually segmented and modified to adapt to the demands of different fluids, such as the microchannels developed by Abate et al (2011) or by Thiele and his colleague (2011) in Fig. 4D.

3.2 Methods of Microchannel Surface Modification

The solid–liquid interfacial force is a nonignorable mesoscale factor for controlling the multiphase flow in microchannels. Its source is the variable molecular composition or microscope structure of the microchannels. In order to attain local modification, several methods have been proposed, and a short summary is provided in this review. The most commonly used method is direct chemical group modification, which is easy to obtain in polydimethylsiloxane (PDMS) or glass microchannels owing to the large amount of silicon hydroxyls on them (Chae et al, 2009). Hydrosilylation reaction (Klasner et al, 2009), UV-mediated graft polymerization (Hu et al, 2004), solvent evaporation film (Kuhn et al, 2011), and chemical vapor deposition (Chen et al, 2006) have been proven to be good methods. Besides these chemical approaches, there are other modification methods. Using the differing contact angles of liquid flowing on patterned anisotropic wetting surfaces, Wang et al (2015b) successfully realized flow direction control of water droplets in a Y-shaped microchannel. A hierarchical anisotropic groove microstructure surface was developed by Kang et al (2013), which was used to operate oil droplets. Besides these static surface modifications, some dynamic modifications those can be erased or regenerated have also been proposed, such as using different kinds of surfactants (Tostado et al, 2011; Xu et al, 2006c) or using electricity (Paik et al, 2003).

4. DYNAMIC INTERFACIAL TENSION IN MULTIPHASE MICROFLOWS

4.1 Generation Mechanism of Dynamic Interfacial Tension

To obtain exact fluid dynamic rules of multiphase microflows, constant interfacial tensions are highly preferred. This is easy to achieve when using purified gas/liquid or liquid/liquid systems without any dissolved substances, such as surfactants or reactants. However, only a few dispersed systems can be operated stably in plastic microfluidic devices without additive, and the reason is the similar wetting properties of both phases on microchannel wall. For example, water/alkanes are ideal systems (Bremond et al, 2008) for PDMS devices. Water/alcohols are infrequent combinations that can obtain stable liquid/liquid dispersed flow in PMMA microchannels (Wang et al, 2013a). Glass microchannels are better since they wet most aqueous solutions, but their weakness is fabrication difficulty (Shen et al, 2006).

In most emulsification experiments, surfactants are necessary to force one phase to wet the microchannel (Baret, 2012). Once this kind of reagent is added to the system, there are more issues in the microflow studies than just fluid dynamic. For example, with the growing of a droplet at the microchannel junction, the fresh interface without adsorbed surfactant is also forming. Therefore, the dynamic adsorption of the surfactant keeps proceeding with the generation of new interface area. Although the surfactant concentration is usually much larger than its critical micelle concentration in an emulsification experiment, the interfacial tension at the droplet break-up moment is hard to reach the lowest value, when droplet surface is fully covered by the surfactant. This is because the adsorption rate of surfactant in the microdroplet generation process is lower than the fresh interface growth rate (Wang et al, 2009). When the droplet generation time is short, such as in several microseconds, the time for surfactant adsorption during the droplet growth process at microchannel junction is not enough and unfinished adsorption of surfactant will happen at the break-up moment. The varied surfactant concentration on droplet surface causes varied interfacial tension at the droplet break-up moment, which is affected by the operating conditions, such as droplet generation time, surfactant concentration, and two-phase velocity (Xu et al, 2012b). This varied interfacial tension further affects the force balance determining the droplet size and makes variable sized droplets. This variable interfacial tension is called dynamic interfacial tension in microfludic study. Besides the study of microfluidic process, the dynamic interfacial tension was observed in some membrane emulsion studies (Geerken et al, 2007; van der Graaf et al, 2004), which has a remarkable effect on droplet break-up process, such as a lag time before pressing out the droplets from the micropores was observed for reducing the interfacial tension. In microfluidic studies, dynamic interfacial tensions, which could be more than three times higher than the interfacial tension at the thermodynamic steady state, were observed in both T-junction and coflowing microchannels (Wang et al, 2009; Xu et al, 2012b).

Essentially, the reason for the appearance of varied interfacial tension is the variable reagent concentration of the interface. When using a surfactant in the droplet generation process, the low adsorption rate of the surfactant is caused by its low mass transfer rate, even for small-molecule surfactants

(Ferri and Stebe, 2000), such as cetyl trimethyl ammonium bromide (CTAB) and Tween-20. The characterized time for surfactant mass transfer (Alvarez et al, 2012) should be much smaller than the droplet generation time, which is less than 100 ms in common microfluidic processes (Fu et al, 2010). Chen et al (2015) used the dynamic adsorption of surfactants to evaluate the mass transfer coefficient during droplet formation in microchannels. By exchanging the continuous phase and the dispersed phase (e.g., *n*-hexane/water with Tween-80 as a surfactant), the mass transfer coefficients inside and outside the droplets were calculated which ranges from 3.3×10^{-4} to 7.1×10^{-3} m/s. In addition to the surfactant mass transfer rate, the surfactant composition is also important. For example, Tostado et al (2012) reported a dynamic interfacial tension phenomenon in a microfluidic experiment with a mixture of SDS-PEG (sodium dodecyl sulfonate-polyethylene glycol) as a surfactant. They found that the macromolecular surfactant PEG had an inhibiting effect on the liquid/liquid interface adsorption of the small-molecule surfactant SDS, which caused varied interfacial tension in the microdroplet generation process.

The effect of surfactants on the variation of interfacial tension is mainly discussed in microfluidic emulsion preparation studies. Except for this emulsification process, mass transfer, and chemical reactions also affect the composition of the interface. However, reports on the extraction and adsorption processes in microchannels are less prevalent in the study of microstructured chemical systems. Several tentative studies have been performed by the authors of this chapter during recent years. In the droplet generation process with mass transfer of phosphoric acid in a working extraction system of methyl isobutyl ketone/water, the dynamic interfacial tensions at the droplet break-up moment were found reducing from 15 to 5 mN/m with the increase of phosphoric acid concentration from 50 to 64 wt% in the water phase, as shown in Fig. 5A. In the study of CO_2 adsorption in microchannel devices with a mono ethanol amine (MEA) solution as the continuous phase, the interfacial tension at the CO_2–N_2 mixture gas break-up moment varied from 48 to 36 mN/m with the variation of CO_2 concentration and the flow rate of the MEA solution, which is exhibited in Fig. 5B. The results in Fig. 5 show that if the mass transfer rate is high, the concentration of mass transfer content on the interface is high and the interfacial tension tends to reduce. Since the concentration of surfactant or mass transfer content may be not uniform, it will cause interfacial tension gradient along the interface too.

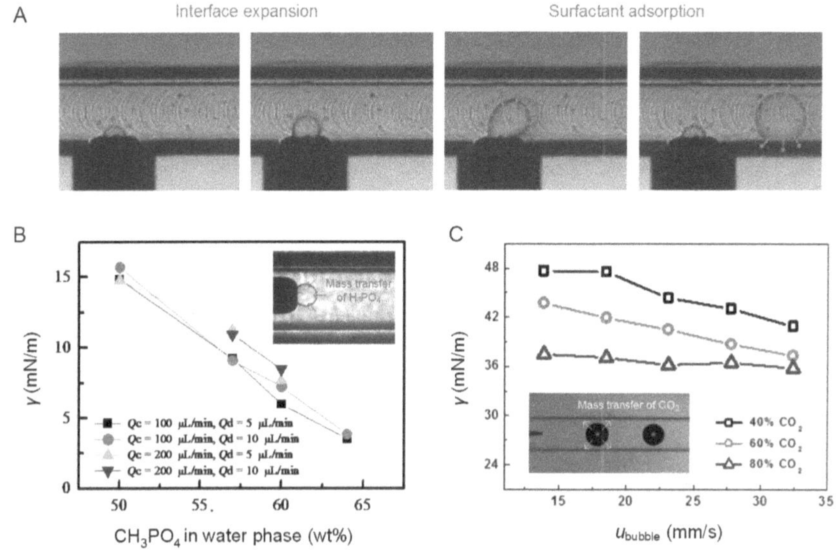

Figure 5 Dynamic interfacial tensions in mass transfer processes. (A) Schematic for variation of surfactant SDS on a droplet surface with its growth at a T-junction. The continuous phase is a Tween-20 aqueous solution and the dispersed phase is hexane in the experiment of those images. (B) Interfacial tensions at the droplet pinch-off moment with the variation of phosphoric acid concentration in the water phase. The continuous phase is a phosphoric acid aqueous solution and the dispersed phase is methyl isobutyl ketone (MIBK). (C) Interfacial tensions at the bubble pinch-off moment with the variation of CO_2 concentration in the gas phase. The continuous phase is a mono ethanol amine (MEA) aqueous solution and the dispersed phase is a CO_2–N_2 mixture. *Panel (A): This figure is adapted from Wang et al (2009) with permission of the American Chemical Society.*

4.2 Measurement of Dynamic Interfacial Tension

The varied interfacial tensions in the microflow processes we discussed above have to be measured online using fluid dynamic models. In recent years, a series of methods have been developed to determine the dynamic interfacial tension, basing on the measurements of droplet size, droplet deformation, and the pressure drop across the droplet. Using the average diameters of droplets generated from joint channels, such as T-junction and cross-junction microchannels, the interfacial tension at the droplet pinch-off moment was obtained. This method should be based on a reliable droplet size model, which contains interfacial tension. In the previous study, the dynamic interfacial tension between liquids at the droplet pinch-off moment in the dripping flow at a T-junction microchannel was determined with the size law of Eq. (3) (Wang et al, 2009). A hexane/Tween-20 aqueous solution was used as the working system and varying droplet diameters were observed when the concentration of Tween-20 was changed. The capillary numbers at the droplet break-up moment were calculated from the measured droplet diameters, then the interfacial tensions from 5 to 16 mN/m were calculated from the capillary number. The measurement error in this process is from the droplet diameter distribution, which relative deviations from the average value can be less than 3%. The total error of interfacial tension could be less than 15% using this method. In the coflowing microchannel, a force balance model given by Eq. (4) was proposed by Xu et al (2008b).

$$\left(\frac{d}{d_{\mathrm{i}}}\right)^2 = k\frac{\gamma}{\mu_{\mathrm{C}}(u_{\mathrm{C}} - u_{\mathrm{D}})}. \tag{4}$$

This model can be used to calculate dynamic interfacial tensions on rupturing droplets. Examples using a hexane/Tween-20 aqueous solution and a hexane/CTAB aqueous solution as the working systems are shown in Fig. 6A (Xu et al, 2012b). More variable interfacial tension is observed in the process preparing smaller droplets that several microns, for the larger specific surface area and shorter droplet generation time to let surfactant adsorb. The measurement error caused by droplet size distribution is less than 2% using this method.

Measurement methods based on droplet deformation use either the shape change ratio or droplet dynamics, when they flow into a microchannel with an expanded or narrowed structure. For example, with the gradual broadening of the flowing chamber, the droplet velocity decreases, which causes round droplets to become spheroid, as shown in Fig. 6B (Hudson et al, 2005). During the ensuing time, the evolution of the deformation ratio of droplet D, defined by Eq. (4), is a function of interfacial tension, which can be used to measure the dynamic interfacial tension.

$$D = \frac{a - b}{a + b} \tag{5}$$

where a and b are the major and minor radii of the spheroid droplet. Cabral and Hudson (2006) suggested that the interfacial tension at the droplet surface in this extensional flow field was a function of deformation ratio D and its derivative along the flow direction $\partial D/\partial x$, droplet velocity u,

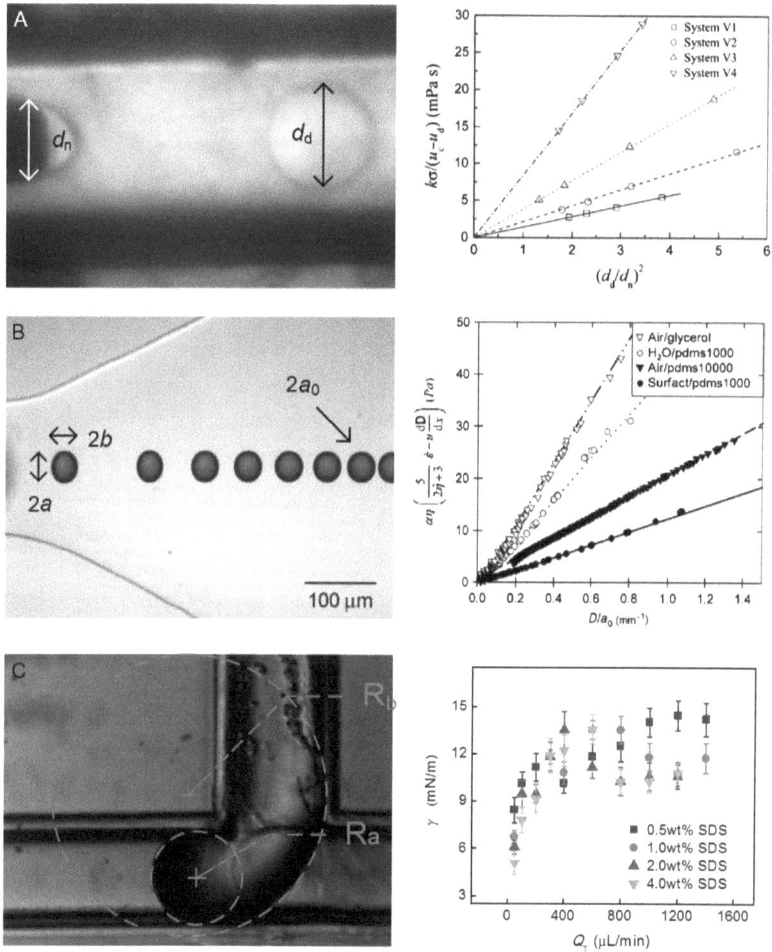

Figure 6 Measurement methods for dynamic interfacial tensions. (A) Measurement of interfacial tension at droplet pinch-off moment in a coflowing microchannel. (B) Measurement of interfacial tension on a flowing droplet surface using a broadening microchannel. (C) Measurement of interfacial tension in the droplet growth process using the pressure drop in a T-shaped microchannel. *Panel (A): This figure is adapted from Xu et al (2008a, 2008b) with permission of the American Chemical Society, panel (B): reprinted with permission from Hudson et al (2005), Copyright 2005, AIP Publishing LLC, and panel (C): with kind permission from Springer Science and Business Media.*

surrounding fluid extension rate ε $(\mathrm{d}u/\mathrm{d}x)$ and two-phase viscosity ratio η (μ_D/μ_C), which could be expressed as

$$\alpha\mu_C\left(\frac{5}{2\eta+3}\varepsilon - u\frac{\partial D}{\partial x}\right) = \gamma\left(\frac{D}{a_0}\right), \tag{6}$$

where a_0 is the equilibrium radius of the droplet and α is a function of η,

$$\alpha = \frac{(2\eta + 3)(19\eta + 16)}{40(\eta + 1)}. \tag{7}$$

According to the basic rules of extensional flow, Eq. (6) is also usable for the process of a droplet flowing into a converging microchannel, where the droplet will be stretched along the flow direction. Hudson et al developed a LabVIEW program to control the tensiometer that they called their microchannel device, and to analyze the data in real-time. The interfacial tension was then calculated from the slope of the data using D/a_0 as the x-axis and $\alpha\mu_C \cdot [(5/(2\eta+3)) \cdot \varepsilon - (u\partial D/\partial x)]$ as the y-axis. This method is suitable for the measurement of dynamic interfacial tension of flowing droplets in a microchannel, such as a mass transfer-induced dynamic interfacial tension in a mineral oil/water system with alcohol and diblock copolymer surfactants and elastic capsules and viscoelastic biological cells (Erk et al, 2012; Martin, 2009; Martin et al, 2011). The measurement errors are between 3% and 15% according to the results of Hudson et al (2005). Brosseau et al (2014) used a similar but simpler method to determine the dynamic interfacial tension of droplet flow in a microchannel. They used the largest deformation ratio of the droplets D_{max} in a suddenly broadening microchannel, and found that D_{max} was a function of the capillary number and a geometrical number $R^* = 2a_0/w$ that relates the droplet radius to the channel width w. Using this correlated equation from previous interfacial tensions, the interfacial tension variation with droplet flowing time was studied to show the adsorption kinetics of a nonionic triblock copolymer on water droplets with about 10% measurement errors.

$$D_{max} = 0.8\mathrm{Ca} \cdot R^{*3.7} \tag{8}$$

The dynamic interfacial tension measurement methods from droplet size and droplet shape analysis are appropriate for microflow processes containing small droplets. Besides these methods, the pressure drop in a microchannel is also a good way to measure the dynamic interfacial tension, especially for liquid/liquid microflows with a liquid plug, whose length is bigger than the width of the microchannel. The pressure drop method is based on the Young–Laplace equation. In the generation of a liquid plug at the T-junction microchannel, the curved plug surface produces a Laplace pressure beside it. Usually the curvature of the plug head, calculated from the width and depth dimensions of a microchannel $(1/R_a + 1/r_a)$ is different from that of the plug tail $(1/R_b + 1/r_b)$ during the plug generation process, as shown in Fig. 6C (Riaud et al, 2013; Wang et al, 2015c). These different curvatures produce a fluctuating pressure drop beside the channel junction, which can be expressed as:

$$\Delta p = \Delta p_{\mathrm{R}} + \gamma\left(\frac{1}{R_{\mathrm{a}}} + \frac{1}{r_{\mathrm{a}}} - \frac{1}{R_{\mathrm{b}}} - \frac{1}{r_{\mathrm{b}}}\right) \tag{9}$$

where Δp_{R} is the pressure drop caused by flow resistance. Glawdel et al (2012) found that in the filling stage of liquid plug growth, which was defined as the time interval before the spontaneous interface constriction took place, the radii of the droplet's head and tail in the depth direction are equal to half of the channel depth $h/2$. This conclusion simplifies Eq. (9), and allows the interfacial tension to be calculated from the measured pressure drop and the interface radii observed with a 2D view field in microscope. Using a self-developed electrical sensor, Wang et al (2015c) measured the dynamic interfacial tensions on the surfaces of growing liquid plugs at the T-junction of a microchannel using a hexane/SDS aqueous solution and a hexane/Tween-20 aqueous solution as the working systems. Using a physical model and computational fluid dynamics (CFD) simulations, similar pressure drops in a T-junction microchannel were calculated by Glawdel and Ren (2012), who measured the dynamic interfacial tensions using the same method with errors less than 20%.

4.3 Effects of Dynamic Interfacial Tension on Flow Evolution

Since interfacial tension plays an important role in determining the sizes and flow patterns of dispersed fluid particles, varying interfacial tension makes the multiphase microflow more complex in microchannels. Variable interfacial tension also causes the droplet/bubble size to be determined by the mass transfer. The surfactant concentration, flow velocity, and droplet/bubble generation time should be considered when predicting the final droplet diameters (Wang et al, 2009; Xu et al, 2012b). Flow patterns are also affected by the dynamic interfacial tension. Shao et al (2012) gave the flow maps of an *n*-butanol/phosphoric acid/water system with the mass transfer of the phosphoric acid from the aqueous phase to the organic phase. They found that the transition line from dripping to jetting flow in the flow maps changed to lower capillary and Webber numbers with increased phosphoric acid concentration in the aqueous phase.

Nonuniform disturbed surfactants further induce the Marangoni effect, which makes the flows of fluid particles different from those in working systems having constant interfacial tensions. For example, Erk et al (2012) measured the circulation velocities in droplets in order to determine the elastic interfacial effect in microflows. A new Marangoni-induced stagnation point was identified, whose distance from the droplet center was small when both the Marangoni and Peclet numbers ($Ma = E/Ca$ and Pe, respectively, where E is the interface elasticity) were large, but became larger when the either of them was small. Ghadiali and Gaver (2003) found that the creation of Marangoni stress influenced the viscous stress along the interface in a cylindrical capillary tube. When the adsorption rate of a surfactant was not close to the steady-state value, the surfactant concentration profile differed substantially from the near equilibrium approximation, which resulted in the thinning of the liquid film between the gas slug and the tube wall.

The tip-streaming of the dispersed phase is another interesting flow phenomenon in microflow processes, caused by a nonuniform surfactant in a liquid/liquid interface. Anna and Mayer (2006) reported the first tip-streaming study in a microfluidic flow-focusing device. By controlling the bulk concentration of a soluble nonionic surfactant ($C_{12}E_8$) in the neighborhood of the critical micelle, a common droplet break-up resulted in not only highly monodispersed droplets in the range of tens of micrometers in size, but also in a new mode, called thread formation, which resembled tip-streaming and yielded tiny droplets in the range of a few micrometers in size. The focusing of the surfactant at the tip of the dispersed fluid that was caused by the sustained shearing flow of the continuous phase was found to be the main reason for the appearance of this phenomenon. Ward et al (2010) further reported an experimental study concerning similar tip-streaming droplets in a microfluidic flow-focusing device. The surfactant was from a chemical reaction of NaOH and oleic acid between water and mineral oil in their study. They found that in the absence of this chemical reaction, the common dripping ruptured droplets were observed at the flow-focusing nozzle, which proceeded to become uniform droplets, but in the presence of this chemical reaction, the shape of the dispersed fluid was strongly modified by the generated surfactant. At a Damkohler number (Da) close to 1, which represented similar rates for the chemical reaction and for the flow of the liquid/liquid interface at the flow-focusing nozzle, the tip-streaming phenomenon occurred easily.

5. CONCLUSIONS AND OUTLOOK

In summary, the importance of interfacial forces between fluid/fluid and fluid/solid systems was introduced in this review. The magnitude of interfacial force can be several orders higher than gravity, and it competes with viscous and inertial forces, which determines the flow patterns and droplet or bubble sizes. Besides the fluid/fluid interfacial tension, which is of much more concern in microfluidics, the wetting property of the fluid/solid interface is also a key parameter that represent the value of additional interfacial force on the fluid from a microchannel wall. In order to adjust the contact angle of the fluid on microchannel wall, chemical and physical modifications, such as surface grafting and surfactant adsorption, have been widely used. Even though the wetting property of the channel wall has been well-controlled, the relatively slow adsorption of surfactant on the fluid/fluid interface can also cause variable flow phenomenon. The dynamic interfacial tension caused by unsaturated surfactant adsorption has been discovered, and its value can be measured using droplet size, droplet shape change, and the pressure drop in the microflow process. The dynamic interfacial tension is important to the microflow process, which makes the flow condition hard to predict. The essence of surfactant-induced dynamic interfacial tension is the dynamic variation of the interface component; therefore, it should be more common in chemical engineering processes with mass transfer or chemical reactions. However, this aspect is still under-reported to date.

As an important meso-scientific issue, the working mechanism of interfacial force in the multiphase microflow process is critical for the development of microstructured chemical systems. Considering the characteristics of chemical engineering processes, future work on the dynamic interfacial phenomenon caused by reagent mass transfer and chemical reactions will be valuable in giving an in-depth understanding of this occurrence. Breakthrough the measurement technology for distributions of surfactant, mass transfer content or chemical reactant on interfaces is very important to reveal the basic rules of dynamic interfacial tension in the multiphase microflows. In addition, the development of flow and concentration field simulation is helpful for understanding of force balance rules in the microfluidic process. Visible experimental methods, including microparticle image velocimetry and laser-induced fluorescence, are good tools. High-accuracy CFD simulations are also necessary.

ACKNOWLEDGMENTS

We gratefully acknowledge the supports from National Natural Science Foundation of China (91334201, U1463208, U1302271), the Foundation for the Author of National Excellent Doctoral Dissertation of the People's Republic of China (FANEDD 201349), and the Tsinghua University Initiative Scientific Research Program (20131089196, 20151080361) on this work.

Hudson SD, Cabral JT, Goodrum WJ, et al: Microfluidic interfacial tensiometry, *Appl Phys Lett* 87:0819058, 2005.

Humphry KJ, Ajdari A, Fernandez-Nieves A, et al: Suppression of instabilities in multiphase flow by geometric confinement, *Phys Rev E* 79:056310, 2009.

Ju SH, Peng P, Wei YQ, et al: Solvent extraction of In^{3+} with microreactor from leachant containing Fe^{2+} and Zn^{2+}, *Green Process Synth* 3:63–68, 2014.

Kang SM, Lee C, Kim HN, et al: Directional oil sliding surfaces with hierarchical anisotropic groove microstructures, *Adv Mater* 25:5756–5761, 2013.

Klasner SA, Metto EC, Roman GT, et al: Synthesis and characterization of a poly(dimethylsiloxane)-poly(ethylene oxide) block copolymer for fabrication of amphiphilic surfaces on microfluidic devices, *Langmuir* 25:10390–10396, 2009.

Kralj JG, Sahoo HR, Jensen KF: Integrated continuous microfluidic liquid-liquid extraction, *Lab Chip* 7:256–263, 2007.

Kuhn S, Hartman RL, Sultana M, et al: Teflon–coated silicon microreactors: impact on segmented liquid-liquid multiphase flows, *Langmuir* 27:6519–6527, 2011.

Lan WJ, Li SW, Lu YC, et al: Controllable preparation of microscale tubes with multiphase co-laminar flow in a double co-axial microdevice, *Lab Chip* 9:3282–3288, 2009.

Li JH, Huang W: *Towards mesoscience: the principle of compromise in competition,* Berlin, Heidelberg, 2014, Springer.

Li W, Nie ZH, Zhang H, et al: Screening of the effect of surface energy of microchannels on microfluidic emulsification, *Langmuir* 23:8010–8014, 2007.

Li JH, Ge W, Wang W, et al: *From multi-scale modeling to meso-science,* Heidelberg, 2013, Springer.

Liu Y, Ismagilov RF: Dynamics of coalescence of plugs with a hydrophilic wetting layer induced by flow in a microfluidic chemistrode, *Langmuir* 25:2854–2859, 2009.

Logtenberg H, Lopez-Martinez MJ, Feringa BL, et al: Multiple flow profiles for two-phase flow in single microfluidic channels through site-selective channel coating, *Lab Chip* 11:2030–2034, 2011.

Lu YC, Xia Y, Luo GS: Phase separation of parallel laminar flow for aqueous two phase systems in branched microchannel, *Microfluid Nanofluid* 10:1079–1086, 2011.

Luo GS, Du L, Wang YJ, et al: Controllable preparation of particles with microfluidics, *Particuology* 9:545–558, 2011.

Malsch D, Kielpinski M, Merthan R, et al: μPIV—analysis of Taylor flow in micro channels, *Chem Eng J* 135:S166–S172, 2008.

Marmottant P, Raven JP: Microfluidics with foams, *Soft Matter* 5:3385–3388, 2009.

Martin JD: Mass transfer and interfacial properties in two-phase microchannel flows, *New J Phys* 11:115005, 2009.

Martin JD, Marhefka JN, Migler KB, et al: Interfacial rheology through microfluidics, *Adv Mater* 23:426–432, 2011.

Mary P, Studer V, Tabeling P: Microfluidic droplet-based liquid-liquid extraction, *Anal Chem* 80:2680–2687, 2008.

Nie ZH, Xu SQ, Seo M, et al: Polymer particles with various shapes and morphologies produced in continuous microfluidic reactors, *J Am Chem Soc* 127:8058–8063, 2005.

Nightingale AM, Phillips TW, Bannock JH, et al: Controlled multistep synthesis in a three-phase droplet reactor, *Nat Commun* 5:3777, 2014.

Nisisako T, Hatsuzawa T: A microfluidic cross-flowing emulsion generator for producing biphasic droplets and anisotropically shaped polymer particles, *Microfluid Nanofluid* 9:427–437, 2010.

Paik P, Pamula VK, Fair RB: Rapid droplet mixers for digital microfluidic systems, *Lab Chip* 3:253–259, 2003.

Pohar A, Lakner M, Plazl I: Parallel flow of immiscible liquids in a microreactor: modeling and experimental study, *Microfluid Nanofluid* 12:307–316, 2012.

Rajesh VM, Buwa VV: Experimental characterization of gas-liquid-liquid flows in T-junction microchannels, *Chem Eng J* 207:832–844, 2012.
Riaud A, Tostado CP, Wang K, et al: A facile pressure drop measurement system and its applications to gas-liquid microflows, *Microfluid Nanofluid* 15:715–724, 2013.
Roberts CC, Rao RR, Loewenberg M, et al: Comparison of monodisperse droplet generation in flow-focusing devices with hydrophilic and hydrophobic surfaces, *Lab Chip* 12:1540–1547, 2012.
Sauret A, Spandagos C, Shum HC: Fluctuation-induced dynamics of multiphase liquid jets with ultra-low interfacial tension, *Lab Chip* 12:3380–3386, 2012.
Serra CA, Chang ZQ: Microfluidic-assisted synthesis of polymer particles, *Chem Eng Technol* 31:1099–1115, 2008.
Shao HW, Lu YC, Wang K, et al: An experimental study of liquid-liquid microflow pattern maps accompanied with mass transfer, *Chin J Chem EnG* 20:18–26, 2012.
Shen H, Fang Q, Fang ZL: A microfluidic chip based sequential injection system with trapped droplet liquid-liquid extraction and chemiluminescence detection, *Lab Chip* 6:1387–1389, 2006.
Shui LL, Eijkel J, van den Berg A: Multiphase flow in micro- and nanochannels, *Sens Actuators B-Chem* 121:263–276, 2007.
Shui LL, Mugele F, van den Berg A, et al: Geometry-controlled droplet generation in head-on microfluidic devices, *Appl Phys Lett* 93:15311315, 2008.
Song H, Ismagilov RF: Millisecond kinetics on a microfluidic chip using nanoliters of reagents, *J Am Chem Soc* 125:14613–14619, 2003.
Song H, Chen DL, Ismagilov RF: Reactions in droplets in microfluidic channels, *Angew Chem Int Ed* 45:7336–7356, 2006.
Tan J, Li SW, Wang K, et al: Gas-liquid flow in T-junction microfluidic devices with a new perpendicular rupturing flow route, *Chem Eng J* 146:428–433, 2009.
Tan J, Lu YC, Xu JH, et al: Modeling investigation of mass transfer of gas-liquid concurrent flow processes, *Sep Purif Technol* 109:77–86, 2013.
Thiele J, Seiffert S: Double emulsions with controlled morphology by microgel scaffolding, *Lab Chip* 11:3188–3192, 2011.
Thorsen T, Roberts RW, Arnold FH, et al: Dynamic pattern formation in a vesicle-generating microfluidic device, *Phys Rev Lett* 86:4163–4166, 2001.
Tice JD, Lyon AD, Ismagilov RF: Effects of viscosity on droplet formation and mixing in microfluidic channels, *Anal Chim Acta* 507:73–77, 2004.
Tonhauser C, Nataello A, Lowe H, et al: Microflow technology in polymer synthesis, *Macromolecules* 45:9551–9570, 2012.
Tostado CP, Xu JH, Luo GS: The effects of hydrophilic surfactant concentration and flow ratio on dynamic wetting in a T-junction microfluidic device, *Chem Eng J* 171:1340–1347, 2011.
Tostado CP, Xu JH, Du AW, et al: Experimental study on dynamic interfacial tension with mixture of SDS-PEG as surfactants in a coflowing microfluidic device, *Langmuir* 28:3120–3128, 2012.
Ushikubo FY, Birribilli FS, Oliveira DRB, et al: Y- and T-junction microfluidic devices: effect of fluids and interface properties and operating conditions, *Microfluid Nanofluid* 17:711–720, 2014.
Utada AS, Fernandez-Nieves A, Stone HA, et al: Dripping to jetting transitions in coflowing liquid streams, *Phys Rev Lett* 99:94502, 2007.
Utada A, Fernandez-Nieves A, Gordillo J, et al: Absolute instability of a liquid jet in a coflowing stream, *Phys Rev Lett* 100:094502, 2008.
van der Graaf S, Schroen C, van der Sman R, et al: Influence of dynamic interfacial tension on droplet formation during membrane emulsification, *J Colloid Interface Sci* 277:456–463, 2004.

van Steijn V, Kleijn CR, Kreutzer MT: Flows around confined bubbles and their importance in triggering pinch–off, *Phys Rev Lett* 103:214501, 2009.
Wang K, Lu YC, Shao HW, et al: Heat-transfer performance of a liquid–liquid micro-dispersed system, *Ind Eng Chem Res* 47:9754–9758, 2008.
Wang K, Lu YC, Xu JH, et al: Determination of dynamic interfacial tension and its effect on droplet formation in the T-shaped microdispersion process, *Langmuir* 25:2153–2158, 2009.
Wang K, Lu YC, Xu JH, et al: Generation of micromonodispersed droplets and bubbles in the capillary embedded T-junction microfluidic devices, *AICHE J* 57:299–306, 2011a.
Wang W, Xie R, Ju XJ, et al: Controllable microfluidic production of multicomponent multiple emulsions, *Lab Chip* 11:1587–1592, 2011b.
Wang K, Lu YC, Tostado CP, et al: Coalescences of microdroplets at a cross–shaped microchannel junction without strictly synchronism control, *Chem Eng J* 227:90–96, 2013a.
Wang K, Lu YC, Qin K, et al: Generating gas-liquid-liquid three-phase microflows in a cross-junction microchannel device, *Chem Eng Technol* 36:1047–1060, 2013b.
Wang K, Xie LS, Lu YC, et al: Generating microbubbles in a co-flowing microfluidic device, *Chem Eng Sci* 100:486–495, 2013c.
Wang XY, Wang K, Riaud A, et al: Experimental study of liquid/liquid second-dispersion process in constrictive microchannels, *Chem Eng J* 254:443–451, 2014.
Wang K, Qin K, Lu YC, et al: Gas/liquid/liquid three-phase flow patterns and bubble/droplet size laws in a double T-junction microchannel, *AICHE J* 61:1722–1734, 2015a.
Wang S, Wang T, Ge P, et al: Controlling flow behavior of water in microfluidics with a chemically patterned anisotropic wetting surface, *Langmuir* 31:4032–4039, 2015b.
Wang XY, Riaud A, Wang K, et al: Pressure drop-based determination of dynamic interfacial tension of droplet generation process in T-junction microchannel, *Microfluid Nanofluid* 18:503–512, 2015c.
Ward T, Faivre M, Stone HA: Drop production and tip-streaming phenomenon in a microfluidic flow-focusing device via an interfacial chemical reaction, *Langmuir* 26:9233–9239, 2010.
Wong H, Radke CJ, Morris S: The motion of long bubbles in polygonal capillaries. Part 1. Thin films, *J Fluid Mech* 292:71, 1995a.
Wong H, Radke CJ, Morris S: The motion of long bubbles in polygonal capillaries. Part 2. Drag, fluid pressure and fluid flow, *J Fluid Mech* 292:95, 1995b.
Worz O, Jackel KP, Richter T, et al: Microreactors—a new efficient tool for reactor development, *Chem Eng Technol* 24:138–142, 2001.
Xu JH, Li SW, Tan J, et al: Preparation of highly monodisperse droplet in a T-junction microfluidic device, *AICHE J* 52:3005–3010, 2006a.
Xu JH, Li SW, Chen GG, et al: Formation of monodisperse microbubbles in a microfluidic device, *AICHE J* 52:2254–2259, 2006b.
Xu JH, Li SW, Tan J, et al: Controllable preparation of monodisperse O/W and W/O emulsions in the same microfluidic device, *Langmuir* 22:7943–7946, 2006c.
Xu JH, Li SW, Tan J, et al: Correlations of droplet formation in T–junction microfluidic devices: from squeezing to dripping, *Microfluid Nanofluid* 5:711–717, 2008a.
Xu JH, Li SW, Lan WJ, et al: Microfluidic approach for rapid interfacial tension measurement, *Langmuir* 24:11287–11292, 2008b.
Xu JH, Chen R, Wang YD, et al: Controllable gas/liquid/liquid double emulsions in co-axial microfluidic devices, *Lab Chip* 12:2029–2036, 2012a.
Xu JH, Dong PF, Zhao H, et al: The dynamic effects of surfactants on droplet formation in coaxial microfluidic devices, *Langmuir* 28:9250–9258, 2012b.
Xu JH, Ge XH, Chen R, et al: Microfluidic preparation and structure evolution of double emulsions with two–phase cores, *RSC Adv* 4:1900–1906, 2014.

Yoshida J, Nagaki A, Iwasaki T, et al: Enhancement of chemical selectivity by microreactors, *Chem Eng Technol* 28:259–266, 2005.

Yu C, Zhang J, Granick S: Selective Janus particle assembly at tipping points of thermally-switched wetting, *Angew Chem Int Ed* 53:4364–4367, 2014.

Yue J, Rebrov EV, Schouten JC: Gas-liquid-liquid three-phase flow pattern and pressure drop in a microfluidic chip: similarities with gas-liquid/liquid-liquid flows, *Lab Chip* 14:1632–1649, 2014.

Zhao CX, Middelberg A: Two-phase microfluidic flows, *Chem Eng Sci* 66:1394–1411, 2011.

Zhao B, Moore JS, Beebe DJ: Surface-directed liquid flow inside microchannels, *Science* 291:1023–1026, 2001.

Zhao YC, Chen GW, Yuan Q: Liquid-liquid two-phase mass transfer in the T-junction microchannels, *AICHE J* 53:3042–3053, 2007.

CHAPTER FOUR

Mesoscale Modeling: Beyond Local Equilibrium Assumption for Multiphase Flow

Wei Wang[1], Yanpei Chen
State Key Laboratory of Multiphase Complex Systems, Institute of Process Engineering, Chinese Academy of Sciences, Beijing, China
[1]Corresponding author: e-mail address: wangwei@ipe.ac.cn

Contents

Abstract

Gas–solid fluidization is a typical nonlinear nonequilibrium system with multiscale structure. In particular, the mesoscale structure in terms of bubbles or clusters, which can be characterized by nonequilibrium features in terms of bimodal velocity distribution, energy nonequipartition, and correlated density fluctuations, is the critical factor.

Advances in Chemical Engineering, Volume 47
ISSN 0065-2377
http://dx.doi.org/10.1016/bs.ache.2015.10.009

Traditional two-fluid model (TFM) and relevant closures depend on local equilibrium and homogeneous distribution assumptions, and fail to predict the dynamic, nonequilibrium phenomena in circulating fluidized beds even with fine-grid resolution. In contrast, the mesoscale modeling, as exemplified by the energy-minimization multiscale (EMMS) model, is consistent with the nonequilibrium features in multiphase flows. Thus, the structure-dependent multifluid model conservation equations with the EMMS-based mesoscale modeling greatly improve the prediction accuracy in terms of flow, mass transfer, and reactions as well as the understanding of flow regime transitions. Such discrepancies raise the question of the applicability of the local equilibrium assumption underlying the TFM and further shed light to the necessity of mesoscale modeling.

1. MULTIPHASE FLOW: NONEQUILIBRIUM SYSTEM WITH MULTISCALE STRUCTURE

Frequently encountered in nature and process industries, multiphase flows may comprise various states of matter, e.g., gas and solid in fluidization; gas and liquid in bubble column; and gas, liquid, and solid in airlift slurry bed (Mudde, 2005). In this article, the term "phase" in multiphase flow is related to the aggregative state of flow, which is normally far from equilibrium states. And it is different from the "phase" for a thermodynamic equilibrium system, where the "phase" is used to refer to a set of equilibrium states that can be demarcated in terms of state variables by a phase boundary on a phase diagram. As a result, it is possible to have a gas–solid flow mixture with more than two phases, which can be classified by size, density of particles, or by the states of dispersion, e.g., polydisperse multiphase flow (Xue and Fox, 2014) and dilute–dense, gas–solid multiphase flow (Hong et al, 2012).

To have a primary idea of multiphase flow, Fig. 1 shows schematically how a gas-fluidized bed varies with increase of superficial gas velocity. Starting from the seemingly homogeneous fixed bed, increase of gas agitation brings flow states from homogeneous expansion with minimum fluidization, to heterogeneous bubbling fluidization, slugging fluidization (if column diameter is comparable to bubble size), turbulent fluidization, circulating fluidization (provided that a separator is equipped to circulate particles entrained from the top outlet), and pneumatic conveying (Grace et al, 2006; Kunii and Levenspiel, 1991). To quantify such transitions under steady state, one may draw the flow regime map, on which the phenomenological description of distinct flow patterns is scaled with the component flow rates, material properties, or some other quantities.

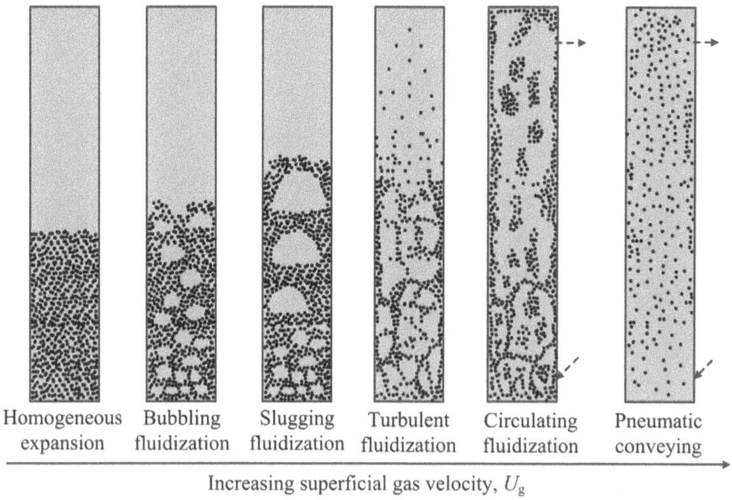

Figure 1 Schematic description of flow regimes in gas–solid fluidization.

Figure 2 shows an example of quantitative flow regime map widely applied in fluidization community, where the space of parameters is separated into material properties, as represented by the Archimedes number, *Ar*, on the abscissa, and operating parameters, as represented on the ordinate in terms of a dimensionless Beranek number Ω in the case of Reh (1971, 1996) or a dimensionless superficial gas velocity in the cases of Grace (1986) and Grace et al (2006). The superficial gas velocity of Grace's map is essentially defined at the reactor scale. In this regard, the effects of dynamic flow structure with variation of solids flux, G_s (and hence various relative/slip velocities) are not considered. In contrast, the Beranek number is a function of relative velocity, thus allowing for different dynamic flow states taken into account. In addition, the relative velocity of the Reh's map can be defined at different scales. For example, in his earlier versions (Reh, 1971), it was defined on the scale of single particle or homogeneous suspension, reflecting an isolated, microscale viewpoint or an averaged viewpoint of fluidization, respectively. In the updated version (Reh, 1996), a fitting function of Froude number, $Fr^*=1$ was included ($Fr^*=3\rho_f\cdot u_{slip}{}^2/(4(\rho_p-\rho_f)\cdot g\cdot d_p)$), where ρ denotes the density, u_{slip} (and v_{rel} in the figure) the slip velocity, d_p the particle diameter, g the gravitational acceleration, and both u and v are used to denote velocity in following sections. This relation was summarized from lots of experimental data to represent the observed fluidization heterogeneity with respect to clusters in circulating fluidized beds (CFBs). So, the cluster scale, which is between the single particle and

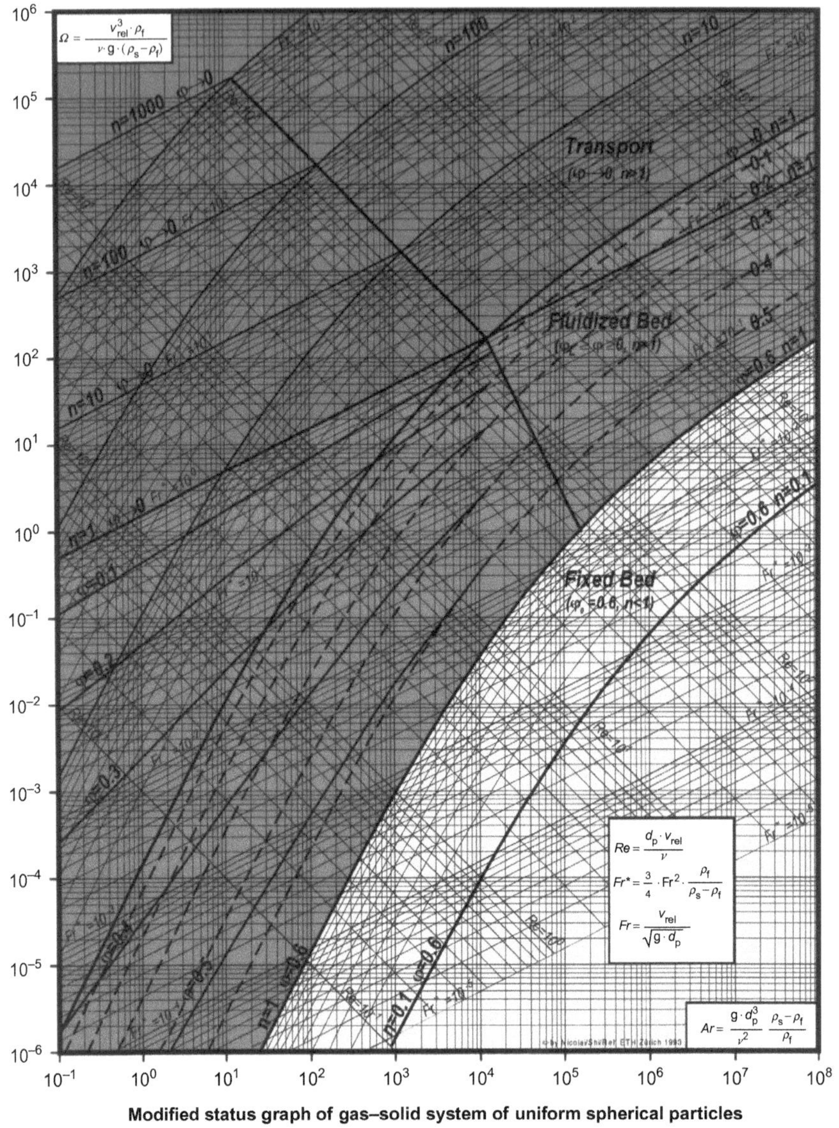

Figure 2 Flow regime map for gas–solid fluidization of Reh (1996). *Adapted from the original chart in his dissertation at University Karlsruhe (1961) and in Reh (1971).*

reactor scales, is also involved in this updated map. However, it is hard to include all different scales into one ordinate of relative velocity. As a result, even for the simplest duct geometries, there exists no universal, dimensionless flow regime map that incorporates the full, parametric dependence of the boundaries on the fluid characteristics (Brennen, 2005). If the possible geometric variation of reactors is further included, the situation will be much more complex (Wang et al, 2008b, 2010). The difficulty encountered in drawing a universal map can be reflected by the obvious blank, undefined space on the flow regime map of Grace et al (2006), which was aimed to cover various sources of experiments conducted on different geometries.

In all, the multiphase flow remains far from being understood because of its dynamic flow structure. The multiphase flow is a typical nonequilibrium system. Here, the equilibrium system refers to a concept coined in molecular thermodynamics and is normally discussed for ideal system without structure and net macroscopic flows of mass and energy (phase change and reactions are excluded here). The macroscopic properties of a molecular equilibrium system do not change with time and its velocity distribution is the Maxwellian. Obviously, a multiphase flow system is time dependent and not ideally homogeneous. In greater detail, as will be discussed in Section 3, the velocity distribution of it is not Maxwellian. Thus, it is always in nonequilibrium state. Such nonequilibrium, multiphase flow system is characterized by not only the key variables concerning net mass/energy flows (say, operating conditions in terms of gas velocity or relative velocity), but more importantly, the complex dependence of these variables on different scales (say, particle, cluster, or reactor scale) that entails a thorough investigation of the multiscale characteristics of multiphase flows.

2. FROM MOLECULAR GAS TO RAPID GRANULAR FLOW: MESOSCALE CHARACTERISTICS

The multiscale feature of nonequilibrium, multiphase flow has been extensively discussed in recent years (Li and Kwauk, 1994; Li et al, 2013; Sundaresan, 2000; van der Hoef et al, 2006, 2008; Wang et al, 2011). A typical example for this is fluidization which is the emphasis of this review.

In a fluidized bed, particles, which are discrete in naked eyes, are fluidized by gas to manifest a fluid-like behavior. Varying with operating conditions and material properties, the gas and particles in a fluidized bed are normally heterogeneously dispersed, resulting in various forms of structures over a wide range of scales with respect to time and space (Li and Kwauk, 1994, 2003; Li et al, 1998). These scales can be defined in a relative manner. If one terms the scale with respect to the smallest space being observed as the microscale, say, a single particle in the discrete element method (DEM) (Tsuji et al, 1993) or a local cell in continuum-based computational fluid dynamics (CFD), and terms the scale with respect to the global reactor or the cross section of it as the macroscale, then the wide span of scales between the micro- and the macroscales can be termed the mesoscale or the intermediate scale. The mesoscale is the bridge between microscale nature and macroscale appearance (Ge et al, 2007). And it bears the dynamic and nonequilibrium information of all flow structures, which may greatly affect the flow, heat/mass transfer, and reactions (Breault, 2006; Li and Kwauk, 2001; Wang et al, 2011). As a result, the mesoscale structures, e.g., bubbles or clusters, are the key to understand the complex multiphase flow behavior (Li et al, 2009, 2013; Wang et al, 2010, 2011).

Fluidization involves both molecular gas and granular flows. The mesoscale structures in terms of bubbles or clusters are determined by both collective motion of large quantity of particles and interstitial gas eddies of large quantity of molecules. So, a statistical view of the relation between a single particle (molecule) and many particles (molecules) is necessary to understand its mesoscale characteristics.

The statistical mechanics was originally proposed for gases. It is the pioneering work of Boltzmann that bridges the gap between the microscopic molecule/atom motion and the macroparameters of gases. In the viewpoint of statistical mechanics (Chapman and Cowling, 1970), the molecules are simplified to be smooth, rigid, elastic spheres. The intermolecular force fields are normally neglected, and hence, the molecules affect each other's motion only at collisions. The path traveled by a moving molecule between two successive collisions is called a free path. In an ordinary gas flow, the density, hydrodynamic velocity, and temperature vary throughout the gas. These inhomogeneities tend to be smoothed out within the mean free path through intermolecular collisions and related transport of mass, momentum, and energy. Therefore, in the kinetic theory of gases, the mean free path is an important indicator of the continuum assumption. For a typical molecular gas (say, CO_2), the mean free path is around 10^{-7} m (Gidaspow, 1994), smaller than the scale of Kolmogorov eddies, which is the smallest dynamically active scale of turbulence (Sagaut, 2001) and also the typical mesoscale for gas flow. The clear separation between the microscale (represented by the mean free path) and the mesoscale (represented by the Kolmogorov eddies) allows to define a series of "scale-independent" parameters, such as flux or mass density. This scale independency indicates that the homogeneous, local equilibrium state can be achieved in a macroscopically nonequilibrium gas flow system, and hence, the hydrodynamic description based on continuum assumption, say, the Navier–Stokes equation for Newtonian fluid, is feasible.

In contrast to molecular gas flow, granular materials and rapid granular flows (or granular gas) are characterized by the dissipative nature, which leads to lack of both spatial and temporal scale separation (Campbell, 1990). For example, in a simply sheared, stationary, monodisperse granular flow, the microtime scale (the mean free time) is close to the macrotime scale which is one over the shear rate. The same is true of the spatial scale in such system, that is, the mean free path is comparable with the system size. In greater detail (Tan and Goldhirsch, 1998), the change of the macroscale velocity within a mean free path is γl, where γ is the shear rate and l is the mean free path, and the fluctuating speed is $\theta^{0.5}$, where θ is the granular temperature, so the ratio between micro- and macroscales is $\gamma l/\theta^{0.5}$, which is proportional to $(1-e^2)^{0.5}$, where e is the restitution coefficient. The smaller e is, the closer the ratio is to 1, and it is harder to separate micro- and macroscales. So, there is no plateau of scales for rapid granular flow (or granular gas) that can be used to define scale-independent parameters. The parameters (Glasser and Goldhirsch, 2001) in granular flow become "scale dependent." This scale dependency does not vanish with the dwindling of the computational cell in CFD simulation. That is the reason why the mesoscale structure is sometimes called microstructure in rapid granular flows (Campbell, 1990; Goldhirsch, 2003), because the micro-, meso-, and even macroscales are hard to be separated.

Huge problems arise due to lack of scale separation. The assumption of fast local equilibrium cannot be employed *a priori* except for the nearly elastic system with slow change of the external parameters. As a result, the hydrodynamic description, which demands that the mean free time is far less than the hydrodynamic scale, cannot be satisfied for granular materials, as indicated in Du et al (1995) by showing the breakdown of hydrodynamics in a one-dimensional system of inelastic particles.

We can expect the scale-dependent problems of fluidization are more complex due to coupling between gas and granular solids. How to quantify the nonequilibrium characteristics related with such mesoscale structures becomes important to tackle the difficulty.

3. NONEQUILIBRIUM FEATURES OF GRANULAR FLOWS

Specifically, at least three typical features are related with nonequilibrium granular flow systems: (i) feasibility of one granular temperature; (ii) non-Gaussian behavior of velocity distribution, and (iii) strong correlated density fluctuations in forms of bubbling or clustering. In the following section, we detail the above three points.

3.1 Is Granular Temperature Existent? Or Lack of Energy Equipartition?

Granular medium is athermal, i.e., it could not be agitated by atomic-level thermal motion. So the granular temperature is fundamentally different with the molecular temperature which is based on the zeroth and the second laws of thermodynamics. Granular temperature is defined as a measure of the velocity fluctuations of particles, first appeared in Ackerman and Shen (1978) and normally reads $\theta = 1/3\langle m(\mathbf{c}-\mathbf{v})^2\rangle$, where $\mathbf{v}$ is the mean velocity of particles and $\mathbf{c}$ is the particle velocity. This concept is paramount to understand the statistical behavior of granular flow and it is widely applied to construct the kinetic and hydrodynamic theories for granular gases (Goldhirsch, 2008). In multiphase flow kinetic theory, the granular temperature is usually defined with $T = 1/3\langle(\mathbf{c}-\mathbf{v})^2\rangle$, which differs from θ by a factor of the particle mass and hence is easier to make use of void fraction in equation (Gidaspow, 1994). Their physical meanings are essentially the same.

However, the granular temperature is a notion borrowed from equilibrium state of molecular gases which is not designed to describe such system. Then it is not surprising that it raises many problems and disputes and requires further investigations (Baldassarri et al, 2005; Evesque, 2002).

First of all, the energy equipartition is violated for granular materials. In other words, the prediction of the "zeroth law" of thermodynamics of equilibrium state, which applies to molecular gases, is not true for granular materials. Here, we provide four situations demonstrating the universal of such unequipartition of granular temperature: (1) Two species of granular mixture driven by vibration do not share the same granular temperature (Feitosa and Menon, 2002; Meer and Reimann, 2006; Wang and Menon, 2008), no matter how long we observe; (2) The granular temperature in a vibrofluidized bed is anisotropic, in the sense that the component of granular temperature in the vibrated direction is not equal to, or, indeed exceeding, those in the other directions, which is not only confirmed in 2D vibrofluidized granular gases but also for 3D granular gases of rod-like grains (Harth et al, 2013); (3) Even confined in the vibrated direction, the positive and negative parts of the granular temperature, or, the magnitudes of fluctuating velocity, are not equal to each other (Chen et al, 2012; van der Meer and Reimann, 2006); and (4) For a granular material in a gas-fluidized bed, the same situation exists such that the granular temperature components are not equal to each other, which is confirmed in experiments (Breault et al, 2005; Gidaspow et al, 2004; Jung et al, 2005), showing that the vertical standard deviation could be four times higher than the lateral one in a bubbling fluidized bed.

Many attempts to define effective granular temperatures (Ono et al, 2002) rather than the granular temperature are mostly still based on the assumption that the local equilibrium is valid (Segre et al, 2001). For example, Edwards (1991) and Edwards and Grinev (1998) identified the dense granular system to microcanonical ensemble and proposed granular temperature based on the so-called Edwards entropy. There are also efforts proposing effective granular temperature through the fluctuation-dissipation theorem (Barrat et al, 2000; Behringer, 2002; Chen and Hou, 2014), though the theorem in itself is based on equilibrium fluctuations. In addition, lack of scale separation in granular matter as mentioned above (Goldhirsch, 1999) makes the measurement of granular temperature scale dependent. As the granular system may not reach local equilibrium in mean free time, the so-called turbulent granular temperature on the scale of clusters is separated from the fluctuation of single particles by Gidaspow and his colleagues (Jung et al, 2005). More discussions on the granular temperature are referred to the work of Goldhirsch (1999).

In short, the definition of the granular temperature is different with the molecular one. It cannot assume the same physical role as the temperature in molecular gases, and cannot represent the detailed information of energy fluctuation for granular gases. Further research is definitely needed on that.

3.2 Non-Gaussian Velocity Distribution

Non-Gaussian velocity distribution is another nonequilibrium feature of granular gases in steady state. The Gaussian velocity distribution or Maxwell–Boltzmann velocity distribution is a fundamental feature for molecular gases. And it is a prerequisite to apply analogously the kinetic theory for molecular gases to inelastic granular gases.

Warr et al (1995) first measured the velocity distribution in vibrofluidized granular gases with high-speed camera, and they reported a Gaussian distribution. Later on, non-Gaussian velocity distributions were revealed in most of granular systems (except for extremely dilute condition; Hou et al, 2008) with advance obtained in measurement technology. Figure 3 shows such a typical velocity distribution (Losert et al, 1999). Now it is widely accepted that the granular velocity distribution has a stretched tail in the high-velocity region compared with the Maxwellian distribution (Losert et al, 1999), i.e., stretched exponential distribution

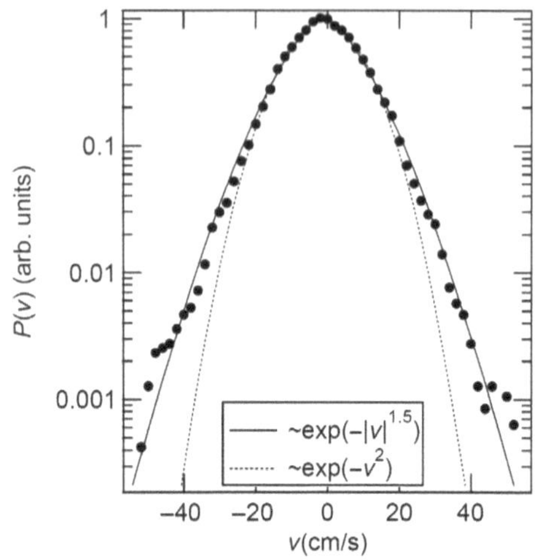

Figure 3 The velocity distribution of granular gases on logarithmic scale. The dots denote the experimental results (Losert et al, 1999), the solid line is exponential fit with $P(v) \sim \exp(|v/\sigma|^{1.5})$, and the dash line is the Gaussian fit.

$P(v) \sim \exp(-|v/\sigma|^{\alpha})$ as reported in both simulations (Brey and Ruiz-Montero, 2003; Brey et al, 1999) and experiments (Kudrolli and Henry, 2000; Losert et al, 1999; Olafsen and Urbach, 1999; Rouyer and Menon, 2000), where the value of α covers a range from 1.5 to 2. Relevant theories (Noije and Ernst, 1998) based on spatially homogeneous white-noise heating (or energy input) obtained $\alpha = 3/2$. However, it should be noted that the heating methods are different in that the boundary heating is usually used in experiments to inject the energy into granular systems, whereas a homogeneous heating is normally used in the theory for the sake of simplicity (Barrat et al, 2005). Further research (van Zon and MacKintosh, 2004, 2005) has found that heating method directly dominates the shape of the velocity distribution. Indeed in a two-layer, vertically vibrated granular system (Baxter and Olafsen, 2003), the distribution of the first layer is found non-Gaussian and the second layer Gaussian, reflecting the close relation of velocity distribution to the boundary.

The above results are based on the analysis of velocity distribution over the whole domain. To probe into the local state, a vibrated cell of a vibrofluidized granular gas in microgravity environment is divided into several strips (Chen et al, 2012, 2013). The results show that local velocity distribution of the velocity component in the vibration direction is asymmetric. In particular, it appears to be a bimodal distribution which comprises

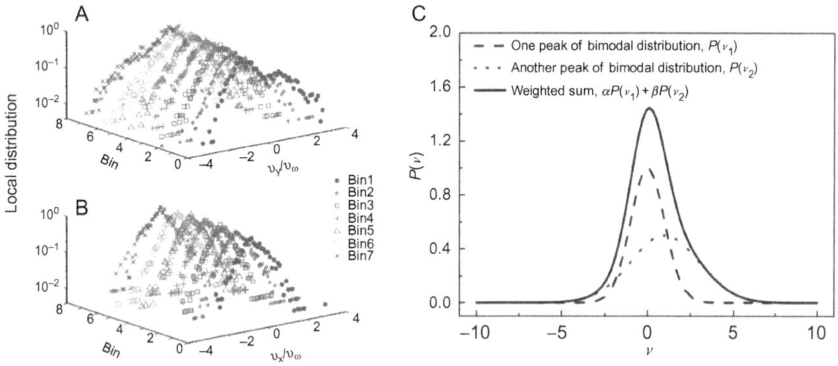

Figure 4 Local distribution functions of (A) v_y and (B) v_x on log-linear scales and (C) the schematic collapse of the bimodal distribution into the exponential. There are seven bins along the vibration direction *y*. The vibration parameters ($A=0.23$ mm, $V_\omega=0.07$ m/s, $\Gamma=21.56$ m/s^2, $f=49$ Hz) (Adapted from Chen et al, 2012).

two peaks in the strips near the boundary, where the energy is supplied (or heating), as illustrated in Fig. 4. The bimodal distribution quickly decays to Maxwell distribution away from the boundary where there is no energy input. The same phenomena are discovered in the DEM simulations (Herbst et al, 2004, 2005). This is believed the reason why the non-Gaussian distribution is only observed in the boundary heating cases (van Zon and MacKintosh, 2004, 2005). In other words, the energy input is critical to the mesoscale structure. It should also be noted that in evenly driven molecular dynamics simulation (Chen et al, 2014), when the granular gas is elastic, the velocity distribution is Gaussian even with a boundary heating. So, the inelasticity and heating (energy input) are the key factors for the non-equilibrium, bimodal velocity distribution. In air-fluidized beds, the heating is more complex, comprising both boundary heating and drag-induced inhomogeneous heating inside the bed. We can expect the non-Gaussian and locally bimodal distributions exist in such cases. And if the mean velocities corresponding to the two peaks are near each other, the bimodal distribution may merge into a stretched exponential distribution, as shown in Fig. 4C.

3.3 Strong Correlated Density Fluctuations

Granular flows are nonequilibrium inherently featuring dissipative, mesoscale structures. To sustain a steady state of granular flows, or, fluidized beds, a continual energy injection is necessary to balance the dissipation. In the past decades, ever since fluidization was recognized as an important field of chemical engineering, inhomogeneous flow structures have been the focus of research. The pioneering work of Wilhelm and Kwauk (1948) classified the whole regime of fluidization into the particulate (homogeneous) and the aggregative (heterogeneous) according to the value of Froude number. At early decades of development, bubbles were recognized as the key to understand heterogeneous fluidization, with focus on the bubble size, shape, rising velocity, and so on (Davidson et al, 1977). It was recognized later that particles are not dispersed uniformly even when the superficial gas velocity is larger than the terminal velocity. With increase of superficial gas velocity, the characteristic flow structure transforms from bubbles to particle clusters with inversion of the continuous phase (Li and Kwauk, 1980; Squires et al, 1985; Yerushalmi et al, 1976), and the flow regime transforms into the CFBs.

In a CFB, clustering may hinder the settling of particles. And bypass of gas around clusters reduces the interphase mass and momentum transfer significantly (Li and Kwauk, 2003). It has been reported that dynamic variation of particle clusters is responsible for the huge discrepancies of five orders of magnitude in mass transfer coefficient and three orders of magnitude in drag coefficient (Breault, 2006; Li and Kwauk, 2001; Wang et al, 2010). Therefore, the particle cluster in terms of its shape, diameter, velocity, and frequency, as well as its effects on heat/mass transfer and reactions, has become a hot topic in fluidization research (Agrawal et al, 2001; Dong et al, 2008a, 2008b; Harris et al, 2002; Holloway and Sundaresan, 2012; Li and Kwauk, 1994). And it is found to trigger the macroscale flow regime transition as represented by the "choking" phenomenon (Wang et al, 2007, 2008a, 2008b; Yang, 2004; Zenz, 1949). Generally, the bubble and cluster are two extremes of the continuous span of flow structures in a fluidized bed. And they are found through bifurcation analysis belonging to the same family of nonuniform solution. Whether instability of the uniform state leads to a cluster or a bubble is based on which "phase" (high solids or low solids) is continuous (Glasser et al, 1998).

Through optical fiber measurement, we now know that the bimodal probability distribution exists over the entire range of solids volume fraction. For a high-velocity CFB, one peak of the bimodal distribution corresponds to the dense "cluster" phase and the other to the dilute "broth" phase (Li and Kwauk, 1994). For a low-velocity, bubbling fluidized bed, this bimodal structure can be characterized by "emulsion" and "bubble" phases (Davidson and Harrison, 1963; Davidson et al, 1977). Lin et al (2001) found that the solids volume fraction of the dense phase exhibits a Gaussian distribution and that of the dilute phase has a log-normal distribution.

From the above discussions, it is clear that the nonequilibrium behavior of granular flows and fluidization is characterized by the nonequipartition of energy, non-Gaussian behavior of velocity distribution, and correlated density fluctuation in forms of bubbling and clustering, and it requires fresh modeling approaches. If we reduce the mesoscale structures into a dilute–dense, structure-based description, we may have updated hydrodynamic modeling that satisfies nonequipartition of energy (two granular temperatures), bimodal velocity distribution (non-Gaussian) and inherent clustering, which are consistent with the nonequilibrium features. That approach seems to be a reasonable simplification to the complex multiphase flow in gas-fluidized beds. Before detailing the relevant methods, in what follows we will first review the classical approaches, in particular, the kinetic theory and two-fluid model (TFM). This constitutes the basis for the following section of the mesoscale modeling.

4. KINETIC THEORY AND HYDRODYNAMIC MODEL

4.1 From Boltzmann Equation to TFM

Boltzmann (Chapman and Cowling, 1970) first gave an integro-differential equation which determines evolution of the velocity distribution function f in phase space for gases as follows:

$$\frac{\partial f}{\partial t} + \mathbf{c} \cdot \frac{\partial f}{\partial \mathbf{r}} + \mathbf{F} \cdot \frac{\partial f}{\partial \mathbf{c}} = \frac{\partial_e f}{\partial t}, \tag{1}$$

where $\mathbf{F}$ denotes the force on a molecule at position $\mathbf{r}$ and time t. On the right hand side (RHS), $\partial_e f/\partial t$ denotes the change rate of f due to collisions. This equation is called Boltzmann equation or Maxwell–Boltzmann equation. It can be derived rigorously from BBGKY hierarchy (Chapman and Cowling, 1970).

The first assumption in deriving the Boltzmann equation is the so-called molecular chaos, or the velocities of colliding particles are uncorrelated. This assumption makes Boltzmann equation irreversible in time. And clusters which reflect strongly correlated density fluctuations cannot be considered in this way. The second assumption is that the gas is dilute and only binary, short-duration collisions are considered. (Technically, the short duration requires the force decaying more rapidly than r^{-2} when the interparticle separation r is large.) This assumption limits the scope of application of Boltzmann equation only at low densities.

When the system is in equilibrium, the velocity distribution function is independent of time. And in the absence of external force field, the Boltzmann equation reduces to $\partial_e f/\partial t = 0$ (Mclennan, 1989). Then solving the Boltzmann's equation under principle of detailed balance, that is, $f'f'_1 = ff_1$, we can obtain the equilibrium distribution function, which is exactly the Maxwell distribution. Multiplying Boltzmann equation by any molecular property $\boldsymbol{\Phi}$, we can obtain the equations of change of any mean molecular properties—the macroscopic balance equations for the moments of the distribution function or the hydrodynamic equations. In these conservation equations, the relationship between the thermodynamic variables and fluxes, namely, the constitutive equations (Mclennan, 1989), such as the constitutive dependence of the stress tensor on the flow properties, can be derived by constructing a particular solution, i.e., the normal solution to the Boltzmann equation using Chapman–Enskog expansion (Chapman and Cowling, 1970) that constitutes the kinetic theories for nonuniform gases.

The seeming analogy between molecular gas and the fluid-like granular flow triggers the wide and successful application of hydrodynamic modeling (Campbell, 1990). By introducing the coefficient of restitution for dissipation, a set of continuum equations of dissipative granular materials can be derived, which constitutes the widely used TFM (Brey et al, 1998; Campbell, 1990; Gidaspow, 1994; Jenkins and Richman, 1985). The TFM is such termed because the solid particles are assumed to obey the local equilibrium with equipartition of energy and hence can be treated as one continuous fluid phase. The statistical properties are simply described by using local values of macroscopic quantities, say, the granular temperature (Du et al, 1995). To close the fluid-like stress/strain rate relation, Savage and Jeffrey (1981) first derived the collision component of the stress tensor with a Maxwellian distribution assumption. However, the above derivations only contain the collisional portion of the stress tensor, and the streaming component vanishes. Lun et al (1984) corrected the assumption of Maxwellian distribution and got more accurate results of the streaming tensor. Gidaspow (1994) summarized the progress obtained in the kinetic theory-based TFM. Energy equipartition was assumed in the above theories until two granular temperatures were included in describing gas-driven binary granular flows (Lu et al, 2001). And recent efforts began to treat the dispersive particles and aggregative clusters as different fluids, which is beyond the conventional TFM framework (Hong et al, 2012; Lu et al, 2008; Wang and Li, 2007).

4.2 Disputes in Two-Fluid Modeling

The TFM has been widely used for simulation of gas–solid fluidization. Based on the local equilibrium assumption, it treats the collective behavior of solid particles as a pseudo-fluid, whose strain–stress relation can be closed with the kinetic theory of granular flow (KTGF). The homogenous drag is further used, which is actually a requirement of the local equilibrium assumption.

In its early development, the TFM with homogeneous drag and kinetic theory is applied successfully to simulate bubbling fluidization and dilute pneumatic transport (Ding and Gidaspow, 1990; Nieuwland et al, 1996; Sinclair and Jackson, 1989). When a dense CFB riser is simulated, however, the situation is quite different, as the high-velocity, dense gas–solids flow is characterized by heterogeneous, dynamic clusters on the mesoscale (Agrawal et al, 2001; Yang et al, 2003), and these mesoscale structures can be smaller than the grid size used in CFD simulations (Schneiderbauer and Pirker, 2014; Wang and Li, 2007).

For simulation of bubbling fluidized beds, some argued if the grid is fine enough to the size of 10 times the particle diameter (or even smaller), then the conventional TFM with closures derived from homogeneous systems may well predict the flow behavior (Agrawal et al, 2001; Parmentier et al, 2012; Wang et al, 2009). Xie et al (2008a, 2008b) found that 2D and 3D simulations coincide with each other for bubbling fluidization of Geldart B particles, but deviation grows for turbulent fluidized bed even when the computational grid is fine enough to reach its grid-independent solution. Lu et al (2011) indicated that the effects of mesoscale structure decay with increasing particle diameter, or, the Archimedes number, so, it is easier to reach grid-independent solution for low-velocity, coarse particle fluidization.

Hot disputes exist as to whether the fine-grid TFM simulation is feasible to capture CFB flow behavior (Benyahia, 2012; Hong et al, 2015; Lu et al, 2009; Syamlal and Pannala, 2011; Wang et al, 2010). For the so-called fast fluidization with Geldart A particles, say, fluid catalytic cracking (FCC) catalyst, Lu et al (2009) pointed out that the fine-grid simulation may improve the results but it is not sufficient to predict correctly the solids flux. Benyahia (2012) refined the grid size further to 1 mm and found that the refinement of grid helps to predict the S-shaped axial profile, though the predicted solids flux is still much higher than experimental data. Hong et al (2015) also pointed out that fine-grid simulation with homogeneous drag fails to predict reasonable solids flux, whereas the solids flux is an important factor to characterize the nonequilibrium state of CFBs, as represented in Fig. 2, the fluidization diagram of Reh (1996), and relevant discussions in Section 1.

Hot disputes also exist in the closure for TFMs. Integration of the normal and shear stresses over the surface of particles gives various interphase forces, in which the drag force represents the time-independent component of the longitudinal force (along direction of relative velocity), besides the time-dependent longitudinal forces (e.g., added mass force, Basset force), transversal forces (or lift forces perpendicular to relative velocity), and the buoyancy irrelevant to slip (Crowe and Michaelides, 2006). Disputes never cease as to the closure of the key component, the effective drag force, when many interacting particles exist in dense multiphase flows. Here, the effective drag force is termed because the symmetrical, steady-state presumptions in defining the drag force for single particles are hard to be satisfied for many-particle systems. And the effective drag force for many particles is actually an averaged interphase force defined on the macroscopically steady state, including probably contributions of all the above components of forces owing to tortuous, asymmetrical flow field between particles and interstitial fluid.

There are several possible approaches to close the effective drag force for multiphase flow systems: (1) Assuming drag force equal to effective gravity (say, gravity of particles minus buoyancy): this force balance relation is normally assumed at the reactor scale under steady state, but it is not valid transiently due to local acceleration of particles; (2) Taking the drag coefficient obtained under the conditions of static array or homogeneous suspension of particles: Ergun (1952) and Richardson and Zaki (1954) drag coefficients from experiments and Koch and Sangani (1999) drag coefficient from simulation are widely cited examples, which actually assumes unchanged structure irrespective of the variation of relative velocity; (3) Deriving the drag coefficient by performing a set of fine-grid simulation over a periodic domain and then from the force balance between gravity and drag force: the filtered drag coefficients of Agrawal et al (2001), Igci et al (2008), and Igci and Sundaresan (2011) are typical examples, which take into account the effects of dynamic structure on the drag coefficient but with constraint of local force balance (also local equilibrium); and (4) Deriving the drag coefficient with the local resolution of the energy-minimization multiscale (EMMS) model: the EMMS/matrix drag (Hong et al, 2012; Lu et al, 2009; Wang and Li, 2007) is a typical example, which considers the effects of dynamic structure with nonequilibrium feature and the detail of which is referred in Section 5.

Of all these approaches, the first one is seldom applied because it contradicts the obvious fact of local acceleration of particles. The second one is widely used in CFD simulation, especially for bubbling fluidized beds, which is expected to be homogeneous in local space and generally satisfy the local equilibrium. Both the third one and the last one have considered the effects of dynamic structure. However, as the third one is obtained under the constraint of force balance over periodic domain, the voidage is the only independent parameter to determine the flow structure properties and thus the structure-dependent drag correction is a function only of voidage. Whereas the EMMS/matrix drag is obtained without such constraint, the structure-dependent drag correction is hence a function of both voidage and slip velocity, which better represents the nonequilibrium features of multiphase flow. More details on the discussion of the functional dependence of drag correction are referred in Wang et al (2010).

In all, the ambiguity of what is appropriate hydrodynamic modeling approach reflects the difficulty of quantification of nonequilibrium systems. In summary to the above sections, various inhomogeneous, mesoscale structures always occupy the central role of research in chemical engineering community ever since the beginning of fluidization as a typical unit operation. In contrast, the past efforts on fluid mechanics modeling largely stay on the hydrodynamic description of fluid-like granular flows, which is based on the seeming analogy between granular gas and molecular gas as well as the local equilibrium assumption (here, discrete simulation methods, such as CFD-DEM (Xu and Yu, 1997) and MP-PIC (Snider, 2001), are not considered). Recent years, however, have witnessed a clear tendency of fusion between the communities of physics, fluid mechanics, and fluidization, and hence a blossom of mesoscale modeling research (Agrawal et al, 2001; Li and Kwauk, 1994; Li et al, 2009; Wang et al, 2010), which can be viewed as a cutting-edge field among traditional chemical engineering, granular matter physics, and fluid mechanics. In what follows, we will discuss particularly the mesoscale modeling and its solution to unravel the nonequilibrium characteristics of multiphase flows with focus on fluidization.

5. MESOSCALE MODELING

Bearing in mind the nonequilibrium features of granular flows with bimodal velocity and density distributions, as presented in Section 3, we proposed a set of structure-dependent multifluid modeling (SFM) methods (Hong et al, 2012, 2013; Liu et al, 2015; Song et al, 2014). The bimodal velocity distribution was incorporated into the Boltzmann equation, thus introducing the mesoscale structure into the conservation equations for the mass, momentum, energy, and species. Based on these conservations, we deduced and analyzed the EMMS model (Li and Kwauk, 1994) for the steady state of fluidized beds and its stability condition. A generalized framework of the multiscale CFD can thus be established with stability-constrained characterization of mesoscale structure, as shown in Fig. 5. In the last section, we will further extend such method to account for the structural effects in Eulerian–Lagrangian approaches (Song et al, 2015).

5.1 Structure-Dependent Hydrodynamics

As discussed above, depending on the combined effects of material properties and operating conditions, the mesoscale structures take the forms of bubbles, clusters, and so on. All these mesoscale structures are found to obey the bimodal distribution, and thus, we can establish the generalized conservation

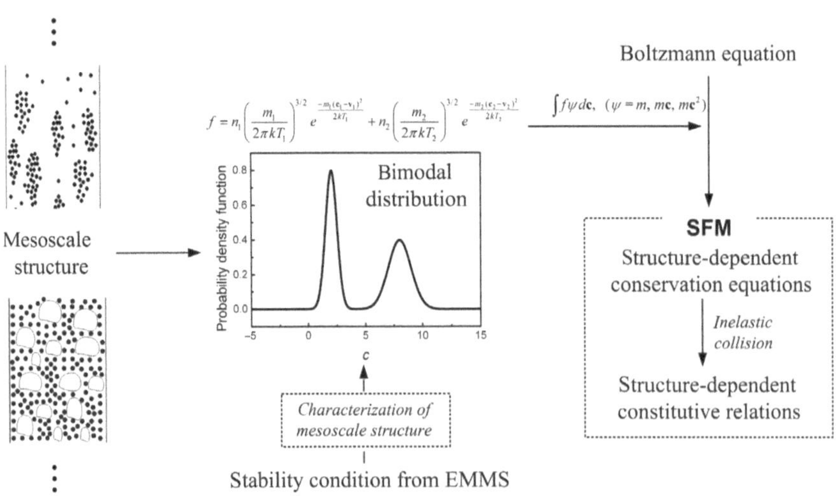

Figure 5 The generalized framework of the multiscale CFD with incorporation of bimodal velocity distribution.

equations with bimodal distribution. The difference induced by the mesoscale structures lies in the closure of the mesoscale interactions.

Corresponding to the bimodal distribution, at superficial gas velocity higher than the transport velocity (Grace et al, 2006), a monodisperse, air-fluidized gas–solid two-phase flow mixture can be classified into dense clusters (denoted by subscript c) and dilute broth (denoted by subscript f) (Li and Kwauk, 1994). The dense clusters are dispersed in the continuous dilute phase. We may refine this broth-cluster structure by defining four continua or structural subelements, as shown in Fig. 6, namely, the dense-phase gas (denoted by subscript gc), the dense-phase solid (denoted by subscript pc), the dilute-phase gas (denoted by subscript gf), and the dilute-phase solid (denoted by subscript pf). The volume fraction of the dense phase is defined by f. The void fractions in the dilute and dense phases are denoted by ε_{gf} and ε_{gc}, respectively. Then the solids volume fraction is $\varepsilon_{pf}=1-\varepsilon_{gf}$ in the dilute phase and $\varepsilon_{pc}=1-\varepsilon_{gc}$ in the dense phase. The velocities with respect to the gas/solids in the dilute and dense phases are $\mathbf{u}_{gf}$, $\mathbf{u}_{pf}$, $\mathbf{u}_{gc}$, and $\mathbf{u}_{pc}$, respectively.

Different from the two-fluid or multifluid model where both the gas and solids are treated as fully interpenetrating continua, the four continua in the SFM are not fully interpenetrating. Thus for certain pair of continua, there may be no interphase forces, denoted by $\mathbf{F}_d$. For example, there is no drag between dilute-phase particles and dense-phase gas, but there are drags between gas and particles in both the dilute and dense phases (for the sake of simplicity, only drag force is discussed in this work, as the other interphase forces, e.g., virtual mass force and lift force, are neglected). Both the dense and dilute phases are assumed to be homogeneous or in local equilibrium state. The gas in the dilute phase surrounds the dense-phase particles and

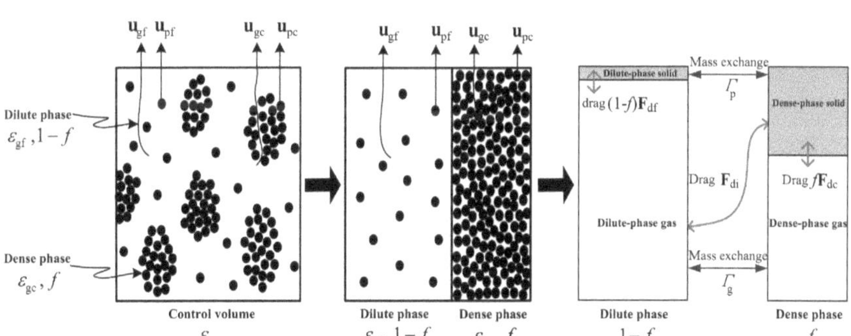

Figure 6 The relation between SFM subelements in a gas–solid riser flow (Hong et al, 2012).

hence exerts a mesoscale drag ($\mathbf{F}_{\text{di}}$) additionally. The gas and particles can further interact through dynamic mass exchange between the dilute and dense phases. This is measured in terms of the rate of interphase mass exchange per unit volume, Γ_{g} and Γ_{p}, for the gas (g) and solid particles (p), respectively. The chemical reactions and mass transfers are also affected by this flow structure and will be discussed in later section.

The conservation equations for four structure-based continua can be derived following the Eulerian spatial averaging method, as follows:

Mass conservation equation for the dense-phase gas:

$$\frac{\partial}{\partial t}\left(f\varepsilon_{\text{gc}}\rho_{\text{g}}\right)+\nabla\cdot\left(f\varepsilon_{\text{gc}}\rho_{\text{g}}\mathbf{u}_{\text{gc}}\right)=\Gamma_{\text{g}}. \tag{2}$$

Mass conservation equation for the dense-phase solid:

$$\frac{\partial}{\partial t}\left(f\varepsilon_{\text{pc}}\rho_{\text{p}}\right)+\nabla\cdot\left(f\varepsilon_{\text{pc}}\rho_{\text{p}}\mathbf{u}_{\text{pc}}\right)=\Gamma_{\text{p}}. \tag{3}$$

Mass conservation equation for the dilute-phase gas:

$$\frac{\partial}{\partial t}\left[(1-f)\varepsilon_{\text{gf}}\rho_{\text{g}}\right]+\nabla\cdot\left[(1-f)\varepsilon_{\text{gf}}\rho_{\text{g}}\mathbf{u}_{\text{gf}}\right]=-\Gamma_{\text{g}}. \tag{4}$$

Mass conservation equation for the dilute-phase solid:

$$\frac{\partial}{\partial t}\left[(1-f)\varepsilon_{\text{pf}}\rho_{\text{p}}\right]+\nabla\cdot\left[(1-f)\varepsilon_{\text{pf}}\rho_{\text{p}}\mathbf{u}_{\text{pf}}\right]=-\Gamma_{\text{p}}. \tag{5}$$

Momentum conservation equation for the dense-phase gas:

$$\begin{aligned}&\frac{\partial}{\partial t}\left(f\varepsilon_{\text{gc}}\rho_{\text{g}}\mathbf{u}_{\text{gc}}\right)+\nabla\cdot\left(f\varepsilon_{\text{gc}}\rho_{\text{g}}\mathbf{u}_{\text{gc}}\mathbf{u}_{\text{gc}}\right)\\&=-f\varepsilon_{\text{gc}}\nabla p+\nabla\cdot\left(f\boldsymbol{\tau}_{\text{gc}}\right)+f\varepsilon_{\text{gc}}\rho_{\text{g}}\mathbf{g}-f\mathbf{F}_{\text{dc}}+\Gamma_{\text{g}}\mathbf{u}_{\text{g}}^{\text{i}}.\end{aligned} \tag{6}$$

Here, the gradient of pressure, p, is localized on the dense phase by $f\varepsilon_{\text{gc}}$, as denoted by the first term on the RHS. This treatment of pressure drop is termed as hydrodynamic model A in literature (Gidaspow, 1994; Song et al, 2014). The fourth term on the RHS, $f\mathbf{F}_{\text{dc}}$, is the gas–solid drag in the dense phase, and the last term is the momentum transfer caused by gas exchange. The stress tensor of the structural subelement for any of the phase k, $\boldsymbol{\tau}_{\text{k}}$, takes the Newtonian form,

$$\boldsymbol{\tau}_{\text{k}}=\varepsilon_{\text{k}}\mu_{\text{k}}\left[\nabla\mathbf{u}_{\text{k}}+(\nabla\mathbf{u}_{\text{k}})^{T}\right]+\varepsilon_{\text{k}}\left(\lambda_{\text{k}}-\frac{2}{3}\mu_{\text{k}}\right)(\nabla\cdot\mathbf{u}_{\text{k}})\mathbf{I}. \tag{7}$$

Momentum conservation equation for the dense-phase solid:

$$\frac{\partial}{\partial t}\left(f\varepsilon_{\mathrm{pc}}\rho_{\mathrm{p}}\mathbf{u}_{\mathrm{pc}}\right)+\nabla\cdot\left(f\varepsilon_{\mathrm{pc}}\rho_{\mathrm{p}}\mathbf{u}_{\mathrm{pc}}\mathbf{u}_{\mathrm{pc}}\right)=-\nabla p_{\mathrm{pc}}-f\varepsilon_{\mathrm{pc}}\nabla p+\nabla\cdot\left(f\boldsymbol{\tau}_{\mathrm{pc}}\right)$$
$$+f\varepsilon_{\mathrm{pc}}\rho_{\mathrm{p}}\mathbf{g}+f\mathbf{F}_{\mathrm{dc}}+\mathbf{F}_{\mathrm{di}}+\Gamma_{\mathrm{p}}\mathbf{u}_{\mathrm{p}}^{i}, \tag{8}$$

where the drag exerted on the dense-phase solid includes that in the dense phase, the fifth term $f\mathbf{F}_{\mathrm{dc}}$ on the RHS, and that between the dilute-phase gas and dense-phase solid, the sixth term $\mathbf{F}_{\mathrm{di}}$. The last term represents the momentum transfer caused by solid exchange. μ denotes the shear viscosity and λ the bulk viscosity.

Momentum conservation equation for the dilute-phase gas:

$$\frac{\partial}{\partial t}\left[(1-f)\varepsilon_{\mathrm{gf}}\rho_{\mathrm{g}}\mathbf{u}_{\mathrm{gf}}\right]+\nabla\cdot\left[(1-f)\varepsilon_{\mathrm{gf}}\rho_{\mathrm{g}}\mathbf{u}_{\mathrm{gf}}\mathbf{u}_{\mathrm{gf}}\right]=-(1-f)\varepsilon_{\mathrm{gf}}\nabla p$$
$$+\nabla\cdot\left[(1-f)\boldsymbol{\tau}_{\mathrm{gf}}\right]+(1-f)\varepsilon_{\mathrm{gf}}\rho_{\mathrm{g}}\mathbf{g}$$
$$-(1-f)\mathbf{F}_{\mathrm{df}}-\mathbf{F}_{\mathrm{di}}-\Gamma_{\mathrm{g}}\mathbf{u}_{\mathrm{g}}^{\mathrm{i}}, \tag{9}$$

where the drag exerted on the dilute-phase gas includes that in the dilute phase, the fourth term $(1-f)\mathbf{F}_{\mathrm{df}}$ on the RHS, and that between the dilute-phase gas and dense-phase solid, the fifth term $\mathbf{F}_{\mathrm{di}}$. The last term is again caused by gas exchange.

Momentum conservation equation for the dilute-phase solid:

$$\frac{\partial}{\partial t}\left[(1-f)\varepsilon_{\mathrm{pf}}\rho_{\mathrm{p}}\mathbf{u}_{\mathrm{pf}}\right]+\nabla\cdot\left[(1-f)\varepsilon_{\mathrm{pf}}\rho_{\mathrm{p}}\mathbf{u}_{\mathrm{pf}}\mathbf{u}_{\mathrm{pf}}\right]=-\nabla p_{\mathrm{pf}}-(1-f)\varepsilon_{\mathrm{pf}}\nabla p$$
$$+\nabla\cdot\left[(1-f)\boldsymbol{\tau}_{\mathrm{pf}}\right]+(1-f)\varepsilon_{\mathrm{pf}}\rho_{\mathrm{p}}\mathbf{g}+(1-f)\mathbf{F}_{\mathrm{df}}-\Gamma_{\mathrm{p}}\mathbf{u}_{\mathrm{p}}^{\mathrm{i}}, \tag{10}$$

where the fifth term on the RHS is the gas–solid drag in the dilute phase. In principle, the full set of above governing equations could be used to solve the independent variables p, f, $\varepsilon_{\mathrm{gf}}$, $\varepsilon_{\mathrm{gc}}$, $\mathbf{u}_{\mathrm{gf}}$, $\mathbf{u}_{\mathrm{gc}}$, $\mathbf{u}_{\mathrm{pf}}$, and $\mathbf{u}_{\mathrm{pc}}$, provided that the dependent parameters $\boldsymbol{\tau}_{\mathrm{pf}}$, $\boldsymbol{\tau}_{\mathrm{pc}}$, p_{pf}, p_{pc}, Γ_{g}, Γ_{p}, $\mathbf{F}_{\mathrm{df}}$, $\mathbf{F}_{\mathrm{dc}}$, and $\mathbf{F}_{\mathrm{di}}$ can be closed. As the four continua are assumed to be homogeneous and in local equilibrium, the KTGF (Gidaspow, 1994) can be used to derive the solid pressure and viscosities in both the dilute and dense phases. And the gas–solid drag ($\mathbf{F}_{\mathrm{df}}$ and $\mathbf{F}_{\mathrm{dc}}$) can be closed using classical drag coefficients for homogeneous suspensions, e.g., the correlations of Wen and Yu (1966) or Ergun (1952). However, the mesoscale drag ($\mathbf{F}_{\mathrm{di}}$) and mass exchange (Γ_{g}, Γ_{p}) terms are hard to be closed due to the dynamic nature of clusters.

It should be further noted that the clusters could be inhomogeneous inside and deviate from the homogeneous assumption, following a nonlinear process. For example, the energy of the homogeneous granular gases decays as $E(t) \sim (1 + t/t_0)^{-2}$ following the Haff (1983), where t is time, $t_0 \propto (1-r^2)^{-1}$. Alvarez-Hamelin and Puglisi (2007) derived energy decay follows t^{-1} in strongly spatially inhomogeneous phase by the method of graph theory. The experimental results of two dense granular gases clusters show that the loss of kinetic energy obeys a power law in time $\Delta E(t) \propto -t^{3/2}$. How to quantify such inhomogeneous clusters needs more efforts in future.

5.1.1 Mesoscale Drag Based on Clusters/Bubbles

For the sake of simplicity, the mesoscale structure can be further assumed to be uniformly dispersed in forms of bubbles or clusters, both satisfying the bimodal distribution. The mesoscale drag force ($\mathbf{F}_{\mathrm{di}}$) can be closed with the same functional of drag but different structure properties (say, bubble diameter or cluster diameter). Accordingly, it is the closure of the mesoscale drag force that may be used to distinguish different SFMs. If the dense phase is assumed to exist in form of clusters with equivalent diameter d_c, uniformly dispersed in the dilute phase, then the mesoscale drag force ($\mathbf{F}_{\mathrm{di}}$) can be closed by

$$\mathbf{F}_{\mathrm{di}} = \frac{3}{4} C_{\mathrm{di}} \frac{f}{d_c} \rho_g \left| \mathbf{U}_{\mathrm{slip,i}} \right| \mathbf{U}_{\mathrm{slip,i}}, \tag{11}$$

where C_d denotes drag coefficient and C_{di} is closed by Wen and Yu (1966) correlation

$$C_{\mathrm{di}} = C_{\mathrm{di0}} (1 - f)^{-4.65}, \tag{12}$$

with $C_{\mathrm{di0}} = 24/Re_i + 3.6/Re_i^{0.313}$ and $Re_i = \rho_g d_c U_{\mathrm{slip,i}}/\mu_g$. Here, the capital U denotes superficial velocity and the interphase superficial slip velocity ($U_{\mathrm{slip,i}}$) is defined by $(U_{gf} - U_{pc}\varepsilon_{gf}/\varepsilon_{pc})(1-f)$. Then, the closure issue shifts to the determination of cluster diameter. Our previous versions normally adopted the original correlation of Li and Kwauk (1994) where the cluster diameter is inversely proportional to the energy consumption rate for suspending and transporting particles, N_{st}. If the dilute phase can be characterized by uniformly dispersed bubbles with diameter of d_b, we can get $\mathbf{F}_{\mathrm{di}}$ as follows:

$$\mathbf{F}_{\mathrm{di}} = \frac{3}{4} C_{\mathrm{db}} \frac{1-f}{d_b} \rho_c \left| \mathbf{U}_{\mathrm{slip,i}} \right| \mathbf{U}_{\mathrm{slip,i}}, \tag{13}$$

where C_{db} reads (Ishii and Zuber, 1979)

$$C_{db} = C_{db0}f^{-0.5}, \tag{14}$$

with $C_{db0} = \begin{cases} 38Re_i^{-1.5} & 0 < Re_i \le 1.8 \\ 2.7 + 24/Re_i & Re_i > 1.8 \end{cases}$

and $Re_i = \rho_c d_b U_{slip,i}/\mu_c$

Here, ρ_c is the mixture density of the dense phase. $U_{slip,i}$ is defined by $f(U_f - U_c)$, where U_f and U_c are mean velocities of the dilute and dense phases, respectively. This definition of mesoscale slip velocity differs a little bit from that in the cluster-based EMMS model, because the continuous phase transforms from the dilute phase to the dense phase. And their quantitative difference is $(1-f)\rho_g U_{gc}/\rho_c$, which is normally negligible for gas–solid systems. Similarly, the closure of $\mathbf{F}_{di}$ switches to the determination of bubble diameter. And it is well documented in literature ever since the classic work of Davidson and Harrison (1963). Compared to cluster diameter, bubble diameter arouses less disputes and hence is easier to characterize. The visual bubbles are normally irregular and in constantly dynamic transformation, which may deviate much from spherical assumption. Thus, the diameter of bubble here can also be viewed as drag-equivalent definition.

5.1.2 Reduction to TFM

To compare the SFM with the conventional TFM, we combine the mass and momentum equations in the SFM for the four structural subelements, yielding the mass conservation for the gas phase,

$$\frac{\partial}{\partial t}\left[f\varepsilon_{gc} + (1-f)\varepsilon_{gf}\right]\rho_g + \nabla\cdot\left[f\varepsilon_{gc}\rho_g\mathbf{u}_{gc} + (1-f)\varepsilon_{gf}\rho_g\mathbf{u}_{gf}\right] = 0, \tag{15}$$

which further reduces to

$$\frac{\partial}{\partial t}\left(\varepsilon_g\rho_g\right) + \nabla\cdot\left(\varepsilon_g\rho_g\mathbf{u}_g\right) = 0 \tag{16}$$

because

$$\varepsilon_k = f\varepsilon_{kc} + (1-f)\varepsilon_{kf} \tag{17}$$

and

$$\mathbf{u}_k = \frac{f\varepsilon_{kc}\mathbf{u}_{kc} + (1-f)\varepsilon_{kf}\mathbf{u}_{kf}}{\varepsilon_k} \tag{18}$$

Here, subscript k denotes g or p. Equation (16) is the same as the continuity equation of the TFM. Likewise, we can derive the same mass conservation equation for the solid phase as follows:

$$\frac{\partial}{\partial t}\left(\varepsilon_{\mathrm{p}}\rho_{\mathrm{p}}\right)+\nabla\cdot\left(\varepsilon_{\mathrm{p}}\rho_{\mathrm{p}}\mathbf{u}_{\mathrm{p}}\right)=0. \tag{19}$$

Combining Eq. (6) with Eq. (9) yields the momentum conservation for the gas phase,

$$\begin{aligned}\frac{\partial}{\partial t}\left(\varepsilon_{\mathrm{g}}\rho_{\mathrm{g}}\mathbf{u}_{\mathrm{g}}\right)+\nabla\cdot\left(\varepsilon_{\mathrm{g}}\rho_{\mathrm{g}}\mathbf{u}_{\mathrm{g}}\mathbf{u}_{\mathrm{g}}\right)=&-\varepsilon_{\mathrm{g}}\nabla p+\nabla\cdot\boldsymbol{\tau}_{\mathrm{ge}}+\varepsilon_{\mathrm{g}}\rho_{\mathrm{g}}\mathbf{g}\\&-\beta_{\mathrm{e}}\left(\mathbf{u}_{\mathrm{g}}-\mathbf{u}_{\mathrm{p}}\right)+\nabla\cdot\boldsymbol{\tau}_{\mathrm{Dg}},\end{aligned} \tag{20}$$

and combining Eq. (8) with Eq. (10) for the solid phase gives

$$\begin{aligned}\frac{\partial}{\partial t}\left(\varepsilon_{\mathrm{p}}\rho_{\mathrm{p}}\mathbf{u}_{\mathrm{p}}\right)+\nabla\cdot\left(\varepsilon_{\mathrm{p}}\rho_{\mathrm{p}}\mathbf{u}_{\mathrm{p}}\mathbf{u}_{\mathrm{p}}\right)=&-\varepsilon_{\mathrm{p}}\nabla p-\nabla p_{\mathrm{pe}}+\nabla\cdot\boldsymbol{\tau}_{\mathrm{pe}}\\&+\varepsilon_{\mathrm{p}}\rho_{\mathrm{p}}\mathbf{g}+\beta_{\mathrm{e}}\left(\mathbf{u}_{\mathrm{g}}-\mathbf{u}_{\mathrm{p}}\right)+\nabla\cdot\boldsymbol{\tau}_{\mathrm{Dp}}.\end{aligned} \tag{21}$$

The whole set of reduced SFM equations is summarized in Table 1. Compared to the hydrodynamic model A of the TFM, the SFM differs in its formulation of stress, drag, and diffusion stress by including the effect of the dilute–dense structure. If the flow is homogeneous in a grid, then the SFM reduces to the conventional TFM. That is, the TFM can be viewed as a simplified case of the more general SFM. It is also worth noting that the drag forces acting on the solid particles inside the dilute and dense phases are not equal to the counteractions on the respective gases (e.g., comparing Eqs. 6 and 8, or Eqs. 9 and 10) because mesoscale interaction works through $\mathbf{F}_{\mathrm{di}}$. However, when combining the equations of the dilute and dense phases, the drag force on the solids counterbalances that on the gas (e.g., comparing Eqs. 20 and 21), satisfying the Newton's third law of motion.

5.1.3 Restoration to Cluster-Based EMMS

The EMMS model (Li and Kwauk, 1994) was originally proposed to describe the global, steady state of fluidization. The mesoscale structure was assumed to take the form of clusters. If we integrate Eq. (15) with the superficial velocity as a constant, we get the mass balance equations of the EMMS model

$$\mathbf{U}_{\mathrm{g}}=f\mathbf{U}_{\mathrm{gc}}+(1-f)\mathbf{U}_{\mathrm{gf}} \tag{22}$$

and

$$\mathbf{U}_{\mathrm{p}}=f\mathbf{U}_{\mathrm{pc}}+(1-f)\mathbf{U}_{\mathrm{pf}}. \tag{23}$$

For an overall description of a steady-state reactor, the dominant factors of the force balance are the drag, gravity, and buoyancy caused by the

Table 1 Summary of the Equations in the Reduced SFM Model

Mass balance
Gas phase
$\frac{\partial}{\partial t}\left(\varepsilon_g\rho_g\right)+\nabla\cdot\left(\varepsilon_g\rho_g\mathbf{u}_g\right)=0$
Solid phase:
$\frac{\partial}{\partial t}\left(\varepsilon_p\rho_p\right)+\nabla\cdot\left(\varepsilon_p\rho_p\mathbf{u}_p\right)=0$
Momentum balance
Gas phase:
$\frac{\partial}{\partial t}\left(\varepsilon_g\rho_g\mathbf{u}_g\right)+\nabla\cdot\left(\varepsilon_g\rho_g\mathbf{u}_g\mathbf{u}_g\right)$
$=-\varepsilon_g\nabla p+\nabla\cdot\boldsymbol{\tau}_{ge}+\varepsilon_g\rho_g\mathbf{g}-\beta_e\left(\mathbf{u}_g-\mathbf{u}_p\right)+\nabla\cdot\boldsymbol{\tau}_{Dg}$
Solid phase:
$\frac{\partial}{\partial t}\left(\varepsilon_p\rho_p\mathbf{u}_p\right)+\nabla\cdot\left(\varepsilon_p\rho_p\mathbf{u}_p\mathbf{u}_p\right)$
$=-\varepsilon_p\nabla p-\nabla p_{pe}+\nabla\cdot\boldsymbol{\tau}_{pe}+\varepsilon_p\rho_p\mathbf{g}+\beta_e\left(\mathbf{u}_g-\mathbf{u}_p\right)+\nabla\cdot\boldsymbol{\tau}_{Dp}$
Structure-dependent drag (EMMS drag)
$\beta_e\left(\mathbf{u}_g-\mathbf{u}_p\right)=f\mathbf{F}_{dc}+\mathbf{F}_{di}+(1-f)\mathbf{F}_{df}$
Heterogeneity index
$H_d=\frac{\beta_e}{\beta_{WY}}$
where β_{WY} refers to homogeneous Wen and Yu (1966) drag
Structure-dependent solid pressure
$p_{pe}=p_{pf}+p_{pc}$
Structure-dependent stress
Gas phase:
$\boldsymbol{\tau}_{ge}=f\boldsymbol{\tau}_{gc}+(1-f)\boldsymbol{\tau}_{gf}$
Solid phase:
$\boldsymbol{\tau}_{pe}=f\boldsymbol{\tau}_{pc}+(1-f)\boldsymbol{\tau}_{pf}$
Diffusion stress
Gas phase:
$\boldsymbol{\tau}_{Dg}=\varepsilon_g\rho_g\mathbf{u}_g\mathbf{u}_g-f\varepsilon_{gc}\rho_g\mathbf{u}_{gc}\mathbf{u}_{gc}-(1-f)\varepsilon_{gf}\rho_g\mathbf{u}_{gf}\mathbf{u}_{gf}$
Solid phase:
$\boldsymbol{\tau}_{Dp}=\varepsilon_p\rho_p\mathbf{u}_p\mathbf{u}_p-f\varepsilon_{pc}\rho_p\mathbf{u}_{pc}\mathbf{u}_{pc}-(1-f)\varepsilon_{pf}\rho_p\mathbf{u}_{pf}\mathbf{u}_{pf}$

pressure drop. Acceleration, stress, and the interphase mass exchange can be neglected. Thus, the dense-phase solid momentum equation reduces to

$$f\mathbf{F}_{dc}+\mathbf{F}_{di}=-f\varepsilon_{pc}\rho_p\mathbf{g}+f\varepsilon_{pc}\nabla p. \tag{24}$$

And the dilute-phase solid momentum equation reduces to

$$\mathbf{F}_{df}=-\varepsilon_{pf}\rho_p\mathbf{g}+\varepsilon_{pf}\nabla p. \tag{25}$$

The dense-phase gas momentum equation reduces to

$$f\mathbf{F}_{\mathrm{dc}} = -f\varepsilon_{\mathrm{gc}}\nabla p + f\varepsilon_{\mathrm{gc}}\rho_{\mathrm{g}}\mathbf{g}. \tag{26}$$

Also, the dilute-phase gas momentum equation can be simplified as

$$(1-f)\mathbf{F}_{\mathrm{df}} + \mathbf{F}_{\mathrm{di}} = -(1-f)\varepsilon_{\mathrm{gf}}\nabla p + (1-f)\varepsilon_{\mathrm{gf}}\rho_{\mathrm{g}}\mathbf{g} \tag{27}$$

Combining Eqs. (26) and (27) gives

$$\frac{\mathbf{F}_{\mathrm{dc}}}{\varepsilon_{\mathrm{gc}}} = \frac{\mathbf{F}_{\mathrm{df}}}{\varepsilon_{\mathrm{gf}}} + \frac{\mathbf{F}_{\mathrm{di}}/\varepsilon_{\mathrm{gf}}}{1-f}. \tag{28}$$

Considering the difference of drag forces between model A and model B (Gidaspow, 1994), Eq. (28) is equivalent to the pressure drop balance equation in the EMMS model (Li and Kwauk, 1994). Likewise, the solid momentum equations can be rewritten as the form of model B, which is the same as the force balance equations for solid particles of the EMMS model (Hong et al, 2013; Song et al, 2014).

5.1.4 Restoration to Bubble-Based EMMS

If the mesoscale structure is assumed to take the form of bubbles, then, for a bubble–emulsion structure, the mass balance equation of gas keeps the same as Eq. (22). If the bubble is assumed to be void of solids, as is the case in Shi et al (2011), then the mass balance equation of solids, Eq. (23), can be further simplified by using $\varepsilon_{\mathrm{pf}}=0$, that is,

$$\mathbf{U}_{\mathrm{p}} = f\mathbf{U}_{\mathrm{pc}}. \tag{29}$$

If one neglects the mesoscale drag force, $\mathbf{F}_{\mathrm{di}}$, in the force balance for the dense-phase solid, as assumed in Shi et al (2011), Eq. (24) reduces to

$$\mathbf{F}_{\mathrm{dc}} = -\varepsilon_{\mathrm{pc}}\varepsilon_{\mathrm{g}}\left(\rho_{\mathrm{p}} - \rho_{\mathrm{g}}\right)\mathbf{g}. \tag{30}$$

Combining Eqs. (24)–(27) gives the pressure gradient by

$$\nabla p = \left(\varepsilon_{\mathrm{p}}\rho_{\mathrm{p}} + \varepsilon_{\mathrm{g}}\rho_{\mathrm{g}}\right)\mathbf{g}. \tag{31}$$

Combining Eqs. (31), (27), and (25), we can reformulate the force balance of bubble as follows:

$$-\mathbf{F}_{\mathrm{di}} = (1-f)\varepsilon_{\mathrm{p}}\left(\rho_{\mathrm{p}} - \rho_{\mathrm{g}}\right)\mathbf{g}. \tag{32}$$

So, Eqs. (22), (29), (30), and (32) recover the balance equations of our bubble-based EMMS model (Shi et al, 2011), but in the form of hydrodynamic model A (Gidaspow, 1994). This version of the bubble-based EMMS model neglects the contribution of the mesoscale drag, $\mathbf{F}_{di}$, in the force balance equation for the dense phase. If we include all components of the drag force, we can directly use Eqs. (22)–(25) and (28), which form the new version of the bubble-based EMMS model, also in forms of hydrodynamic model A.

From the above derivation, we can see that the SFM may revert to the cluster-based EMMS or bubble-based EMMS hydrodynamic models under different assumptions of structure parameters. Thus, SFM unifies these models. Indeed, the mesoscale structure has always been the critical part of the EMMS model. We may expect more variants of the EMMS model will be proposed because of different descriptions of the mesoscale structure (Hong et al, 2013).

5.2 Structure-Dependent Energy Analysis

The equations presented in Section 5.1 are consistent with the dilute–dense, bimodal structure. The quantification of the mesoscale structure in these equations can be empirical, using experimental correlations for the cluster/bubble diameter or, more fundamentally, from theoretical analysis. As discussed above, the mesoscale structure in dissipative granular flows heavily depends on the heating or the rate of energy input besides the inelastic collisions. According to the principle of compromise in competition (Li and Huang, 2014; Li and Kwauk, 1994), this dependency is determined by the minimization of energy consumption rate for suspending and transporting particles in a fluidized bed. It is the stability condition that distinguishes the EMMS model from the conventional hydrodynamic models and it can be used to close the mesoscale structures of SFM, thus constituting the EMMS-based multifluid model (EFM) (Hong et al, 2012). The CFD simulation based on this model is the so-called multiscale CFD (Wang and Li, 2007; Wang et al, 2010). To better understand the energy consumption and dissipation with respect to the dilute–dense, bimodal structure, in what follows we will conduct an energy analysis rooted in the SFM conservation equations. The energy balance based on the gas and solids phases is presented consequently. And their relations with the EMMS energy decomposition and stability condition are discussed (Song et al, 2014).

5.2.1 Energy Conservation of the Gas Phase

As presented in Section 5.1, the gas–solid flow in a local cell can be divided into four parts: dilute-phase gas, dilute-phase solid, dense-phase gas, and dense-phase solid, among which there are mass and momentum exchanges. Further, as a definition, the acceleration of phase k is given by

$$\mathbf{a}_{\mathrm{k}} \equiv \frac{D\mathbf{u}_{\mathrm{k}}}{Dt} = \frac{\partial \mathbf{u}_{\mathrm{k}}}{\partial t} + \mathbf{u}_{\mathrm{k}} \cdot \nabla \mathbf{u}_{\mathrm{k}}. \quad (33)$$

Combining mass conservation equations for gas and Eq. (33), the left hand side (LHS) of gas momentum equations for dilute and dense gases can be rewritten as

$$\frac{\partial}{\partial t}\left(f\varepsilon_{\mathrm{gc}}\rho_{\mathrm{g}}\mathbf{u}_{\mathrm{gc}}\right) + \nabla \cdot \left(f\varepsilon_{\mathrm{gc}}\rho_{\mathrm{g}}\mathbf{u}_{\mathrm{gc}}\mathbf{u}_{\mathrm{gc}}\right) = f\varepsilon_{\mathrm{gc}}\rho_{\mathrm{g}}\mathbf{a}_{\mathrm{gc}} + \Gamma_{\mathrm{g}}\mathbf{u}_{\mathrm{gc}}, \quad (34)$$

$$\begin{aligned}\frac{\partial}{\partial t}\left[(1-f)\varepsilon_{\mathrm{gf}}\rho_{\mathrm{g}}\mathbf{u}_{\mathrm{gf}}\right] + \nabla \cdot \left[(1-f)\varepsilon_{\mathrm{gf}}\rho_{\mathrm{g}}\mathbf{u}_{\mathrm{gf}}\mathbf{u}_{\mathrm{gf}}\right] \\ = (1-f)\varepsilon_{\mathrm{gf}}\rho_{\mathrm{g}}\mathbf{a}_{\mathrm{gf}} - \Gamma_{\mathrm{g}}\mathbf{u}_{\mathrm{gf}}.\end{aligned} \quad (35)$$

Substituting Eqs. (34) and (35) into gas momentum equations, Eqs. (6) and (9), we get

$$f\varepsilon_{\mathrm{gc}}\rho_{\mathrm{g}}\mathbf{a}_{\mathrm{gc}} = -f\varepsilon_{\mathrm{gc}}\nabla p + \nabla \cdot \left(f\boldsymbol{\tau}_{\mathrm{gc}}\right) + f\varepsilon_{\mathrm{gc}}\rho_{\mathrm{g}}\mathbf{g} - f\mathbf{F}_{\mathrm{dc}} + \Gamma_{\mathrm{g}}\left(\mathbf{u}_{\mathrm{g}}^{\mathrm{i}} - \mathbf{u}_{\mathrm{gc}}\right) \quad (36)$$

$$\begin{aligned}(1-f)\varepsilon_{\mathrm{gf}}\rho_{\mathrm{g}}\mathbf{a}_{\mathrm{gf}} = &-(1-f)\varepsilon_{\mathrm{gf}}\nabla p + \nabla \cdot \left[(1-f)\boldsymbol{\tau}_{\mathrm{gf}}\right] + (1-f)\varepsilon_{\mathrm{gf}}\rho_{\mathrm{g}}\mathbf{g} \\ &-(1-f)\mathbf{F}_{\mathrm{df}} - \mathbf{F}_{\mathrm{di}} - \Gamma_{\mathrm{g}}\left(\mathbf{u}_{\mathrm{g}}^{\mathrm{i}} - \mathbf{u}_{\mathrm{gf}}\right).\end{aligned} \quad (37)$$

The energy conservation of the entire gas phase can be obtained as a dot product of the momentum equations and relevant velocities. Then, Eq. $(36) \cdot \mathbf{u}_{\mathrm{gc}} + \mathrm{Eq.}(37) \cdot \mathbf{u}_{\mathrm{gf}}$ gives

$$\begin{aligned}f\varepsilon_{\mathrm{gc}}\rho_{\mathrm{g}}\mathbf{u}_{\mathrm{gc}} \cdot \mathbf{a}_{\mathrm{gc}} + (1-f)\varepsilon_{\mathrm{gf}}\rho_{\mathrm{g}}\mathbf{u}_{\mathrm{gf}} \cdot \mathbf{a}_{\mathrm{gf}} = &-\left[f\varepsilon_{\mathrm{gc}}\mathbf{u}_{\mathrm{gc}} + (1-f)\varepsilon_{\mathrm{gf}}\mathbf{u}_{\mathrm{gf}}\right] \cdot \nabla p \\ &+ \rho_{\mathrm{g}}\left[f\varepsilon_{\mathrm{gc}}\mathbf{u}_{\mathrm{gc}} + (1-f)\varepsilon_{\mathrm{gf}}\mathbf{u}_{\mathrm{gf}}\right] \cdot \mathbf{g} \\ &+ \left\{\nabla \cdot \left(f\boldsymbol{\tau}_{\mathrm{gc}}\right) \cdot \mathbf{u}_{\mathrm{gc}} + \nabla \cdot \left[(1-f)\boldsymbol{\tau}_{\mathrm{gf}}\right] \cdot \mathbf{u}_{\mathrm{gf}}\right\} \\ &- \left[f\mathbf{F}_{\mathrm{dc}} \cdot \mathbf{u}_{\mathrm{gc}} + (1-f)\mathbf{F}_{\mathrm{df}} \cdot \mathbf{u}_{\mathrm{gf}} + \mathbf{F}_{\mathrm{di}} \cdot \mathbf{u}_{\mathrm{gf}}\right] \\ &+ \left[\Gamma_{\mathrm{g}}\left(\mathbf{u}_{\mathrm{g}}^{\mathrm{i}} - \mathbf{u}_{\mathrm{gc}}\right) \cdot \mathbf{u}_{\mathrm{gc}} - \Gamma_{\mathrm{g}}\left(\mathbf{u}_{\mathrm{g}}^{\mathrm{i}} - \mathbf{u}_{\mathrm{gf}}\right) \cdot \mathbf{u}_{\mathrm{gf}}\right].\end{aligned} \quad (38)$$

If we define the kinetic energy of the gas phase per unit volume by

$$E_{\mathrm{k,g}} = \frac{1}{2}f\varepsilon_{\mathrm{gc}}\rho_{\mathrm{g}}\mathbf{u}_{\mathrm{gc}} \cdot \mathbf{u}_{\mathrm{gc}} + \frac{1}{2}(1-f)\varepsilon_{\mathrm{gf}}\rho_{\mathrm{g}}\mathbf{u}_{\mathrm{gf}} \cdot \mathbf{u}_{\mathrm{gf}}, \quad (39)$$

then the rate of change of $E_{\mathrm{k,g}}$ is

$$W_{\mathrm{k,g}} = \frac{DE_{\mathrm{k,g}}}{Dt} = f\varepsilon_{\mathrm{gc}}\rho_{\mathrm{g}}\mathbf{u}_{\mathrm{gc}} \cdot \frac{D\mathbf{u}_{\mathrm{gc}}}{Dt} + (1-f)\varepsilon_{\mathrm{gf}}\rho_{\mathrm{g}}\mathbf{u}_{\mathrm{gf}} \cdot \frac{D\mathbf{u}_{\mathrm{gf}}}{Dt} \\ = f\varepsilon_{\mathrm{gc}}\rho_{\mathrm{g}}\mathbf{u}_{\mathrm{gc}} \cdot \mathbf{a}_{\mathrm{gc}} + (1-f)\varepsilon_{\mathrm{gf}}\rho_{\mathrm{g}}\mathbf{u}_{\mathrm{gf}} \cdot \mathbf{a}_{\mathrm{gf}}. \tag{40}$$

So the LHS of Eq. (38) stands for the rate of change of the kinetic energy of the gas per unit volume. Further,

$$f\varepsilon_{\mathrm{gc}}\mathbf{u}_{\mathrm{gc}} + (1-f)\varepsilon_{\mathrm{gf}}\mathbf{u}_{\mathrm{gf}} = f\mathbf{U}_{\mathrm{gc}} + (1-f)\mathbf{U}_{\mathrm{gf}} = \mathbf{U}_{\mathrm{g}}, \tag{41}$$

then the first term of the RHS of Eq. (38) reduces to

$$-\left[f\varepsilon_{\mathrm{gc}}\mathbf{u}_{\mathrm{gc}} + (1-f)\varepsilon_{\mathrm{gf}}\mathbf{u}_{\mathrm{gf}}\right] \cdot \nabla p = -\mathbf{U}_{\mathrm{g}} \cdot \nabla p, \tag{42}$$

which stands for the net power of the gas pressure exerted on unit volume, i.e., $W_{\mathrm{T,g}}$. The second term of the RHS of Eq. (38) can be expressed as

$$\rho_{\mathrm{g}}\left[f\varepsilon_{\mathrm{gc}}\mathbf{u}_{\mathrm{gc}} + (1-f)\varepsilon_{\mathrm{gf}}\mathbf{u}_{\mathrm{gf}}\right] \cdot \mathbf{g} = \rho_{\mathrm{g}} \cdot \mathbf{U}_{\mathrm{g}} \cdot \mathbf{g}, \tag{43}$$

which means the variation rate of gas gravity potential per unit volume. The third term of the RHS of Eq. (38) is related to the viscous force. It can be divided into the power done by viscous force, $W_{\tau,\mathrm{g}}$, and the power dissipated, that is,

$$\nabla \cdot \left(f\boldsymbol{\tau}_{\mathrm{gc}}\right) \cdot \mathbf{u}_{\mathrm{gc}} + \nabla \cdot \left[(1-f)\boldsymbol{\tau}_{\mathrm{gf}}\right] \cdot \mathbf{u}_{\mathrm{gf}} = \nabla \cdot \left[f\boldsymbol{\tau}_{\mathrm{gc}} \cdot \mathbf{u}_{\mathrm{gc}} + (1-f)\boldsymbol{\tau}_{\mathrm{gf}} \cdot \mathbf{u}_{\mathrm{gf}}\right] \\ - \left[f\boldsymbol{\tau}_{\mathrm{gc}} : \nabla\mathbf{u}_{\mathrm{gc}} + (1-f)\boldsymbol{\tau}_{\mathrm{gf}} : \nabla\mathbf{u}_{\mathrm{gf}}\right] \\ = W_{\tau,\mathrm{g}} - \left[f\boldsymbol{\tau}_{\mathrm{gc}} : \nabla\mathbf{u}_{\mathrm{gc}} + (1-f)\boldsymbol{\tau}_{\mathrm{gf}} : \nabla\mathbf{u}_{\mathrm{gf}}\right]. \tag{44}$$

The fourth term of the RHS of Eq. (38) is related to the structure-dependent drag forces, which can be further divided into the rate of energy dissipation for suspending particles per unit volume, W_{s}, and the rate of energy consumption for transporting particles per unit volume, W_{t}, both of which are driven by the drag forces, that is,

$$W_{\mathrm{st}} = W_{\mathrm{s}} + W_{\mathrm{t}} = f\mathbf{F}_{\mathrm{dc}} \cdot \mathbf{u}_{\mathrm{gc}} + (1-f)\mathbf{F}_{\mathrm{df}} \cdot \mathbf{u}_{\mathrm{gf}} + \mathbf{F}_{\mathrm{di}} \cdot \mathbf{u}_{\mathrm{gf}}, \tag{45}$$

where

$$W_{\mathrm{s}} = f\mathbf{F}_{\mathrm{dc}} \cdot \mathbf{u}_{\mathrm{slip,c}} + (1-f)\mathbf{F}_{\mathrm{df}} \cdot \mathbf{u}_{\mathrm{slip,f}} + \mathbf{F}_{\mathrm{di}} \cdot \mathbf{u}_{\mathrm{slip,f}}, \tag{46}$$

$$W_{\mathrm{t}} = f\mathbf{F}_{\mathrm{dc}} \cdot \mathbf{u}_{\mathrm{pc}} + (1-f)\mathbf{F}_{\mathrm{df}} \cdot \mathbf{u}_{\mathrm{pf}} + \mathbf{F}_{\mathrm{di}} \cdot \mathbf{u}_{\mathrm{pc}}. \tag{47}$$

The last term of the RHS of Eq. (38) stands for the rate of energy dissipation related to gas exchange between the dilute and dense phases. Combining it with the energy dissipation terms related to the viscous force, we get the rate of entire energy dissipation of the gas phase, $W_{d,g}$. So the energy equation of the SFM for gas–solid fluidization, Eq. (38), can be rewritten as follows:

$$W_{T,g} + W_{\tau,g} = W_{k,g} - \rho_g \mathbf{U}_g \cdot \mathbf{g} + W_{st} + W_{d,g}, \quad (48)$$

where

$$W_{T,g} = -\nabla p \cdot \mathbf{U}_g. \quad (49)$$

$$W_{\tau,g} = \nabla \cdot \left[f \boldsymbol{\tau}_{gc} \cdot \mathbf{u}_{gc} + (1-f) \boldsymbol{\tau}_{gf} \cdot \mathbf{u}_{gf} \right] \quad (50)$$

$$W_{d,g} = \left[f \boldsymbol{\tau}_{gc} : \nabla \mathbf{u}_{gc} + (1-f) \boldsymbol{\tau}_{gf} : \nabla \mathbf{u}_{gf} \right] - \Gamma_g \left(\mathbf{u}_g^i - \mathbf{u}_{gc} \right) \cdot \mathbf{u}_{gc} + \Gamma_g \left(\mathbf{u}_g^i - \mathbf{u}_{gf} \right) \cdot \mathbf{u}_{gf} \quad (51)$$

5.2.2 Energy Conservation of the Solid Phase

The conservation equations of the solid phase are similar to those of the gas phase. We can analyze the composition of energy consumption and get the energy equation for the solid phase, by multiplying the momentum equations with their relevant velocities, as follows:

$$W_{T,p} + W_{sp,p} + W_{\tau,p} + W_t = W_{k,p} - \rho_p \mathbf{U}_p \cdot \mathbf{g} + W_{d,p}. \quad (52)$$

Here, $W_{sp,p}$ stands for the power of the solid pressure exerted on unit volume, which reads

$$W_{sp,p} = -\nabla p_{pc} \cdot \mathbf{u}_{pc} - \nabla p_{pf} \cdot \mathbf{u}_{pf}. \quad (53)$$

And definitions of the variables in Eq. (52) are all similar with those defined in the gas phase, except that the parameters of the gas phase should be replaced with corresponding solid parameters. In detail, those variables can be calculated from

$$W_{T,p} = -\nabla p \cdot \mathbf{U}_p \quad (54)$$

$$W_{\tau,p} = \nabla \cdot \left[f \boldsymbol{\tau}_{pc} \cdot \mathbf{u}_{pc} + (1-f) \boldsymbol{\tau}_{pf} \cdot \mathbf{u}_{pf} \right] \quad (55)$$

$$W_{k,p} = f \varepsilon_{pc} \rho_p \mathbf{u}_{pc} \cdot \mathbf{a}_{pc} + (1-f) \varepsilon_{pf} \rho_p \mathbf{u}_{pf} \cdot \mathbf{a}_{pf} \quad (56)$$

$$W_{d,p} = \left[f \boldsymbol{\tau}_{pc} : \nabla \mathbf{u}_{pc} + (1-f) \boldsymbol{\tau}_{pf} : \nabla \mathbf{u}_{pf} \right] - \Gamma_p \left(\mathbf{u}_p^i - \mathbf{u}_{pc} \right) \cdot \mathbf{u}_{pc} + \Gamma_p \left(\mathbf{u}_p^i - \mathbf{u}_{pf} \right) \cdot \mathbf{u}_{pf} \quad (57)$$

From Eq. (52), it can be seen that the solid pressure does have some effects on the energy consumption and dissipation, which explains in part why the solid pressure affects the state of the system as shown in the literature of Liu et al (2014). And it should be noted that, by comparing Eqs. (48) and (52), W_t assumes two different roles: for the gas phase it is a rate of energy consumption, while for the solid phase it is a power exerted by the gas phase. With Eqs. (56) and (57), we can also see that if the solid particles are carried by the gas uniformly, then the variation of the kinetic energy of solids and the energy dissipation due to particle collision are negligible. Considering the contributions of $W_{T,p}$, $W_{sp,p}$, and $W_{\tau,p}$ are also much smaller than the power consumption for transporting particles, W_t, we can further reduce Eq. (52) to

$$W_t = -\rho_p \mathbf{U}_p \cdot \mathbf{g}, \tag{58}$$

which reverts to the simplified definition of W_t in the original EMMS model (Li and Kwauk, 1994). Furthermore, combining the energy conservation equations for both gas and solid phases, we get the total energy balance equation:

$$W_{T,g} + W_{T,p} + W_{sp,p} + W_{\tau,g} + W_{\tau,p} = W_{k,g} + W_{k,p} - \left(\rho_g \mathbf{U}_g + \rho_p \mathbf{U}_p\right) \cdot \mathbf{g} + \left(W_{d,g} + W_{d,p} + W_s\right). \tag{59}$$

That means the LHS, the total power input per unit volume, which is generated by the pressure gradient, viscous stress, and solid pressure, is consumed to change the kinetic energy (the first two terms on the RHS) and the gravity potential (the third term) of both gas and solid phases, besides the last three terms in bracket which are dissipated into heat.

5.2.3 Restoration to Energy Terms in EMMS

Starting from the energy conservation equation of the gas phase, we may restore the energy balance defined in the EMMS model. Note that $\mathbf{F}_{dc}$ is defined in the whole dense phase, while $\mathbf{u}_{gc}$ applies only in the gas part of the dense phase, then, for the averaged energy consumption for suspending and transporting particles in any cell,

$$\overline{\mathbf{F}_{dc} \cdot \mathbf{u}_{gc}} = \overline{\mathbf{F}_{dc} \cdot \left(\mathbf{u}_{gc}\delta\right)} = \frac{\int_{\Omega_c} \mathbf{F}_{dc} \cdot \mathbf{u}_{gc} \delta \mathrm{d}V}{\int_{\Omega_c} \mathrm{d}V}, \tag{60}$$

where

$$\delta = \begin{cases} 0, & \text{particle phase} \\ 1, & \text{gas phase} \end{cases} \tag{61}$$

And $\mathbf{F}_{df} \cdot \mathbf{u}_{gf}$ and $\mathbf{F}_{di} \cdot \mathbf{u}_{gf}$ can be treated likewise. Thus, Eq. (45) can be rewritten as follows:

$$\begin{aligned}
\overline{W_{st}} &= f\overline{\mathbf{F}_{dc} \cdot \mathbf{u}_{gc}} + (1-f)\overline{\mathbf{F}_{df} \cdot \mathbf{u}_{gf}} + \overline{\mathbf{F}_{di} \cdot \mathbf{u}_{gf}} \\
&= f\frac{\int_{\Omega_c} \mathbf{F}_{dc} \cdot \mathbf{u}_{gc}\delta_c \mathrm{d}V}{\int_{\Omega_c} \mathrm{d}V} + (1-f)\frac{\int_{\Omega_f} \mathbf{F}_{df} \cdot \mathbf{u}_{gf}\delta_f \mathrm{d}V}{\int_{\Omega_f} \mathrm{d}V} + \frac{\int_{\Omega} \mathbf{F}_{di} \cdot \mathbf{u}_{gf}\delta_f \mathrm{d}V}{\int_{\Omega} \mathrm{d}V} \\
&= f\mathbf{F}_{dc} \cdot \frac{\int_{\Omega_c} \mathbf{u}_{gc}\delta_c \mathrm{d}V}{\int_{\Omega_c} \mathrm{d}V} + (1-f)\mathbf{F}_{df} \cdot \frac{\int_{\Omega_f} \mathbf{u}_{gf}\delta_f \mathrm{d}V}{\int_{\Omega_f} \mathrm{d}V} + \mathbf{F}_{di} \cdot \frac{\int_{\Omega} \mathbf{u}_{gf}\delta_f \mathrm{d}V}{\int_{\Omega} \mathrm{d}V} \\
&= f\mathbf{F}_{dc} \cdot \mathbf{U}_{gc} + (1-f)\mathbf{F}_{df} \cdot \mathbf{U}_{gf} + (1-f)\mathbf{F}_{di} \cdot \mathbf{U}_{gf},
\end{aligned} \tag{62}$$

which is the same as that defined in the EMMS model. For a riser exemplified in Li and Kwauk (1994), if it is operated under steady state, then the kinetic energy of gas is constant. Further we neglect the wall friction, and then the viscous force does no work to the whole system, so the mean $W_{k,g}$ and $W_{T,g}$ vanish. For a steady-state system, the pressure gradient is approximately equal to the buoyant gravity of particles, then for a typical flow state (say, $\varepsilon_g = 0.894, \rho_g = 1.2$ kg/m^3, and $\rho_p = 930$ kg/m^3), $\rho_p U_g g / W_{T,g} \sim 0.01$. Thus, compared with $W_{T,g}$, the power of gas gravity potential is negligible. Using the force balance equations for particles in the dense and dilute phases in the traditional EMMS model (Li and Kwauk, 1994), we can see that the drag forces for different phases have the same magnitude as the effective gravity of solids, so the mean W_{st} has the same magnitude as mean $W_{T,g}$. Then compared with mean $W_{T,g}$, mean W_{st}, and $W_{d,g}$, the other terms in Eq. (48) are negligible, and Eq. (48) reduces to

$$\overline{W_{T,g}} = \overline{W_{st}} + \overline{W_{d,g}}. \tag{63}$$

If we rewrite Eq. (63) with respect to unit mass of particles by dividing it with $\varepsilon_p \rho_p$, we get

$$\overline{N_{T,g}} = \overline{N_{st}} + \overline{N_{d,g}}, \tag{64}$$

which is the same as the energy partition in the EMMS model.

As discussed in Wang et al (2011), to reflect the bilateral coupling on the mesoscale modeling, certain trans-scale criterion such as the stability condition in EMMS should be needed to untangle the correlation among the micro-, meso-, and macroscales. The analysis of energy consumption and dissipation in local space of a fluidized bed is hence very important to understand the variation and the scale-dependent constraint of mesoscale structures. It may shed light for us to search for a generalized EMMS model whose structure parameters are correlated with mesoscale energy consumption or dissipation terms. And the steady-state results of it may revert to the original EMMS formulations. Some researchers have tried to directly use certain stability condition to determine the flow structures within a computational cell (Shuai et al, 2011, 2014) or rewrite the force balance model by introducing interparticle forces within clusters or agglomerates (Tebianian et al, 2014). However, Li et al (2004) have pointed out that the stability condition defined in the EMMS model applies only to macro- and mesostates, not to micro motion. For a given set of operating conditions, the gas velocity and solid flux are constant. In the original EMMS model, the global state of a fluidized bed under steady state can thus be determined with a set of conservation equations and a stability condition. The stability condition can be obtained by applying the principle of "compromise in competition" (Li and Huang, 2014). That is, the energy consumption for suspending and transporting particles per unit mass of particles has a tendency to reach its minimum value, $N_{st} \rightarrow \min$, through a compromise between the minimum tendencies of the voidage and the energy consumption for suspending and transporting particles per unit volume (Ge and Li, 2002). That extremum stability condition is reasonable because the flow behavior has steady-state values at the reactor level. However, at a local point, such as in a CFD cell, both gas and solid velocities are in dynamic variation, which will cause W_{st} and $W_{d,g}$ fluctuating around their mean values. So whether N_{st} and $N_{d,g}$ can assume their extreme tendencies at local points is questionable. In other words, the dilute–dense structure always exists and hence the SFM and its energy analysis apply in any control volume, whereas the extremum tendency exists only on the meso- or larger scales. If one tends to close the SFM directly with the EMMS stability condition, the grid size should be large enough to satisfy such requirement (Li et al, 2004). More efforts are deserved to understand the scale dependence of such stability condition in fluidized beds.

5.2.4 Macroscale Stability Constraint for Mesoscale Structure

In the above processes, the cluster diameter is one of the key parameters for EMMS prediction. There are some empirical or semi-empirical expressions proposed for the cluster diameter (say, the review of Harris et al, 2002). Most of these cluster diameters are functions of local voidage or cross-sectionally averaged voidage. Li and Kwauk (1994) proposed to correlate the cluster diameter with $N_{st} \rightarrow \min$, that is,

$$N_{st} = \frac{fF_{dc} \cdot U_{gc} + (1-f)F_{df} \cdot U_{gf} + F_{di} \cdot U_{gf}(1-f)}{(1-\varepsilon_g)\rho_p} \rightarrow \min. \quad (65)$$

For fast fluidization, it can be rewritten as

$$N_{st} = \frac{\rho_p - \rho_g}{\rho_p}\left[U_g - \frac{\varepsilon_{gf} - \varepsilon_g}{1-\varepsilon_g} f(1-f)U_{gf}\right] \cdot g \rightarrow \min. \quad (66)$$

For bubbling fluidization where solids entrainment is negligible, it can be reformulated as in Shi et al (2011), as follows:

$$N_{st} = \frac{fF_{dc} \cdot U_{gc} + (1-f)F_{df} \cdot U_{gf}}{(1-\varepsilon_g)\rho_p} + \frac{(1-f)\varepsilon_{gf} U_g \cdot g}{\varepsilon_g} \rightarrow \min. \quad (67)$$

These two stability conditions are actually the same in physics and the only difference lies in the energy consumption term related to mesoscale drag because of their different characterization of the mesoscale structure. For general purpose, Ge and Li (2002) indicated that the stability condition is better defined by a dimensionless ratio, $N_{st}/N_T \rightarrow \min$, where N_T is the total mass-specific energy consumption. Equations (65)–(67) are presented without boldface because they are defined at the reactor scale.

As discussed above, the stability condition is expected to reach extremum in sufficiently large space (e.g., cross section of a fluidized bed) instead of local cell. The energy to sustain mesoscale structures in a fluidized bed comes largely from the mean relative motion between gas and particles on the macroscale. Furthermore, the dynamic evolution of mesoscale structure and its energy transfer is subject to both macroscale operating conditions and the conservation laws in microscale computational cells. As a result, a two-step scheme was proposed to fulfill the coupling between EMMS and hydrodynamic conservation equations, called EMMS/matrix (Lu et al, 2009; Wang and Li, 2007). At the macroscale (reactor), the bi-objective optimization method in terms of $N_{st} \rightarrow \min$ was first used to resolve the mesoscale parameters, say, d_c and ε_{gc}. These mesoscale parameters were then incorporated into the microscale hydrodynamic equations in the second step. Based on this two-step scheme, the EMMS-based drag correction can be calculated and normally scaled with the so-called heterogeneity index, H_d (Benyahia, 2012; Lu et al, 2009; Schneiderbauer and Pirker, 2014; Wang and Li, 2007). Such an EMMS-based drag correction has proved extensively to enable improving the accuracy and efficiency by allowing use of coarser grid scheme (to mention but a few of these validation, Igci and Sundaresan, 2011; Lu et al, 2011; Lungu et al, 2014; Nikolopoulos et al, 2010; Schneiderbauer and Pirker, 2014; Wang and Li, 2007; Wang et al, 2010). In our previous review, we have shown the profiles of typical cluster-based EMMS drag corrections (Wang et al, 2010, 2011). Here, Fig. 7 gives a typical drag correction in terms of heterogeneity index based on bubbles.

In literature, some modeling efforts of incorporating a subgrid, structure-dependent drag model are misinterpreted to be based on the TFM. However, as presented in above sections, the structure-dependent drag is consistent with structure-dependent conservations, and once it is used, the governing equations or the hydrodynamic model should be the filtered TFM (if the filtered drag is used) (Agrawal et al, 2001; Igci et al, 2008) or the

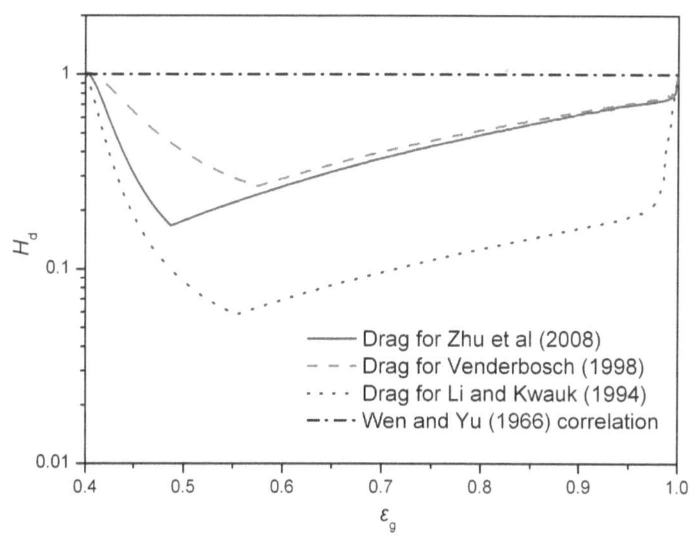

Figure 7 Variation of H_d with voidage ε_g for three operating conditions: a bubbling fluidized bed with $U_g = 0.2$ m/s (Zhu et al, 2008), a turbulent fluidized bed with $U_g = 0.6$ m/s (Venderbosch, 1998), and a fast fluidized bed with $U_g = 1.52$ m/s and $G_s = 14.3$ kg/m^2 s (Li and Kwauk, 1994). The Wen and Yu drag (1966) is used to scale the other drags (Hong et al, 2013).

SFM (if the EMMS drag is used) (Hong et al, 2012, 2013) or certain other models equivalent.

5.2.5 Extension to Cluster-Void Approach and Generalized EFM

The mesoscale structure varies with the operating conditions and material properties. In a fluidized bed with a given superficial gas velocity, locally the structure may manifest void or clusters (Lin et al, 2001) due to inhomogeneous distribution of gas. In that case, solo assumption of cluster or bubble may deviate from the reality. Ullah et al (2014) tried to resolve this issue by considering two mixtures existing alternately with respect to time and space, i.e., "gas and particles" at microscale and "dense cluster and dilute void" at mesoscale. As a first approximation, the void phase is assumed free of particles.

A combination of the EMMS/matrix model for clusters and the EMMS/bubbling model for bubbles was presented in this work. To close this set of equations, they adopted a simple criterion to quantify the phase inversion phenomenon, which is if $f > 1-f$, then the system has dispersed bubbles in continuous dense phase, else it has continuous gas phase with dispersed clusters. Furthermore, clusters will exist only when input solids flux is greater than zero. The traditional stability condition of EMMS was, however, adopted to determine these structures, irrespective of bubble or cluster.

How to characterize cluster or bubble is a critical issue for the EMMS-based modeling. As an alternative to the current stability-constrained cluster diameter, Wang et al (2008a) related the cluster diameter to the solids volume fraction by introducing an equation to correlate the dense-phase acceleration, dilute-phase acceleration, and mean solids concentration in clusters. In the long run, we still need elaborate analysis of how the energy dissipation and consumption is related with mesoscale structure. In the realm of the nonequilibrium thermodynamics, both minimum entropy production (Jaynes, 1980; Prigogine, 1955) and maximum entropy production principles (Martyushev and Seleznev, 2006) have been proposed to determine the steady state of nonequilibrium systems. Leaving aside the extremum criterion, as analyzed in the above section, the variable in the stability condition of the EMMS, N_{st}, involves not only the entropy production term due to dissipation but also the efficient work done by gas to transporting particles, as shown in Eqs. (45)–(47). Unification of these seemingly different criteria for nonequilibrium systems, in particular, multiphase flows, is strongly needed. And in practice, it may provide a unified approach for predicting flow behavior over the whole range of flow regimes, freeing ourselves from phenomenological cluster/bubble characterization.

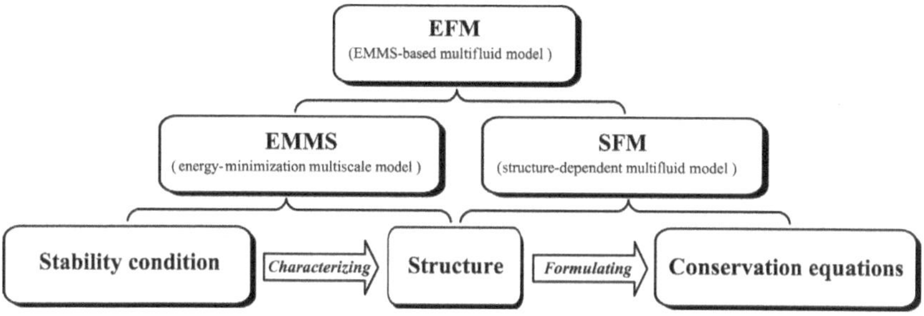

Figure 8 Framework of the EMMS-based multifluid model (EFM) (Hong et al, 2012).

Looking back on our past efforts in coupling EMMS-based models and continuum models (Lu et al, 2009; Shi et al, 2011; Wang and Li, 2007; Wang et al, 2008b; Yang et al, 2003), we can see that to be consistent with the intrinsic mesoscale structures in gas–solid two-phase flow, besides the conservation equations, the mesoscale structure as well as its relevant stability condition should be taken into account, as shown in Fig. 8. That is, the conservation equations should be formulated with full consideration of the involved structures. If the flow is comparatively homogeneous within a grid, as is the case in bubbling fluidized beds of coarse particles, the only structure one should tackle is gas and solids and then the traditional TFM can be derived. To characterize or close the structural terms, one may introduce an extra constraint which is the stability condition of $N_{st} \rightarrow \min$ as defined in the original EMMS model. Thus, the EMMS-based EFM is defined with the stability-constrained SFM. For the other multiphase systems, we may expect different formulations of the stability conditions (Ge et al, 2007) and hence more generalized forms of EFM.

5.3 Kinetic Theory with Nonequipartition Energies

Besides the EMMS-based drag closures, as shown in Eq. (21) and Table 1, the solids stress needs structure-dependent closure that is beyond the current KTGFs. Bearing in mind the bimodal velocity distribution as shown in Fig. 5, we can calculate it with two Maxwellian distributions for the dilute and dense phases, respectively. The mass exchange between the dilute and dense phases is further neglected because of the mathematical difficulties in dealing with phase change in kinetic theory. Thus, the closure of the structure-dependent stress transforms to the kinetic theory of binary particles or dilute–dense structure with unequal mean velocities and nonequipartition energies (i.e., different granular temperatures).

In literature, the dense gas kinetic theory of Chapman and Cowling (1970) has been employed to quantify rapid flow of granular mixture systems for decades (Huan et al, 2004; Serero et al, 2006; Yang et al, 2002), in which granular mixtures are assumed to be smooth, nearly elastic, and spherical grains. Generally, two approaches can be classified, as elaborated by Galvin et al (2005): the first one is derived via systematic expansion, such as Chapman–Enskog method, and the second one is on the base of a hypothesized velocity distribution. All the above derivations supposed that the mean velocities of two species are identical, which is obviously not the case in gas-fluidized systems.

In this work, for a binary granular mixture composed of hard, smooth, inelastic spheres of species α and β with mass m_i and diameter d_i, $i=\alpha, \beta$, due to suffering inhomogeneous external energy input, such as the different drag forces exerted on fluidized particles, two species are assumed to have their own temperatures and mean velocities, respectively. It should be noted that the binary granular mixture here is a generalized concept and it may refer to a mixture composed of particles in the dilute broth and dense cluster. Here, we consider the most general case for binary collisions. Let i and j represent either species, α or β. For an inelastic collision between two particles in species i and j, with velocity $\mathbf{c}_i$ and $\mathbf{c}_j$, according to the kinetic theory (Chapman and Cowling, 1970), the number of binary collisions between species i and j at position $\mathbf{r}$ per unit time per unit volume has the form:

$$N_{ij} = \int_{\mathbf{c}_{ij}\cdot\mathbf{k}>0} f_{ij}^{(2)}\left(\mathbf{c}_i, \mathbf{r}, \mathbf{c}_j, \mathbf{r} + d_{ij}\mathbf{k}\right)\left(\mathbf{c}_{ij}\cdot\mathbf{k}\right) d_{ij}^2 d\mathbf{k} d\mathbf{c}_i d\mathbf{c}_j, \tag{68}$$

where $f_{ij}^{(2)}$ is the pair distribution function, $\mathbf{k}=\mathbf{r}_{ij}/r_{ij}$ is the unit vector directing from the center of particle with velocity $\mathbf{c}_i$ to that of particle with velocity $\mathbf{c}_j$ upon contact, specifying the geometry of the impact. Following Chapman and Cowling (1970), the assumptions of chaos allow us to write the pair distribution function as a product of two single-velocity distributions:

$$f_{ij}^{(2)}\left(\mathbf{c}_i, \mathbf{r}_i, \mathbf{c}_j, \mathbf{r}_j\right) = \frac{1}{8\pi^3} g_{ij} n_i n_j \left(\frac{m_i m_j}{\theta_i \theta_j}\right)^{3/2} \exp\left[-\frac{m_i(\mathbf{c}_i - \mathbf{v}_i)^2}{2\theta_i} - \frac{m_j\left(\mathbf{c}_j - \mathbf{v}_j\right)^2}{2\theta_j}\right]. \tag{69}$$

Using this pair distribution, the collision frequency becomes

$$N_{ij} = \frac{1}{8\pi^2} d_{ij}^2 g_{ij} n_i n_j \left(\frac{m_i m_j}{\theta_i \theta_j}\right)^{3/2} \iint c_{ij} \exp\left[-\frac{m_i(\mathbf{c}_i - \mathbf{v}_i)^2}{2\theta_i} - \frac{m_j\left(\mathbf{c}_j - \mathbf{v}_j\right)^2}{2\theta_j}\right] \times d\mathbf{c}_i d\mathbf{c}_j, \tag{70}$$

where g_{ij} is the radial distribution function between spheres of species i and j, n_i is the particle number density, $\mathbf{v}_i$ is the mean velocity of species i, and θ_i is the granular temperature of species i. Due to the symmetry of collisions, the number of collisions of species i is expected to equal the number of collisions of species j, i.e., $N_{ij}=N_{ji}$.

For species i, let ψ_i be any function of particle velocity $\mathbf{c}_i$. Multiplying both sides of Boltzmann equation by $\psi_i d\mathbf{c}_i$ and integrating them throughout the velocity space, we could obtain the transport equation for the quantity ψ_i as follows:

$$\frac{\partial n_i\langle\psi_i\rangle}{\partial t}+\nabla\cdot\left(n_i\langle\mathbf{c}_i\psi_i\rangle+\sum_{j=\alpha,\beta}\mathbf{P}_{c,ij}\right)=n_i\left\langle\mathbf{F}_i\cdot\frac{\partial}{\partial\mathbf{c}_i}\psi_i\right\rangle+\sum_{j=\alpha,\beta}\mathbf{N}_{c,ij}, \quad (71)$$

where $\mathbf{P}_{c,ij}$ is the collisional part of the stress tensor and $N_{c,ij}$ the collisional source term for particles in species i during the collision with particles in species j. Here, we use the expressions of $\mathbf{P}_{c,ij}$ and $N_{c,ij}$ given by Jenkins and Mancini (1987), which reads

$$\begin{aligned}\mathbf{P}_{c,ij}= & -\frac{1}{2}d_{ij}^3\int_{\mathbf{k}\cdot\mathbf{c}_{ij}>0}\left(\psi_{i'}-\psi_i\right)\left(\mathbf{k}\cdot\mathbf{c}_{ij}\right)\\ & \times\mathbf{k}f_{ij}^{(2)}\left(\mathbf{c}_i,\mathbf{r}-\frac{1}{2}d_{ij}\mathbf{k},\,\mathbf{c}_j,\mathbf{r}+\frac{1}{2}d_{ij}\mathbf{k}\right)d\mathbf{k}\,d\mathbf{c}_i\,d\mathbf{c}_j,\end{aligned} \quad (72)$$

$$\begin{aligned}N_{c,ij}= & \, d_{ij}^2\int_{\mathbf{k}\cdot\mathbf{c}_{ij}>0}\left(\psi_{i'}-\psi_i\right)\left(\mathbf{k}\cdot\mathbf{c}_{ij}\right)f_{ij}^{(2)}\left(\mathbf{c}_i,\mathbf{r}-\frac{1}{2}d_{ij}\mathbf{k},\,\mathbf{c}_j,\mathbf{r}+\frac{1}{2}d_{ij}\mathbf{k}\right)\\ & \times d\mathbf{k}\,d\mathbf{c}_i\,d\mathbf{c}_j.\end{aligned} \quad (73)$$

Setting $\psi_i=m_i$, $m_i\mathbf{c}_i$, and $1/2m_i\mathbf{c}_i^2$, we get the transport equation of mass, momentum, and granular temperature, respectively, as follows:

$$\frac{\partial(n_i m_i)}{\partial t}+\nabla\cdot(n_i m_i\mathbf{v}_i)=0, \quad (74)$$

$$\begin{aligned}\frac{\partial(n_i m_i\mathbf{v}_i)}{\partial t}+\nabla\cdot(n_i m_i\mathbf{v}_i\mathbf{v}_i)= & -\nabla\cdot\left[\sum_{j=\alpha,\beta}\mathbf{P}_{c,ij}(m_i\mathbf{C}_i)+\mathbf{P}_{k,i}\right]\\ & +n_i m_i\mathbf{F}_i+\sum_{j=\alpha,\beta}\mathbf{N}_{c,ij}(m_i\mathbf{C}_i),\end{aligned} \quad (75)$$

$$\frac{3}{2}\left[\frac{\partial}{\partial t}(n_i\theta_i)+\nabla\cdot(n_i\mathbf{v}_i\theta_i)\right]=\left[\mathbf{P}_{k,i}+\sum_{j=\alpha,\beta}\mathbf{P}_{c,ij}(m_i\mathbf{C}_i)\right]:\nabla\mathbf{v}_i-\nabla\left(\mathbf{q}_{k,i}+\mathbf{q}_{c,i}\right)$$
$$+n_i m_i\langle\mathbf{F}\cdot\mathbf{C}_i\rangle$$
$$+\sum_{j=\alpha,\beta}\mathbf{N}_{c,ij}\left(\frac{1}{2}m_i\mathbf{C}_i^2\right),\tag{76}$$

where $\mathbf{P}_{k,i}$ is the kinetic part of the stress tensor for species i,

$$\mathbf{P}_{k,i}=\int m_i\mathbf{C}_i\mathbf{C}_i f_i d\mathbf{c}_i,\tag{77}$$

$\mathbf{q}_{c,i}$ represents $\left(\sum_{j=\alpha,\beta}\mathbf{P}_{c,ij}\left(\frac{1}{2}m_i\mathbf{C}_i^2\right)\right)$, and $\mathbf{q}_{k,i}$ is the kinetic part of energy flux,

$$\mathbf{q}_{k,i}=\int\frac{1}{2}m_i\mathbf{C}_i\mathbf{C}_i^2 f_i d\mathbf{c}_i.\tag{78}$$

Calculated from the first-order approximation to the distribution, $\mathbf{P}_{k,i}$ becomes

$$\mathbf{P}_{k,i}=n_i\theta_i\hat{\mathbf{I}}-\frac{2\mu_{i,\text{dil}}}{g_{ii}}\left(1+\frac{2\pi}{15}d_i^3 n_i g_{ii}(1+e_i)\right)\nabla^s v_i,\tag{79}$$

where $\mu_{i,\text{dil}}$ is the viscosity for dilute suspensions. The second part of the stress tensor describes the momentum transfer caused by collisions, that is,

$$\mathbf{P}_{c,ij}(m_i\mathbf{c}_i)=\sum_j\left[\frac{1}{2}d_{ij}^3 g_{ij}\int_{\mathbf{k}\cdot\mathbf{c}_{ij}>0}m_i\left(C_i'-C_i\right)(\mathbf{k}\cdot\mathbf{c}_{ij})\mathbf{k}\cdot f_i f_j d\mathbf{k}d\mathbf{c}_i d\mathbf{c}_j\right.$$
$$\left.-\frac{1}{4}g_{ij}d_{ij}^4\int_{\mathbf{k}\cdot\mathbf{c}_{ij}>0}m_i\left(C_i'-C_i\right)(\mathbf{k}\cdot\mathbf{c}_{ij})\mathbf{k}\cdot f_i f_j\nabla\ln\frac{f_i}{f_j}d\mathbf{k}d\mathbf{c}_i d\mathbf{c}_j\right]$$
$$=\sum_j\left(\mathbf{P}_{c,ij}^1+\mathbf{P}_{c,ij}^2\right).\tag{80}$$

Consequently, we acquire the total particle pressure $\mathbf{p}_i$ defined from the normal term and the coefficient of viscosity μ_i defined from the shear term of the sum of $\mathbf{P}_{k,i}$ and $\mathbf{P}_{c,ij}$, as follows:

$$\begin{aligned}\mathbf{p}_i =& n_i\theta_i + \sum_j \frac{m_i m_j}{m_0}\frac{1}{48\pi^2}\left(e_{ij}+1\right)d_{ij}^3 g_{ij} n_i n_j \left(\frac{m_i m_j}{\theta_i\theta_j}\right)^{3/2} \\ &\times \frac{1}{A^{3/2}D^{5/2}}\left(1+\frac{5B^2}{2AD}+\frac{35B^4}{8A^2D^2}+\cdots\right),\end{aligned} \tag{81}$$

$$\begin{aligned}\mu_i =& \frac{\mu_{i,\,dil}}{g_{ii}}\left(1+\frac{2\pi}{15}d_i^3 n_i g_{ii}(1+e_i)\right) \\ &\times\left(1+\frac{2\pi}{15}\frac{n_j m_j}{m_0}d_i^3 g_{ii}\left(1+e_{ij}\right)+\frac{2\pi}{15}d_i^3 n_i g_{ii}(1+e_i)\right) \\ &+\sum_j \frac{1}{960\pi^2}\frac{m_i m_j}{m_0}\left(e_{ij}+1\right) \\ &\times d_{ij}^4 g_{ij} n_i n_j \left(\frac{m_i m_j}{\theta_i\theta_j}\right)^{3/2}\left(\frac{m_i m_j}{m_0}\frac{1}{\theta_i}R_1-\frac{2Bm_i}{3\,\theta_i}R_6\right).\end{aligned} \tag{82}$$

The collisional contribution to the energy flux is determined by

$$\begin{aligned}&\mathbf{P}_{ij}\left(\frac{1}{2}m_i\mathbf{C}_i^2\right) \\ &= -\frac{m_i}{4}d_{ij}^3\int_{\mathbf{k}\cdot\mathbf{c}_{ij}>0}\left(\mathbf{C}_{i'}^2-\mathbf{C}_i^2\right)\left(\mathbf{k}\cdot\mathbf{c}_{ij}\right)\mathbf{k}f_{ij}^{(2)}\left(\mathbf{c}_i,\mathbf{r}-\frac{d_{ij}}{2}\mathbf{k},\,\mathbf{c}_j,\mathbf{r}+\frac{d_{ij}}{2}\mathbf{k}\right)d\mathbf{k}\,d\mathbf{c}_i\,d\mathbf{c}_j \\ &= \mathbf{q}_{c,\,ij}^1+\mathbf{q}_{c,\,ij}^2.\end{aligned} \tag{83}$$

And collision part of energy source term can be written as

$$\begin{aligned}&\mathbf{N}_{ij}\left(\frac{1}{2}m_i\mathbf{C}_i^2\right) = d_{ij}^2\int_{\mathbf{k}\cdot\mathbf{c}_{ij}>0}\frac{m_i}{2}\left(\mathbf{C}_{i'}^2-\mathbf{C}_i^2\right)\left(\mathbf{k}\cdot\mathbf{c}_{ij}\right) \\ &\times f_{ij}^{(2)}\left(\mathbf{c}_i,\mathbf{r}-\frac{1}{2}d_{ij}\mathbf{k},\,\mathbf{c}_j,\mathbf{r}+\frac{1}{2}d_{ij}\mathbf{k}\right)d\mathbf{k}\,d\mathbf{c}_i\,d\mathbf{c}_j = \gamma_{c,\,ij}^1+\gamma_{c,\,ij}^2,\end{aligned} \tag{84}$$

where

$$\begin{aligned}\gamma_{c,\,ij}^1 &= d_{ij}^2 g_{ij}\int_{\mathbf{k}\cdot\mathbf{c}_{ij}>0}\left(\frac{1}{2}m_i\mathbf{C}_{i'}^2-\frac{1}{2}m_i\mathbf{C}_i^2\right)\left(\mathbf{k}\cdot\mathbf{c}_{ij}\right)f_i f_j\,d\mathbf{k}\,d\mathbf{c}_i\,d\mathbf{c}_j, \\ \gamma_{c,\,ij}^2 &= \frac{1}{2}d_{ij}^3 g_{ij}\int_{\mathbf{k}\cdot\mathbf{c}_{ij}>0}\left(\frac{1}{2}m_i\mathbf{C}_{i'}^2-\frac{1}{2}m_i\mathbf{C}_i^2\right)\left(\mathbf{k}\cdot\mathbf{c}_{ij}\right)f_i f_j\nabla\ln\frac{f_i}{f_j}\,d\mathbf{k}\,d\mathbf{c}_i\,d\mathbf{c}_j.\end{aligned} \tag{85}$$

Figure 9 plots 3D surfaces of scaled collision frequency N_{ij} in terms of granular temperature of two species and comparison with previous results

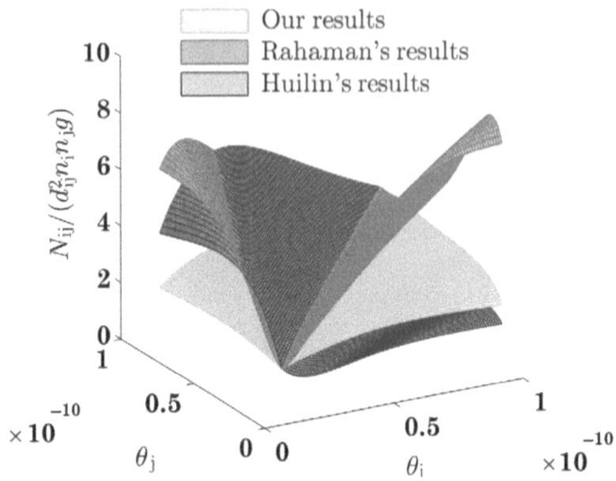

Figure 9 Surfaces of scaled collision frequency in terms of two granular temperatures θ_i and θ_j, where $m_i = 2.79 \times 10^{-10}$ kg, $m_j = 6.62 \times 10^{-10}$ kg, and N_{ij} is scaled by $d_{ij}^2 n_i n_j g$.

(Lu et al, 2001; Rahaman et al, 2003). N_{ij} is scaled by $d_{ij}{}^2 n_i n_j g$ considering that N_{ij} varies greatly with the Enskog factor. Three surfaces intersect at line $\theta_i = \theta_j$, about which both our results and Rahaman's results are symmetric. That is consistent with theoretical analysis as discussed above, while Huilin's result fails. Furthermore, we can observe that our surfaces are higher than the Huilin's prediction and lower than Rahaman's when $\theta_i < \theta_j$. This is expected as the Huilin's profile can only be applied in energy equipartition system, which may underestimate the collision frequencies, whereas Rahaman's hypothesis implies all the collisions happen in a 2D plane, which obviously overestimates the collision frequency.

Figure 10 plots the solids shear viscosity and its comparison with literature results. Different sources of data are scale by the Enskog factor to be viewed in the same figure. Unlike Rahaman's result where the viscosity has a minimum at $\theta_i = \theta_j$, our result shows that the μ_i are still mainly determined by θ_i and the maximum of viscosity appears at the maximum of θ_i, which demonstrates that the dilute viscosity increases with the increase of the granular temperature.

To investigate the influence of mean velocities, Fig. 11 plots the surface of energy dissipation part $\gamma^1_{c,ij}$ in terms of two mean velocities. It can be found that when the mean velocities are not equal, $\gamma^1_{c,ij}$ varies greatly, reaching its peak when $\mathbf{u}_i$ is maximum and $\mathbf{u}_j$ is minimum. So the difference of two mean velocities should not be ignored. In an air-fluidized bed, the mean velocities of clusters and dispersed particles are usually not equal to

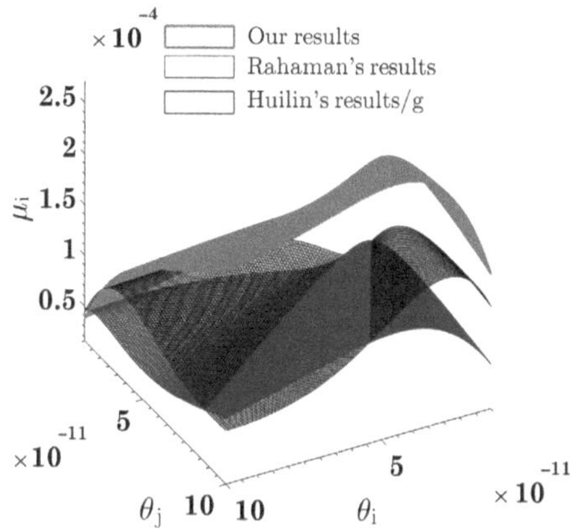

Figure 10 Surfaces of scaled viscosity as a function of granular temperatures, where $m_i = 2.79 \times 10^{-10}$ kg, $m_j = 6.62 \times 10^{-10}$ kg, $d_i = d_j = 5 \times 10^{-4}$ m, $\varepsilon_i = \varepsilon_j = 0.25$, and $e_i = e_{ij} = 0.9$.

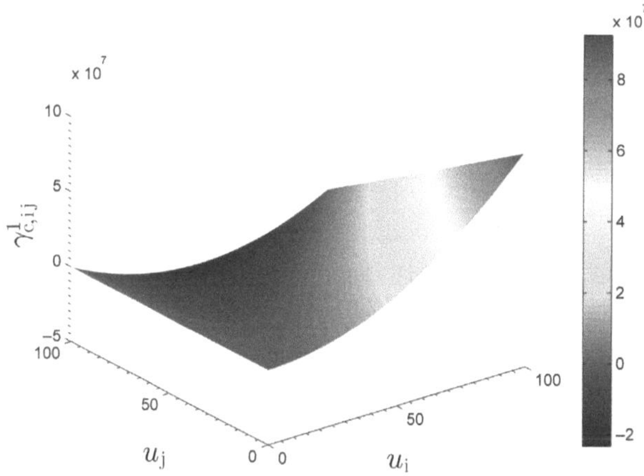

Figure 11 Surfaces of energy source $\gamma^1_{c,ij}$ as a function of mean velocities, where $\varepsilon_i = \varepsilon_j = 0.25$, $d_i = d_j = 5 \times 10^{-4}$ m, $\theta_i = \theta_j = 3.9 \times 10^{-11}$ kg/m^2 s^2, $m_i = 2.79 \times 10^{-10}$ kg, $m_j = 6.62 \times 10^{-10}$ kg, and $e_{ij} = 0.9$.

each other. Our model is sensitive to this factor. And more elaborate validation needs further efforts.

It should be emphasized again, that, for a gas-fluidized system, the results based on the two Maxwellian distributions, which correspond to the EMMS-featured dilute and dense phases, respectively, are the same as those on the bimodal distribution. In greater detail, the total velocity distribution reads

$$f(\mathbf{c}_\mathrm{c}, \mathbf{r}_\mathrm{c}, \mathbf{c}_\mathrm{f}, \mathbf{r}_\mathrm{f}) = f\frac{\varepsilon_{\mathrm{pc}}\rho_\mathrm{p}}{m}\left(\frac{m}{2\pi\theta_\mathrm{c}}\right)^{3/2}\exp\left[-\frac{m\left(\mathbf{c}_\mathrm{c}-\mathbf{v}_{\mathrm{pc}}\right)^2}{2\theta_\mathrm{c}}\right] + (1-f)\frac{\varepsilon_{\mathrm{pf}}\rho_\mathrm{p}}{m}\left(\frac{m}{2\pi\theta_\mathrm{f}}\right)^{3/2}\exp\left[-\frac{m\left(\mathbf{c}_\mathrm{f}-\mathbf{v}_{\mathrm{pf}}\right)^2}{2\theta_\mathrm{f}}\right] \tag{86}$$

Now the whole set of SFM hydrodynamic equations with EMMS mesoscale modeling is established, where the drag and stress closures are based on the bimodal distribution. However, the mass exchange (or phase change) between the dilute and dense phases is still a big challenge to the kinetic theory derivation. More elaborate efforts are needed on this topic.

5.4 Structure-Dependent Mass Transfer and Reactions

Although the effects of mesoscale structure on the flow have been extensively investigated in recent years, the mass transfer study receives much less attention. There are three possible reasons for this situation (Liu et al, 2015; Wang et al, 2011):

First, mass transfer acts with reactions and momentum transfer, complicating its dependency on structure so that it is hard to measure. Second, the importance of mass transfer on the overall reaction rate is often underestimated. For fine particles reacting with surrounding gas under high slip velocity, it is normally assumed that the fluid–particle mass transfer rate is very high because of the large specific area such that the overall reaction rate is controlled by the slower step, the surface reaction. However, the mesoscale structure may reduce the mass transfer rate in the same way that it affects the drag coefficient, making realistic Damköhler number $Da \sim 1$ (defined on the scale of CFD cell). That is, the reactive behavior is actually controlled by both surface reaction and mass transfer. The last and the most easily overlooked reason lies in that the classic concept of intrinsic reaction rate (Levenspiel, 1999), which is based on the assumption that the reaction and transport phenomena are separable, is oversimplified. For example, the observed catalytic reaction mechanism may be affected by the cavity size of molecular sieves (Li et al, 2015), and there is evidence that elementary kinetic reactions (Turns, 2000) may be inseparable from catalyst (particle) concentration. Thus, the so-called intrinsic reaction rate on the scale of a single particle is actually an oversimplification. Indeed, although there are lots of reaction kinetics for various types of reactions reported in literature, they are hard to be called intrinsic ones. As a result, the exact effects of mass transfer on the overall reaction rate at the scale of global reactors are difficult to quantify (Breault, 2006; Dong et al, 2008a, 2008b; Holloway and Sundaresan, 2012).

The successful application and generalization of the SFM model triggers our attempts to establish a structure-dependent mass transfer model which may unify our past efforts on mass transfer modeling. Similar to what has been presented in Section 5.1, we introduce the nonequilibrium, bimodal distribution into the modeling of mass transfer and reactions in a similar way, as detailed in the following.

5.4.1 Structure-Dependent Mass Transfer/Reaction Model

In a gas–solid CFB with heterogeneous reactions and mass transfer, in line with the structural characteristics of the SFM model (Hong et al, 2012), as shown in Fig. 12, the mass transfer and reaction in any local space can be divided into components of the dense cluster (denoted by subscript c), the dilute broth (denoted by subscript f), and in-between (denoted by subscript i), respectively. And these terms can be represented by R_l (l = gc, gf, gi, sc, sf, si). Both the dense and dilute phases are assumed homogenous and continuous inside, and the dense phase is further assumed suspended uniformly in the dilute phase in forms of clusters of particles. Then the mass transfer terms can be described with Ranz–Marshall-like relations for uniform suspension of particles (Halder and Basu, 1988). In particular, the mesoscale interaction over the cluster will be treated as is for a "big particle" with hydrodynamic equivalent diameter of d_c. Due to dynamic nature of clusters, there are mass exchanges between the dilute and dense phases with rate of Γ_k (k = g, s), pointing outward from the dilute to the dense phase.

Likewise, the above physical process can be translated into a set of structure-dependent conservation equations for mass, momentum, and

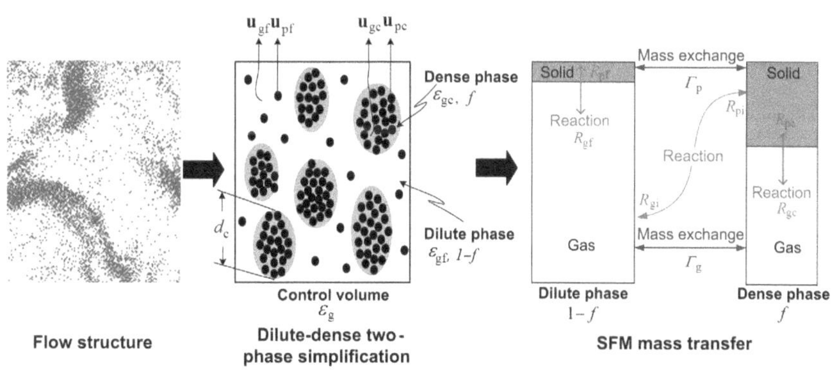

Figure 12 Schematic drawing of the structure-dependent multifluid model for mass transfer and reactions (Liu et al, 2015).

species of dilute and dense phases (here the hydrodynamic model B (Gidaspow, 1994) is used to fully comply with the EMMS description), as detailed in the following.
Continuity equation for the dense-phase gas:

$$\frac{\partial}{\partial t}\left(f\varepsilon_{\mathrm{gc}}\rho_{\mathrm{g}}\right)+\nabla\cdot\left(f\varepsilon_{\mathrm{gc}}\rho_{\mathrm{g}}\mathbf{u}_{\mathrm{gc}}\right)=\Gamma_{\mathrm{g}}+R_{\mathrm{gc}}. \tag{87}$$

Continuity equation for the dense-phase solid:

$$\frac{\partial}{\partial t}\left(f\varepsilon_{\mathrm{pc}}\rho_{\mathrm{p}}\right)+\nabla\cdot\left(f\varepsilon_{\mathrm{pc}}\rho_{\mathrm{p}}\mathbf{u}_{\mathrm{pc}}\right)=\Gamma_{\mathrm{p}}+R_{\mathrm{pc}}+R_{\mathrm{pi}}. \tag{88}$$

Continuity equation for the dilute-phase gas:

$$\frac{\partial}{\partial t}\left[(1-f)\varepsilon_{\mathrm{gf}}\rho_{\mathrm{g}}\right]+\nabla\cdot\left[(1-f)\varepsilon_{\mathrm{gf}}\rho_{\mathrm{g}}\mathbf{u}_{\mathrm{gf}}\right]=-\Gamma_{\mathrm{g}}+R_{\mathrm{gf}}+R_{\mathrm{gi}}. \tag{89}$$

Continuity equation for the dilute-phase solid:

$$\frac{\partial}{\partial t}\left[(1-f)\varepsilon_{\mathrm{pf}}\rho_{\mathrm{p}}\right]+\nabla\cdot\left[(1-f)\varepsilon_{\mathrm{pf}}\rho_{\mathrm{p}}\mathbf{u}_{\mathrm{pf}}\right]=-\Gamma_{\mathrm{p}}+R_{\mathrm{pf}}. \tag{90}$$

To better describe the momentum transfer due to reactions, here we distinguish the source being consumed, $R_{\mathrm{l}}^{\mathrm{r}}$, from that being generated, $R_{\mathrm{l}}^{\mathrm{p}}$, that is,

$$R_{\mathrm{l}}=R_{\mathrm{l}}^{\mathrm{p}}-R_{\mathrm{l}}^{\mathrm{r}}. \tag{91}$$

Momentum conservation equation for the dense-phase gas:

$$\begin{aligned}&\frac{\partial}{\partial t}\left(f\varepsilon_{\mathrm{gc}}\rho_{\mathrm{g}}\mathbf{u}_{\mathrm{gc}}\right)+\nabla\cdot\left(f\varepsilon_{\mathrm{gc}}\rho_{\mathrm{g}}\mathbf{u}_{\mathrm{gc}}\mathbf{u}_{\mathrm{gc}}\right)=-f\nabla p\\&\quad+\nabla\cdot\left(f\boldsymbol{\tau}_{\mathrm{gc}}\right)+f\rho_{\mathrm{g}}\mathbf{g}-f\mathbf{F}_{\mathrm{dc}}+\Gamma_{\mathrm{g}}\mathbf{u}_{\mathrm{gx}}+\left(R_{\mathrm{gc}}^{\mathrm{p}}\mathbf{u}_{\mathrm{gc}}^{\mathrm{p}}-R_{\mathrm{gc}}^{\mathrm{r}}\mathbf{u}_{\mathrm{gc}}\right).\end{aligned} \tag{92}$$

Momentum conservation equation for the dense-phase solid:

$$\begin{aligned}\frac{\partial}{\partial t}\left(f\varepsilon_{\mathrm{pc}}\rho_{\mathrm{p}}\mathbf{u}_{\mathrm{pc}}\right)&+\nabla\cdot\left(f\varepsilon_{\mathrm{pc}}\rho_{\mathrm{p}}\mathbf{u}_{\mathrm{pc}}\mathbf{u}_{\mathrm{pc}}\right)=-\nabla p_{\mathrm{pc}}+\nabla\cdot\left(f\boldsymbol{\tau}_{\mathrm{pc}}\right)\\&+f\varepsilon_{\mathrm{pc}}\left(\rho_{\mathrm{p}}-\rho_{\mathrm{g}}\right)\mathbf{g}+f\mathbf{F}_{\mathrm{dc}}+\mathbf{F}_{\mathrm{di}}+\Gamma_{\mathrm{p}}\mathbf{u}_{\mathrm{px}}\\&+\left(R_{\mathrm{pc}}^{\mathrm{p}}\mathbf{u}_{\mathrm{pc}}^{\mathrm{p}}-R_{\mathrm{pc}}^{\mathrm{r}}\mathbf{u}_{\mathrm{pc}}\right)+\left(R_{\mathrm{pi}}^{\mathrm{p}}\mathbf{u}_{\mathrm{pi}}-R_{\mathrm{pi}}^{\mathrm{r}}\mathbf{u}_{\mathrm{pc}}\right).\end{aligned} \tag{93}$$

Momentum conservation equation for the dilute-phase gas:

$$\begin{aligned}\frac{\partial}{\partial t}\left[(1-f)\varepsilon_{\mathrm{gf}}\rho_{\mathrm{g}}\mathbf{u}_{\mathrm{gf}}\right] &+\nabla\cdot\left[(1-f)\varepsilon_{\mathrm{gf}}\rho_{\mathrm{g}}\mathbf{u}_{\mathrm{gf}}\mathbf{u}_{\mathrm{gf}}\right]=-(1-f)\nabla p\\ &+\nabla\cdot\left[(1-f)\boldsymbol{\tau}_{\mathrm{gf}}\right]+(1-f)\rho_{\mathrm{g}}\mathbf{g}\\ &-(1-f)\mathbf{F}_{\mathrm{df}}-\mathbf{F}_{\mathrm{di}}-\Gamma_{\mathrm{g}}\mathbf{u}_{\mathrm{gx}}+\left(R_{\mathrm{gf}}^{\mathrm{p}}\mathbf{u}_{\mathrm{gf}}^{\mathrm{p}}-R_{\mathrm{gf}}^{\mathrm{r}}\mathbf{u}_{\mathrm{gf}}\right)\\ &+\left(R_{\mathrm{gi}}^{\mathrm{p}}\mathbf{u}_{\mathrm{pi}}-R_{\mathrm{gi}}^{\mathrm{r}}\mathbf{u}_{\mathrm{gf}}\right).\end{aligned} \tag{94}$$

Momentum conservation equation for the dilute-phase solid:

$$\begin{aligned}\frac{\partial}{\partial t}\left[(1-f)\varepsilon_{\mathrm{pf}}\rho_{\mathrm{p}}\mathbf{u}_{\mathrm{pf}}\right]+\nabla\cdot\left[(1-f)\varepsilon_{\mathrm{pf}}\rho_{\mathrm{p}}\mathbf{u}_{\mathrm{pf}}\mathbf{u}_{\mathrm{pf}}\right]&=-\nabla p_{\mathrm{pf}}+\nabla\cdot\left[(1-f)\boldsymbol{\tau}_{\mathrm{pf}}\right]\\ &+(1-f)\varepsilon_{\mathrm{pf}}\left(\rho_{\mathrm{p}}-\rho_{\mathrm{g}}\right)\mathbf{g}\\ &+(1-f)\mathbf{F}_{\mathrm{df}}\\ &-\Gamma_{\mathrm{p}}\mathbf{u}_{\mathrm{px}}+\left(R_{\mathrm{pf}}^{\mathrm{p}}\mathbf{u}_{\mathrm{pf}}^{\mathrm{p}}-R_{\mathrm{pf}}^{\mathrm{r}}\mathbf{u}_{\mathrm{pf}}\right).\end{aligned} \tag{95}$$

The mass exchanges of gas and solids between the dilute and dense phases also cause momentum transfer, $\Gamma_{\mathrm{g}}\mathbf{u}_{\mathrm{gx}}$ and $\Gamma_{\mathrm{s}}\mathbf{u}_{\mathrm{sx}}$, where $\mathbf{u}_{\mathrm{gx}}$ and $\mathbf{u}_{\mathrm{sx}}$ are the velocities of the carrier phase, or the phases where the gas or solids comes from. Conservation equation of species in the dense-phase gas:

$$\frac{\partial}{\partial t}\left(f\varepsilon_{\mathrm{gc}}\rho_{\mathrm{g}}y_{\mathrm{jc}}\right)+\nabla\cdot\left(f\varepsilon_{\mathrm{gc}}\rho_{\mathrm{g}}y_{\mathrm{jc}}\mathbf{u}_{\mathrm{gc}}-f\varepsilon_{\mathrm{gc}}D_{\mathrm{mj}}\rho_{\mathrm{g}}\nabla y_{\mathrm{jc}}\right)=\Gamma_{\mathrm{g}}y_{\mathrm{jh}}+R_{\mathrm{jc}}. \tag{96}$$

Conservation equation of species in the dilute-phase gas:

$$\begin{aligned}&\frac{\partial}{\partial t}\left[(1-f)\varepsilon_{\mathrm{gf}}\rho_{\mathrm{g}}y_{\mathrm{jf}}\right]+\nabla\cdot\left[(1-f)\varepsilon_{\mathrm{gf}}\rho_{\mathrm{g}}y_{\mathrm{jf}}\mathbf{u}_{\mathrm{gf}}-(1-f)\varepsilon_{\mathrm{gf}}D_{\mathrm{mj}}\rho_{\mathrm{g}}\nabla y_{\mathrm{jf}}\right]\\ &=-\Gamma_{\mathrm{g}}y_{\mathrm{jh}}+R_{\mathrm{jf}}+R_{\mathrm{ji}}.\end{aligned} \tag{97}$$

Here y_{jc} and y_{jf} refer to the mass fractions of species j in the dense and dilute gas phases, and y_{jh} is the mass fraction of species j in the phase where Γ_{g} comes from.
Conservation equation of species in the dense-phase solid:

$$\begin{aligned}&\frac{\partial}{\partial t}\left(f\varepsilon_{\mathrm{pc}}\rho_{\mathrm{p}}x_{\mathrm{wc}}\right)+\nabla\cdot\left(f\varepsilon_{\mathrm{pc}}\rho_{\mathrm{p}}x_{\mathrm{wc}}\mathbf{u}_{\mathrm{pc}}-f\varepsilon_{\mathrm{pc}}D_{\mathrm{mw}}\rho_{\mathrm{p}}\nabla x_{\mathrm{wc}}\right)\\ &=\Gamma_{\mathrm{p}}x_{\mathrm{wh}}+R_{\mathrm{wc}}+R_{\mathrm{wi}}.\end{aligned} \tag{98}$$

Conservation equation of species in the dilute-phase solid:

$$\frac{\partial}{\partial t}\left[(1-f)\varepsilon_{\mathrm{pf}}\rho_{\mathrm{p}}x_{\mathrm{wf}}\right]+\nabla\cdot\left[(1-f)\varepsilon_{\mathrm{pf}}\rho_{\mathrm{p}}x_{\mathrm{wf}}\mathbf{u}_{\mathrm{pf}}-(1-f)\varepsilon_{\mathrm{pf}}D_{\mathrm{mw}}\rho_{\mathrm{p}}\nabla x_{\mathrm{wf}}\right]$$
$$=-\Gamma_{\mathrm{p}}x_{\mathrm{wh}}+R_{\mathrm{wf}}. \tag{99}$$

Here, x_{wc} and x_{wf} refer to the mass fractions of species w in the solids of the dense and the dilute phases, and x_{wh} is the mass fraction of species w in the phase where Γ_{s} comes from. And the source term of the species j or w caused by the heterogeneous reaction in the dilute and dense phase in the above species conservation equations are R_{jc}, R_{jf}, R_{ji} R_{wc}, R_{wf}, and R_{wi}.

5.4.2 Reduction to TFM with Structure-Dependent Closures

Direct solution of this SFM for mass transfer and reactions is possible but much more demanding than our normally used TFM. To simplify its solution, as discussed above and in Hong et al (2012), we may reduce it by combining the conservation equations of the dilute and dense phases. The combined equations read

$$\frac{\partial}{\partial t}\left(\varepsilon_{\mathrm{g}}\rho_{\mathrm{g}}\right)+\nabla\cdot\left(\varepsilon_{\mathrm{g}}\rho_{\mathrm{g}}\mathbf{u}_{\mathrm{g}}\right)=R_{\mathrm{ge}}. \tag{100}$$

$$\frac{\partial}{\partial t}\left(\varepsilon_{\mathrm{p}}\rho_{\mathrm{p}}\right)+\nabla\cdot\left(\varepsilon_{\mathrm{p}}\rho_{\mathrm{p}}\mathbf{u}_{\mathrm{p}}\right)=R_{\mathrm{pe}}. \tag{101}$$

The main difference between Eqs. (100) and (101) and the normally used mass conservation equations lies in its structure-dependent mass source term on the RHS, i.e.,

$$R_{\mathrm{ke}}=R_{\mathrm{kc}}+R_{\mathrm{kf}}+R_{\mathrm{ki}}. \tag{102}$$

Likewise, we can get the momentum conservation and species conservation for the gas and solid phases, as follows:

$$\frac{\partial}{\partial t}\left(\varepsilon_{\mathrm{g}}\rho_{\mathrm{g}}\mathbf{u}_{\mathrm{g}}\right)+\nabla\cdot\left(\varepsilon_{\mathrm{g}}\rho_{\mathrm{g}}\mathbf{u}_{\mathrm{g}}\mathbf{u}_{\mathrm{g}}\right)$$
$$=-\nabla p+\nabla\cdot\boldsymbol{\tau}_{\mathrm{ge}}+\rho_{\mathrm{g}}\mathbf{g}-\beta_{\mathrm{Be}}\left(\mathbf{u}_{\mathrm{g}}-\mathbf{u}_{\mathrm{p}}\right)+\left(R_{\mathrm{ge}}^{\mathrm{p}}\mathbf{u}_{\mathrm{ge}}^{\mathrm{p}}-R_{\mathrm{ge}}^{\mathrm{r}}\mathbf{u}_{\mathrm{ge}}\right)+\nabla\cdot\boldsymbol{\tau}_{\mathrm{Dg}}, \tag{103}$$

$$\begin{aligned}\frac{\partial}{\partial t}\left(\varepsilon_{\mathrm{p}}\rho_{\mathrm{p}}\mathbf{u}_{\mathrm{p}}\right)+\nabla\cdot\left(\varepsilon_{\mathrm{p}}\rho_{\mathrm{p}}\mathbf{u}_{\mathrm{p}}\mathbf{u}_{\mathrm{p}}\right) &= -\nabla p_{\mathrm{pe}}+\nabla\cdot\boldsymbol{\tau}_{\mathrm{pe}}+\varepsilon_{\mathrm{p}}\left(\rho_{\mathrm{p}}-\rho_{\mathrm{g}}\right)\mathbf{g}\\ &\quad+\beta_{\mathrm{Be}}\left(\mathbf{u}_{\mathrm{g}}-\mathbf{u}_{\mathrm{p}}\right)+\left(R_{\mathrm{pe}}^{\mathrm{p}}\mathbf{u}_{\mathrm{pe}}^{\mathrm{p}}-R_{\mathrm{pe}}^{\mathrm{r}}\mathbf{u}_{\mathrm{pe}}\right)\\ &\quad+\nabla\cdot\boldsymbol{\tau}_{\mathrm{Dp}}.\end{aligned} \tag{104}$$

$$\frac{\partial}{\partial t}\left(\varepsilon_g\rho_g y_j\right) + \nabla\cdot\left(\varepsilon_g\rho_g y_j \mathbf{u}_g - \varepsilon_g D_m \rho_g \nabla y_j\right) = R_j + S_j^u + S_j^d \tag{105}$$

$$\frac{\partial}{\partial t}\left(\varepsilon_p\rho_p x_w\right) + \nabla\cdot\left(\varepsilon_p\rho_p x_w \mathbf{u}_p - \varepsilon_p D_m \rho_p \nabla x_w\right) = R_w + S_w^u + S_w^d, \tag{106}$$

where

$$R_q = R_{qc} + R_{qf} + R_{qi}. \tag{107}$$

Here, q represents species j or w in gas or solid phase. In all, we reduce our model to a set of TFM-like equations. The structure-dependent terms for diffusion, stress, drag force, and mass sources distinguish our model from the TFM. The normally used mass transfer and reaction equations of the TFM model can be viewed as a special case of this new model without subgrid structure.

5.4.3 Local Balance Between Mass Transfer and Reaction

The relations between the dilute/dense concentrations and the mean concentration can be resolved with sequential analysis of the balance between mass transfer and reactions, as is in Levenspiel (1999). Thereon, the structure-dependent correction for mass transfer and reaction rate can be determined. Here we exemplify the solution with reference to a reactive system where a gas species is catalyzed by solid particles. The gas concentrations are therefore of our major concern. In greater details, the mean molar concentration of species j (C_j) in a control volume relates with the dense-phase and dilute-phase concentrations by

$$C_j = \left(f\cdot\varepsilon_{gc}\cdot C_{jc} + (1-f)\cdot\varepsilon_{gf}\cdot C_{jf}\right)/\varepsilon_g. \tag{108}$$

As shown in Fig. 13, the gas–solid heterogeneous reactions in a CFB riser can be divided into the following steps:

(1) Diffusion of gaseous reactant j through the bulk gas in the dilute phase to the surface of solid particles in the dilute phase and the surface of clusters in the dense phase;

(2) Diffusion of reactant j through the surface of clusters into the bulk gas of the dense phase and to the surface of solid particles;

(3) Penetration and diffusion of reactant j through the porous framework of solid particles in the dilute and dense phases where adsorption and reaction take place;

(4) Adsorption of reactant j inside the particles of the dilute and dense phases;

(5) Surface reaction from reactant j to product p in the dilute and dense phases;

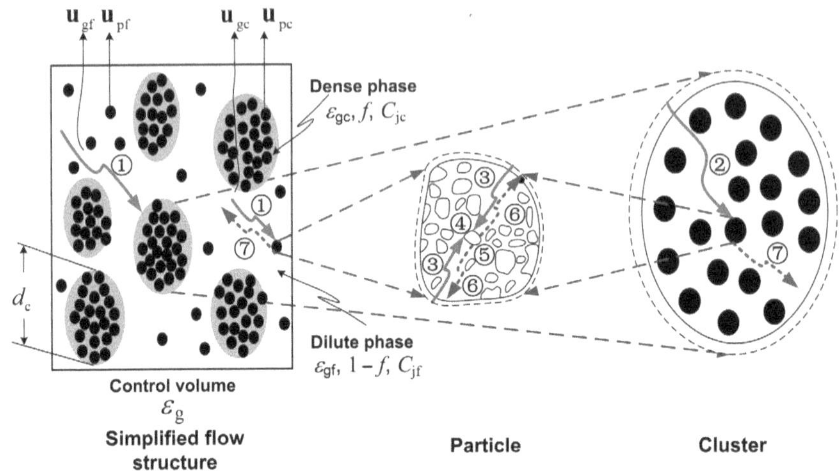

Figure 13 The sequential balance between surface reaction and mass transfer in a riser (Liu et al, 2015).

(6) Desorption of product p inside the particles of the dilute and dense phase;

(7) Diffusion of product p from inside the particles back to the gas of the dilute and dense phases sequentially.

If CFB is maintained at a steady state, then the mass diffused to the particles or clusters is supposed to be equal to what is consumed in reactions. For the sake of simplicity, the reaction is assumed to be first order and happens only at the gas–solid interface. Thus, the balance equations for species concentration over the dilute and dense phases and their interface are as follows:
For the particles in the dilute phase:

$$k_{df}\frac{6(1-f)\varepsilon_{pf}}{d_p}\left(C_{jf}-C_{jf}^0\right)=k_r(1-f)\varepsilon_{pf}C_{jf}^0. \tag{109}$$

For the particles in the dense phase:

$$k_{dc}\frac{6f\varepsilon_{pc}}{d_p}\left(C_{jc}-C_{jc}^0\right)=k_r f\varepsilon_{pc}C_{jc}^0. \tag{110}$$

Further, by assuming local balance between reaction in clusters and the mass transfer between the dilute-phase gas and the surface of cluster, we get

$$k_{di}\frac{6f}{d_c}\varepsilon_{gc}\left(C_{jf}-C_{jc}\right)+k_{di}\frac{6f}{d_c}\varepsilon_{pc}\left(C_{jf}-C_{jc}^0\right)=k_r f\varepsilon_{pc}C_{jc}^0, \tag{111}$$

where C^0 represents the concentration at the particle surface, k_{dn} (n = c, f, i) represents the mass transfer coefficient of species j from the bulk gas to catalyst surface in each phase. For the sake of simplicity, here we assume that the area fraction of the solids over the cluster interface is also ε_{pc}, and the concentrations of species j over the cluster interface are the same as those inside. Further, the Sherwood number relation for homogeneous system is used for mass transfer correlation (Halder and Basu, 1988); we can obtain the relations between C_{jc}, C_{jf}, C_{jc}^0, C_{jf}^0, and C_j, thereby the overall reaction rate can be determined by summing those from the dilute and dense phases,

$$r = r_f + r_c + r_i = k_r(1-f)\varepsilon_{pf} C_{jf}^0 + k_r f \varepsilon_{pc} C_{jc}^0. \tag{112}$$

In contrast, the reaction rate without considering the dilute–dense flow structure reads

$$r_0 = k_r f \varepsilon_p C_j^0. \tag{113}$$

Likewise, C_j^0 can be calculated from the equilibrium between the mass transfer and surface reaction, that is,

$$k_d \frac{6\varepsilon_p}{d_p}\left(C_j - C_j^0\right) = k_r \varepsilon_p C_j^0, \tag{114}$$

Similar to Wang and Li (2007), the heterogeneity index for structure-dependent reaction can be defined as follows:

$$H_r = \frac{r}{r_0}. \tag{115}$$

If there is no chemical reaction but mass exchange between phases, such as evaporation, sublimation, and condensation, we can replace the reaction rate with certain mass exchange rate r_m, and similarly define the heterogeneity index for structure-dependent mass transfer. For example, assume that the concentration of transferred species at particle surface is saturated; further, the diffusion rate from the particles to the gas in the dense phase is equal to that from the gas in the dense phase to the gas in the dilute phase and we get

$$k_{dc}\frac{6f\varepsilon_{pc}}{d_p}\left(C_j^0 - C_{jc}\right) = k_{di}\frac{6f}{d_c}\varepsilon_{gc}\left(C_{jc} - C_{jf}\right) + k_{di}\frac{6f}{d_c}\varepsilon_{pc}\left(C_j^0 - C_{jf}\right). \tag{116}$$

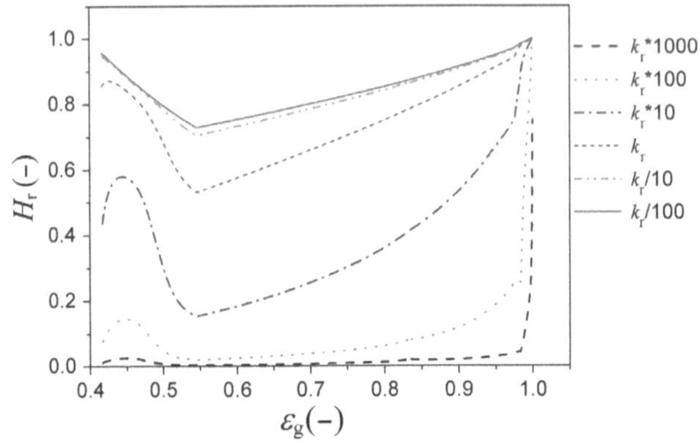

Figure 14 Variation of H_r against voidage under different reaction rate ($U_g = 3.8$ m/s, $G_s = 106$ kg/m^2 s, $k_r = 57.21$ s^{-1}) (Liu et al, 2015).

The relations between C_{jc}, C_{jf}, and C_j can thus be calculated by solving Eqs. (108) and (116). Then we can get the heterogeneity index for mass transfer as follows:

$$H_m = \frac{r_m}{r_{m0}}. \tag{117}$$

To have some idea of the magnitude of the effect of structure on the mass transfer and reactions, Fig. 14 shows the heterogeneity index for reactions. In the investigated range of voidage, H_r varies between zero and unity. At the two ends of voidage, H_r approaches unity, where the flow structure reduces to homogeneous states. In the rest range of voidage, the nonuniform flow structure causes additional mass diffusion resistance and hence reduction of overall reaction rate. The structural effect decreases with the decrease of reaction rate, since lower reaction is easier to become the control process, or the bottleneck. For higher reaction rates, e.g., the case of 1000 times k_r, the reduction ratio could be as low as 1×10^{-3}. That infers one may greatly overestimate the conversion rate when using a homogeneous model in a CFB reactor, especially with fast reactions such as gasification or combustion. Our structure-dependent mass model can be expected to give more reasonable prediction of the reactive behavior for such cases.

5.5 Mesoscale Modeling for Coarse-Grained Approaches

Besides the Eulerian–Eulerian approaches (say, TFM and SFM) discussed above, various Eulerian–Lagrangian approaches have been proposed and developed rapidly in recent years, say, CFD-DEM (Tsuji et al, 1993; Xu and Yu, 1997), MP-PIC (Snider, 2001), and DPM (Hoomans et al, 1996). Though the solid phase is resolved on the particle scale (for CFD-DEM) or parcel scale (for MP-PIC), the solution of the fluid phase still rests on the cell or grid scale, as discussed in Xu et al (2007) and Song et al (2015). In addition, the positions of particles are just used to calculate the motion of particles (Deen et al, 2007; Zhu et al, 2007), leaving rich information of these positions unused for calculating the heterogeneity of particle distribution. When calculating the drag force between particles and fluid, the mean fluid velocity on the grid node or its interpolation on the particles is used (Helland et al, 2002; Hoomans et al, 2000; Li et al, 2012b; Xu et al, 2000). Such treatment can be applied in principle if the fluid velocity is slowly varying. However, that is hard to be satisfied for fluidized systems that are far from equilibrium. As a result, the gap between the particle-resolved and fluid-resolved scales may lead to unresolved, subgrid structures. Absence of such local flow heterogeneity in the model may lead to inaccurate prediction of particle motions.

Compared to the rapidly growing recognition of the importance of mesoscale modeling in TFM (Agrawal et al, 2001; Li and Kwauk, 1994; Lu et al, 2009; Wang and Li, 2007; Yang et al, 2003), the subgrid modeling of drag force receives less attention in various DPMs (for brevity, hereafter in this review DPMs refers to Eulerian–Lagrangian approaches) except certain pioneering work (Benyahia and Sundaresan, 2012; Radl and Sundaresan, 2014; Xu et al, 2007). In practice, the EMMS drag (Li and Kwauk, 1994; Wang and Li, 2007) derived for the TFM can be directly used in DPMs (Li et al, 2012a, 2012b). However, as the information of particle positions was not fully utilized in this way, the slip velocity on each particle was still calculated by using linear interpolation of gas velocities on the grid node.

In this section, we try to propose a new method to calculate the drag force with consideration of subgrid, mesoscale structure by using the local information of particle positions. In principle, this method is compatible with all kinds of DPMs mentioned above. In order to save computing time, we choose the coarse-grained, MP-PIC method (Andrews and Orourke, 1996; Snider, 2001) as a representative to test this method.

5.5.1 MP-PIC Method

The MP-PIC method was first developed by Andrews and O'Rourke (1996) and Snider (2001) for dense particulate flows. In this method, the gas phase motion is solved by using the Eulerian equations adopted in

TFM. And the particle phase motion is solved by tracking discrete parcels, each representing a number of particles with the same properties and following the Newton's law of motion. In the MP-PIC method, the collisional interaction between particles is replaced by using the normal stress of solids (Snider, 2001), which is calculated on the grid points for gas phase and interpolated to the positions of parcels. The gas continuity equation is the same as Eq. (16), whereas the gas momentum equation reads

$$\frac{\partial}{\partial t}\left(\varepsilon_g \rho_g \mathbf{u}_g\right) + \nabla \cdot \left(\varepsilon_g \rho_g \mathbf{u}_g \mathbf{u}_g\right) = -\varepsilon_g \nabla p + \nabla \cdot \left\{\mu_g\left[\nabla \mathbf{u}_g + \left(\nabla \mathbf{u}_g\right)^T\right] - \frac{2}{3}\mu_g \nabla \cdot \mathbf{u}_g \mathbf{I}\right\} + \varepsilon_g \rho_g \mathbf{g} - \sum_{p=1}^{n_T} n_p \frac{V_p \rho_p}{V_{cell}} \mathbf{F}_{d,p} \tag{118}$$

where n_T and n_p are the parcel number in a fluid cell and particle number in a parcel, respectively, V_p and V_{cell} stand for the volume of a particle and the volume of a cell, respectively, and $F_{d,p}$ is the drag force per unit mass of a parcel. The parcel motion equations are as follows:

$$\frac{d\mathbf{x}_p}{dt} = \mathbf{u}_p, \tag{119}$$

$$\frac{d\mathbf{u}_p}{dt} = -\frac{\nabla p}{\rho_p} - \frac{\nabla p_p}{\varepsilon_p \rho_p} + \mathbf{g} + \mathbf{F}_{d,p}, \tag{120}$$

5.5.2 Subgrid EMMS/DP Drag

For any two parcels i and j in a cell, the surrounding flow fields and hence drag force on them may be different. As a first approximation to quantify such local distribution within a cell for each parcel, we introduce the local porosity to replace the use of mean porosity of cell and use the smoothed particle hydrodynamics method to calculate the local porosity, ε_{local}^{i}, around parcel i by using a weighting function with consideration of pairwise distance between parcel i and the others (Morris et al, 1997). With such a normalized kernel, the local porosity ε_{local}^{i} is just affected by the parcels inside the circle. As in Xiong et al (2011), the smoothing length is chosen equal to half of the grid size.

Besides the spatial positions of parcels, we further include the effects of subgrid heterogeneity of gas flow field. Similar to the treatment in the EMMS model (Li and Kwauk, 1994), we simplify the mixture of parcels and fluid in a cell into several phases: one dilute phase that is assumed to be free of particles and n_T dense phases, each containing one parcel that comprises n_k real particles. Such a parcel can be viewed as a kind of porous cluster of particles, where the cluster diameter is d_c^k and the solid concentration of the cluster is ε_{pc}^k. The volume fraction of the dense phase k is f_k. It should be noted that the above parameters for the parcel or cluster (ε_{pc}^k and d_c^k) are determined by the structure inside the parcel, whereas the local voidage of the parcel, ε_{local}^i, reflects the interstructure between different parcels. The velocities of the gas in the dilute and kth dense phases are represented by U_f and $U_{c,k}$, while the velocity of the solid phase in the kth dense phase is represented by $U_{pc,k}$, respectively.

The closure of the parcel parameters actually represents the coarse graining scheme on the mesoscale. In this work, we adopt the method proposed by Wang et al (2008a, 2008b) to calculate the solid concentration of the cluster ε_{pc}^k. The cluster diameter d_c^k of the dense phase k is calculated by using the EMMS/matrix method as detailed in Wang and Li (2007). And f_k can be determined by using its definition

$$f_k = \frac{n_k V_p}{\varepsilon_{pc}^k V_{cell}} \tag{121}$$

Then the total volume fraction of the dense phase in a cell is given by

$$f_{total} = \sum_{k=1}^{n_T} f_k \tag{122}$$

When a parcel is near the boundary of a cell, the searching area may exceed it. So we normalize the dense-phase volume fraction with the mean dense-phase fraction in cell f_{cell}, that is,

$$f_k = f_k \cdot \frac{f_{cell}}{f_{total}} \tag{123}$$

Here, the mean dense-phase fraction in cell is determined by the mean solids fraction in cell, that is,

$$f_{cell} = \frac{n_T n_k V_p}{\varepsilon_{pc} V_{cell}}, \tag{124}$$

$$\varepsilon_{pc} = f(\varepsilon_p). \tag{125}$$

As assumed in the EMMS model (Li and Kwauk, 1994) and verified in our structure-dependent analysis (Hong et al, 2012; Song et al, 2014), we further assume that the pressure drop balance between the dilute phase and each dense phase is achieved. Because the dilute phase is assumed free of particles, this balance can be expressed as

$$\frac{f_k}{1-f_{\text{cell}}} C_{\text{di}}^k \frac{1}{d_{\text{c}}^k} \rho_{\text{g}} \left(U_{\text{slip,i}}^k \right)^2 = C_{\text{dc}}^k \frac{1-\varepsilon_{\text{gc}}^k}{d_{\text{p}}} \rho_{\text{g}} \left(U_{\text{slip,c}}^k \right)^2 \tag{126}$$

Here C_{dc}^k and $U_{\text{slip,c}}^k$ represent the drag coefficient and superficial slip velocity for particles inside the dense phase k, respectively. With these relations, one can determine all the superficial slip velocities of each dense phase and the superficial fluid velocities of the dilute and dense phases by using binary search method. Then, each particle in the dense phase k experiences two drag components, one suffering from the fluid in the dense phase and the other from the dilute phase as a whole and equally distributed over each particle in the dense phase. So, the drag force acting on each particle in the dense phase is

$$F_{\text{d}}^k = C_{\text{dc}}^k \frac{\pi d_{\text{p}}^2}{4} \frac{\rho_{\text{g}}}{2} \left(U_{\text{slip,c}}^k \right)^2 + \frac{m_{\text{i}} V_{\text{cell}} C_{\text{di}}^k \dfrac{\pi \left(d_{\text{c}}^k \right)^2}{4} \dfrac{\rho_{\text{g}}}{2} \left(U_{\text{slip,i}}^k \right)^2}{n_k} \tag{127}$$

To reduce the computational time, we find the fitting function between the structure-dependent drag and the heterogeneity of particle distribution by traversing the randomly generated values of mean solids volume fraction and the mean gas and solids velocities in their permissible ranges. Thus, the drag force can be correlated with the corresponding local solids concentration and slip velocity. Figure 15 shows the fitting function of the normalized heterogeneity index obtained in this work. The drag obtained using this method is called the EMMS/DP drag, as it is expected to be applicable to different coarse-grained, generalized discrete particle approaches (Song et al, 2015).

6. COMPARISON BETWEEN METHODS WITH/WITHOUT MESOSCALE MODELING

In previous sections, we have shown the mesoscale modeling based on the bimodal structure. In what follows, we will show the advantage of such approaches over the ones based on local equilibrium and homogeneous closures, in particular, TFMs. The comparison starts from a simple one-dimensional force balance analysis, aiming to shed light on which kind

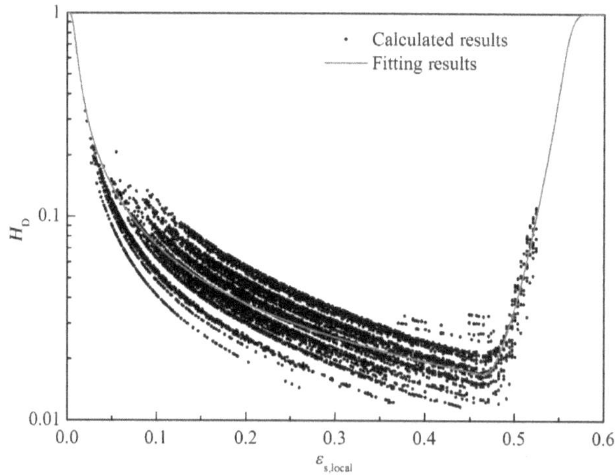

Figure 15 Variation of H_D against local solids concentration (Song et al, 2015).

of drag can predict the bistable flow state. Then, CFD simulation with and without mesoscale modeling of drag is compared, in particular for the fluidization of fine particles classified as Geldart A (Geldart, 1973). The effects of mesoscale modeling on both Eulerian–Eulerian (TFM vs. EFM) and Eulerian–Lagrangian (MP-PIC) approaches and reactive simulation are discussed. Finally, the flow regime map with mesoscale modeling is highlighted to help understand the choking and flooding phenomena.

6.1 Bistable State Analysis Based on 1D Force Balance

From the typical curve for the heterogeneity index, H_d, as shown in Figs. 7 and 15, we can see that this drag factor is higher near the two ends of voidage and lower in the middle section. Thus, it is easy to imagine that if drawing a horizontal line, one may have two intersections on the curve. However, such two intersections do not correspond to the interesting bistable state because they are under different voidages, hence different homogenous drag values and different operating conditions, and most importantly, such intersections do not reflect the force balance between gravity and drag.

The drift flux diagram well suits the target of investigating whether there are two stable states under the specific conditions (Wallis, 1969). The simplest hydrodynamic analysis of two-phase flow systems is a lumped force balance where the most influential factors such as drag force and gravity are equilibrated. Such balance is normally devoid of the particle–particle interaction, wall effects and mixing effects, etc. The drift flux is closely related to the slip velocity as detailed in Wallis (1969) and Ullah et al (2013a). For a unit cross section, the drift flux for gas phase can be defined as the volumetric flux of the gas phase relative to a surface moving at the average velocity, and can then be written as

$$J_{gs} = U_g - \varepsilon_g(U_g + U_p) = \varepsilon_g(1 - \varepsilon_g)u_{slip} \tag{128}$$

The slip velocity can be determined from the force balance

$$\frac{\beta u_{slip}}{\varepsilon_g} = (1 - \varepsilon_g)(\rho_p - \rho_g)g. \tag{129}$$

Once the slip velocity is known, the drift flux can be obtained by using Eq. (128), which is termed the *hydrodynamic curve*. On the other hand, for the given operating conditions with specified superficial velocities of particles and fluid, the drift flux can also be directly obtained through the *operating line* given by

$$J_{gs} = (1 - \varepsilon_g)U_g - \varepsilon_g U_p, \tag{130}$$

which is a straight line (as a function of voidage) connecting $J_{gs} = U_g$ for $\varepsilon_g = 0$ and $J_{gs} = -U_p$ for $\varepsilon_g = 1$. The vertical axis at $\varepsilon_g = 0$ then represents the gas velocity. The axis at other end of voidage, $\varepsilon_g = 1$, represents the reversed solids velocity. Equating Eqs. (128) and (130), the intersection of the operating line and the hydrodynamic curve predicts the possible voidage for the system under investigation under the selected operating conditions. More than one intersection between the operating line and hydrodynamic curve means coexistence of multiple voidage values at the same operating conditions.

It is clear that by incorporating different drag, with or without mesoscale modeling of the subgrid structure, we can have different prediction of the force-balanced states. Indeed, if we incorporate the homogeneous drag, e.g., Wen and Yu (1966), as shown in Fig. 16, only one intersection can be found on the two curves of the drift flux diagram for concurrent-up riser flow, no matter how the operating conditions are varied. If we use the EMMS drag instead, we can find two intersections under certain set of operating conditions, which corresponds to the dilute–dense coexisting flow state, or the choking transition. Refining grids to obtain the so-called filtered drag and using it also in the force balance, we find it is also possible to have two intersections, though the value may differ a lot from the prediction of using the EMMS drag. In greater detail of the analysis (Ullah et al, 2013a), we can find that the use of the EMMS drag can substantially change the *hydrodynamic curve* from convex to concave, thus allowing for two solutions

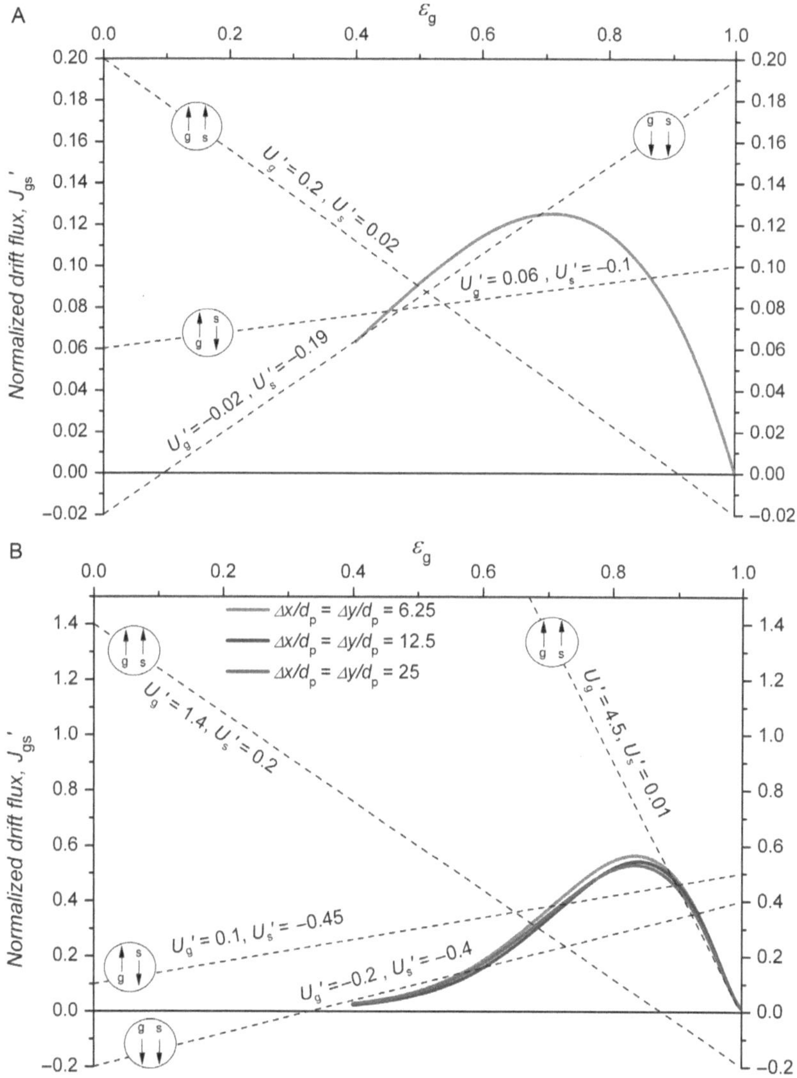

Figure 16 Dimensionless drift flux for air-FCC particles normalized by terminal velocity of single particle (the dashed operating lines are marked with corresponding dimensionless superficial gas velocity and solids velocity; the solid lines correspond to the hydrodynamic curve; $\Delta x/d_p$ denotes the dimensionless grid size; (A) with Wen and Yu drag; (B) with EMMS drag under different grid sizes) (Ullah et al, 2013a).

under one specific operating conditions, i.e., *operating line*. Using sufficiently fine grid may help to achieve such change from convexity to concavity, but the obviously quantitative difference reminds us that more efforts are needed to fully understand the feasibility of fine-grid simulation based on TFM.

That is the reason why we conduct the comparison based on CFD simulation in the next section.

6.2 CFD Simulation with/without Mesoscale Modeling

6.2.1 Comparison Between Fine-Grid TFM and EFM

To give a comprehensive evaluation of the TFM modeling of gas-fluidized beds, in what follows we will test its performance under fine-grid resolution. The particle property discussed here is restricted into Geldart A ($d_p = 54$ μm, $\rho_p = 930$ kg/m^3). In the case that fine-grid resolution was unaffordable, we performed coarse-grid simulations with subgrid drag laws derived either from fine-grid two-fluid modeling or the EMMS model.

Fine-grid simulation here is to get the real solution, or the grid-independent solution of TFM. Such a grid-dependency test is a prerequisite to discuss the feasibility of TFM for simulation of fluidized beds, or else the errors induced by coarse-grid resolution may blur the physical reality and mislead our judgment. In this regard, the physical law and numerical measures/algorithms are two inseparable aspects for testing of a model at least currently.

The bubbling fluidized bed of Mckeen and Pugsley (2003), the turbulent fluidized bed of Venderbosch (1998), and the CFB riser of Li and Kwauk (1994) were simulated. Initially, the particles were uniformly distributed across the riser with $\varepsilon_p = 0.09$. The TFM of Fluent® 6.3.26 was used, on which the drag force can be modified through user-defined functions. Thus, the SFM can also be implemented. The solids stress was closed with the algebraic form of the KTGF to save computational time. A combination of Wen and Yu (1966) and Ergun (1952) correlations was selected to be the homogeneous drag. The gas was assumed to flow uniformly into the bottom inlet and leave from the top outlet, where the atmospheric pressure boundary was prescribed. To ensure constant solids inventory inside the bed, the entrained solids were recycled into the solid inlet at the bottom. The no-slip boundary condition was prescribed for the gas phase, whereas the partial-slip boundary condition developed by Johnson and Jackson (1987) was used for the solid phase. The time-averaged analysis was carried out for 10 s after an elapse of initial period of developing.

Figure 17 shows the effects of size of square grids (ΔG) on the time-averaged solids distribution in the bubbling, turbulent, and circulating fluidized beds, sequentially from left to right. The bubbling bed expansion height decreases gradually with increasing grid resolution and approaches the experimental data (about 0.6 m). From the results with grid sizes of

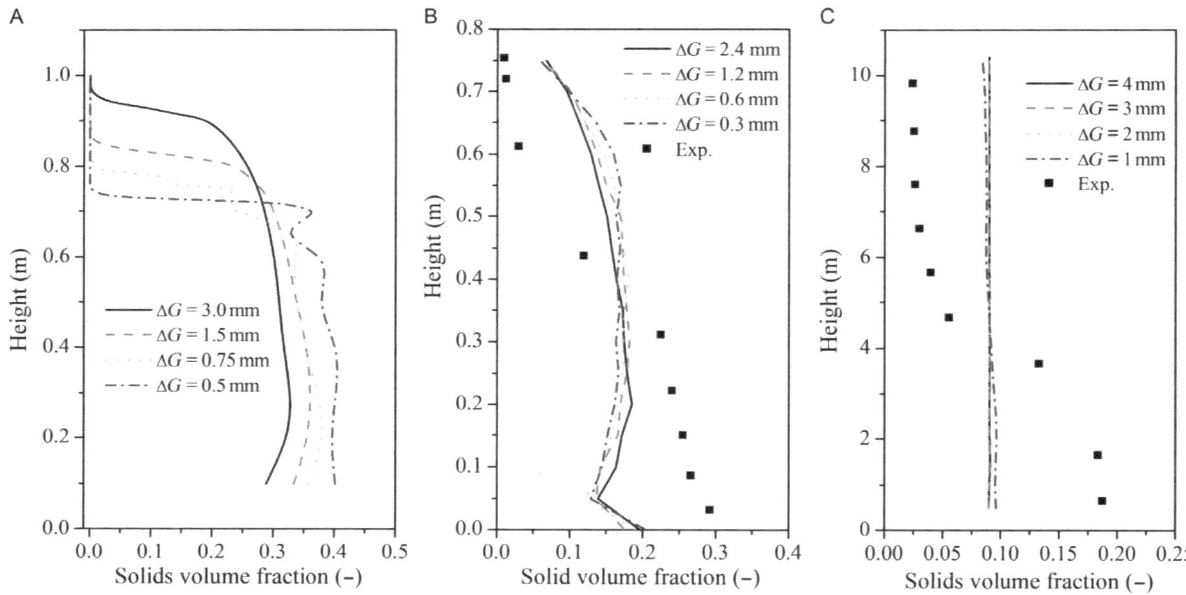

Figure 17 Predicted time-averaged distribution of solids volume fraction in (A) bubbling fluidized bed, (B) turbulent fluidized bed, and (C) CFB riser (Hong et al, 2015).

0.75 and 0.5 mm, the grid independence has not been reached, though the smallest grid size corresponds to less than 10 times the particle diameter. So, the grid size needs to be further refined to give reasonable prediction. In contrast, the axial profiles of turbulent fluidized bed depend less on the grid resolution, but all predictions deviate obviously from the experimental measurement. For the CFB riser, the predicted profiles of solids volume fraction are close to each other but deviate obviously from the S-shaped curve of data even for the case with the smallest grid size (1 mm) corresponding to 16 times the particle diameter. This result is generally in agreement with what was reported by Lu et al (2009). In general, deviation grows with the increase of superficial gas velocity for these three cases, which is qualitatively in agreement with Xie et al (2008a, 2008b).

The time-mean circulating solids flux is an important parameter to characterize the hydrodynamic performance of a CFB riser. Its value can be obtained by monitoring the solids flow rate at the outlet. The predicted solids flux is around 120 kg/m^2 s when the grid size is equal to 1.0 mm, roughly eight times the experimental data (14.3 kg/m^2 s). It is worth noting that the prediction of Benyahia (2012) under the similar settings is around 130 kg/m^2 s, which is comparable to our result of fine-grid simulation but still much higher than experimental data.

One may question whether the grid resolution in this work is high enough. In addition, the errors that may be induced by algebraic formulation of the granular temperature equation and 2D settings also need to be excluded. As further refinement of grid size or using the PDE granular temperature equation in fine-grid simulations leads to convergence difficulties, and direct 3D simulation with fine-grid scheme is not affordable even for high-performance computing, so, we cannot make a direct comparison for this issue. As a remedy, we performed coarse-grid simulation with subgrid modeling. The subgrid models can be derived from fully resolved fine-grid simulation (Agrawal et al, 2001; Igci et al, 2008) or from the EMMS model (Wang and Li, 2007; Lu et al, 2009; Hong et al, 2012). In literature, these two approaches are termed the filtered two-fluid modeling (Igci and Sundaresan, 2011; Igci et al, 2008) and EMMS-based multifluid modeling (Hong et al, 2012, 2013), respectively. In fact, Igci and Sundaresan (2011) found that the statistical results obtained by solving the filtered models agree well with those from highly resolved simulations with full set of the kinetic theory model. Thus, if the fine-grid TFM with closures derived from homogeneous systems is sufficient to predict the two-phase flow behavior of a fluidized bed, then the relevant filtered two-fluid modeling should be also able

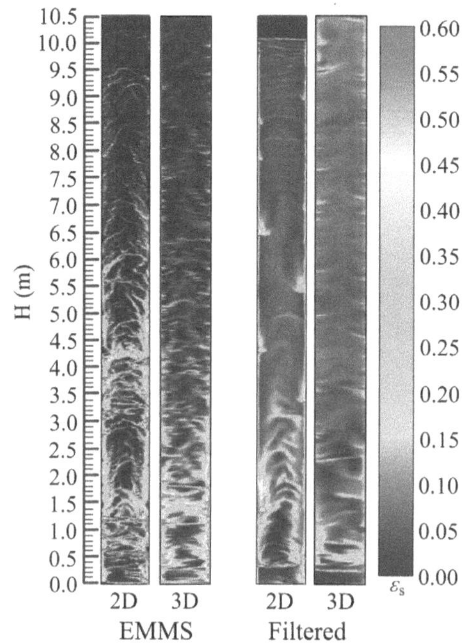

Figure 18 Comparison of solids distribution predicted by using the EMMS and filtered drag models in 2D and 3D riser flow cases (Hong et al, 2015).

to predict the flow behavior properly. On the basis of that, we can view the filtered two-fluid modeling as a reasonable substitute to the direct use of fine-grid TFM modeling. And it compensates the lack of highly resolved simulation with grid size smaller than 10 times the particle diameter. The detailed numerical settings are referred in literature (Hong et al, 2015).

Figure 18 presents typical snapshots of solids distribution on the vertical plane for using both filtered drag and EMMS drag models. 2D results are also provided for comparison. Dense clusters are captured with both drags especially near the bottom zone. And the EMMS-based approach gives more pronounced segregation of particles.

Figure 19 shows the axial profiles of time-mean, cross-sectionally averaged distribution of solids volume fraction. The S-shaped axial profile can be predicted in both 2D and 3D cases when the EMMS drag model is used. And the 2D EMMS results seem to give better prediction compared with experimental data. When the filtered approach is used, it is not easy to make a clear statement as to whether 2D or 3D results are better than the other. Both profiles are not like the S-shaped and their solids concentrations in the upper region are higher than experiment. Similar tendency can also be found in the

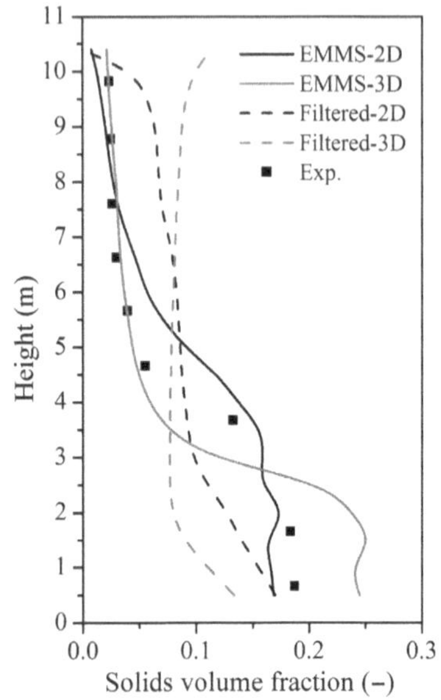

Figure 19 Comparison of axial profiles of solids volume fraction predicted with the EMMS and filtered drag models in 2D and 3D geometries (Hong et al, 2015).

literature (Igci et al, 2012). This implies that the filtered approach overpredicts the solids entrainment for this riser. The same statement can be addressed by comparing the solids flux. The solids fluxes predicted with using the EMMS drag (around 15 kg/m^2 s) agree well with the experimental data. The filtered approach, however, overpredicts the solids flux (around 110 kg/m^2 s). It is also interesting to note that both filtered approach and fine-grid simulations of ours and Benyahia (2012) predict almost the same value of solids flux. Such convergence of prediction means that these fine-grid simulations or filtered approaches have exhibited the real and grid-independent solution of the TFM. However, the predicted solids flux of the TFM with homogeneous drag is roughly eight times higher than experimental data.

In literature, there are reports concerning the effects of the inlet/outlet settings, backflow at the outlet, aspect ratio of grid, and boundary conditions for the solid phase. These factors are found to have big impact on the simulation results under certain specific settings (Li and Benyahia, 2013; Li et al, 2010). To exclude the effects of these factors, we also performed a series of simulations by adjusting these factors. We found that the different inlet/outlet configurations obviously affect the solids volume fraction nearby, but the effects decay rapidly with distance. High aspect ratio of the grid, say, $\Delta y/\Delta x > 8$, leads to increase of computational errors and the predicted profiles deviate more from the experimentally observed S-shaped curve. The effects of boundary conditions are very complex. And we found that the value of specularity coefficient affects clearly the axial profile of solids volume fraction and the value of solids flux in the case when the subgrid EMMS drag was used. In spite of this, using the EMMS drag still has a better performance than using the other approaches generally.

In all, the TFM is actually derived with the assumption that the local equilibrium can be achieved for both gas and solid phases. And that assumption is consistent with the use of homogeneous drag and the kinetic theory based on the assumption of molecular chaos. However, the fluidized system does not always satisfy the local equilibrium assumption. For a bubbling fluidized bed, or a CFB with coarse particles (say, Geldart B, D particles), which seems to be close to the requirement of local equilibriums states, the fine-grid TFM simulation with homogeneous drag is a plausible approach. However, the flow state seems to deviate more from the local equilibrium with increase of gas velocity and then the fine-grid TFM with homogeneous drag closure is not sufficient to resolve all the mesoscale structures in high-velocity fluidized beds and hence the solids flux predicted is not reliable. In contrast, the EMMS-based modeling depends on bimodal distribution, which is closer to reality but far from local equilibrium, and hence enables reasonable prediction of both axial profiles and solids flux. Obviously, the current understanding of fluidization and nonequilibrium physics is still far from maturity so that we are hardly able to quantify precisely the mesoscale structure and to what extent the local flow behavior of a fluidized bed deviates from the equilibrium state. Certain structure-dependent conservation equations and closure laws, such as the EMMS drag or something equivalent, need more efforts, as recently explored in Hong et al (2012, 2013) and Song et al (2014). Our understanding of the mesoscale structure is still at the preliminary state. Mesoscale modeling definitely deserves more efforts.

6.2.2 MP-PIC Simulations with/without Mesoscale Modeling

The experiments reported by Horio et al (1988) were used to validate the EMMS/DP drag model for Eulerian–Lagrangian approaches. The riser was simplified with 2D, rectangular domain. In the simulations, air enters the riser from the bottom with a uniform superficial velocity, the top of the domain is set to be a pressure outlet for the gas phase, and the entrained particles are directly returned to the riser from the bottom to maintain the

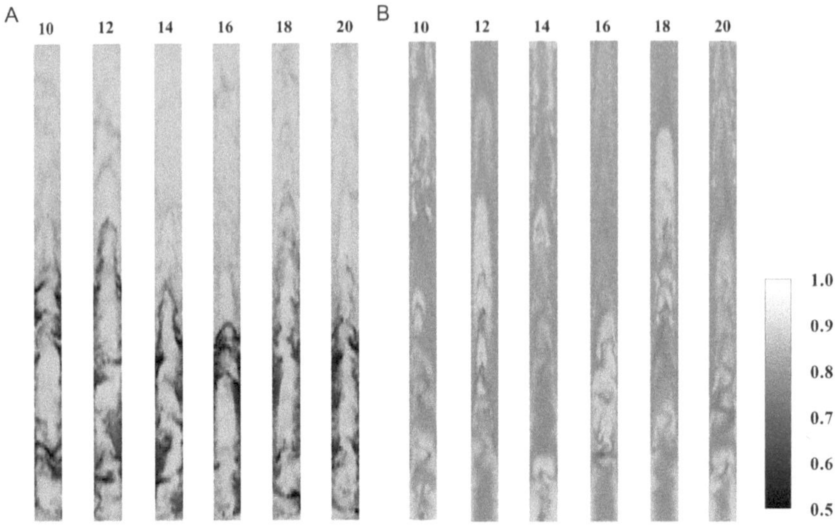

Figure 20 Snapshots of voidage distribution at different instants (from No. 10 to 20s), calculated with (A): EMMS/DP drag and (B): homogeneous drag (Song et al, 2015).

bed inventory constant. The left and right of the domain are set to be nonslip wall boundaries for the gas phase. The open source code MFIX is taken as the platform (Syamlal et al, 1993). The time step varies from 10^{-6} to 10^{-3} s to maintain the convergence of the program.

Figure 20 shows the snapshots of voidage distribution at different instants. It can be seen that the results with the EMMS/DP drag show densely heterogeneous clusters, forming the core-annular distribution in the radial direction and the axially S-shaped distribution. In contrast, the results using the homogeneous drag show nearly uniform distribution under the current resolution of grid. In our previous test of grid-size effects (Li et al, 2012b), the grid resolution here (30×200) is found sufficient to reach grid-independent results. Such different tendencies of the axial distribution of voidage can be evidenced further in Fig. 21B, which shows an S-shaped distribution of voidage for the case with the EMMS/DP drag but a much more uniform distribution for the case with the homogeneous drag. Figure 21A shows the solid flux calculated using both drag models. It can be seen that the solid fluxes are stable at around 4 s with both drag models. The mean value of the solid flux with the EMMS/DP drag is close to the experimental data of 11.7 kg/m^2 s, while the one with homogeneous drag is much higher than it. Similar findings have been reported in many previous simulations based on TFM. These results show that the same situation does exist for both the

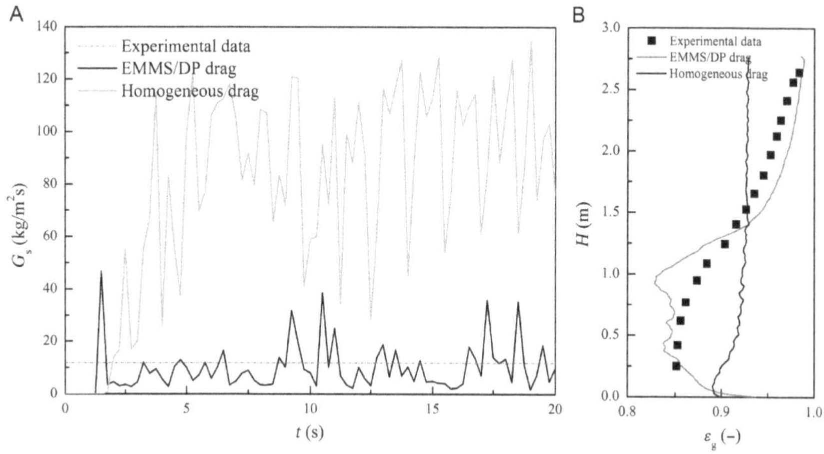

Figure 21 Solid flux (A) and axial distribution of voidage (B) calculated with the EMMS/DP and homogeneous drags (Song et al, 2015).

TFM and the MP-PIC modeling that is partly continuum based. That is, subgrid modeling is needed for both Eulerian–Eulerian and Eulerian–Lagrangian approaches.

6.3 Reactive Simulation with Mesoscale Modeling

An ambient CFB reactor for ozone decomposition was simulated to verify the structure-dependent mass transfer and reaction model (Liu et al, 2015). The CFB consists of a 76 mm I.D. riser with height of 10.2 m, cyclones, a downcomer, and solids recycling and feeding devices. To save the computing time, our simulation only includes the riser and the solid feeding pipe. The Eulerian multiphase flow model or TFM of Fluent® was selected to be the platform. The structure-dependent drag and mass transfer models were adopted through UDFs. According to the experimentally observed profile of solids volume fraction data (Li et al, 2012a), we set the packing height of solids in the riser to be 0.85 m. The simulation ran for an elapse of 20 s to reach quasi-stable flow state. Afterward, the reactions and mass transfer were added and kept running for another 25 s.

Figure 22 shows a snapshot of solids distribution as well as the axial and radial profiles of solids volume fraction. Experimental data and simulation results are collected along four azimuthal directions to reflect the inlet-induced, asymmetrical flow in the riser. The asymmetry decays with the riser height. Generally, the prediction is in reasonable agreement with experimental data.

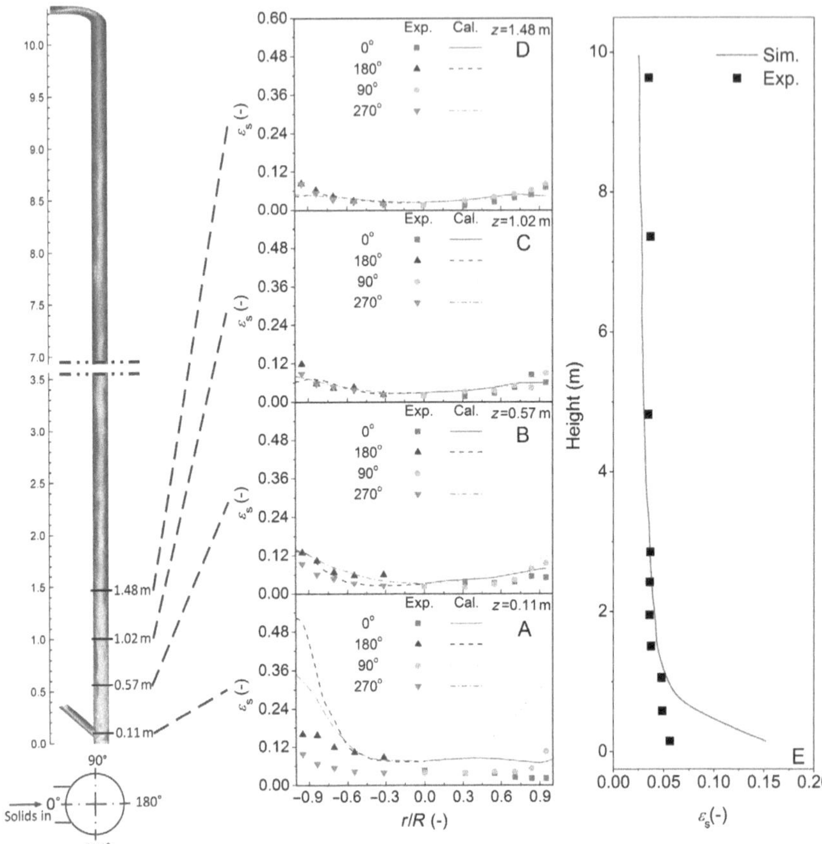

Figure 22 Snapshot of solids volume fraction distribution and time-averaged profiles of solids volume fraction against experimental data ($U_g = 3.0$ m/s, $G_s = 100$ kg/m^2 s) (Liu et al, 2015).

Figure 23 shows the radial profiles of dimensionless ozone concentrations at different heights. Generally, the results agree with experimental data. The asymmetric profiles along the directions of 0° and 180° at $z = 0.11$ m are captured. With the increase of height, the asymmetry weakens. As the ozone decomposition is closely related with the presence of solid catalyst, the ozone concentration also shows the core-annulus radial distribution. At the top section, the prediction is obviously lower than the measured data.

If we use the TFM with homogeneous drag and mass transfer coefficient, as indicated in our previous work and review (Dong et al, 2008b; Wang et al, 2011), the reaction rate will be greatly overpredicted, resulting in quick, nearly complete conversion of ozone near the distributor. Such an overestimation of reaction rate is similar to the situation of drag overprediction

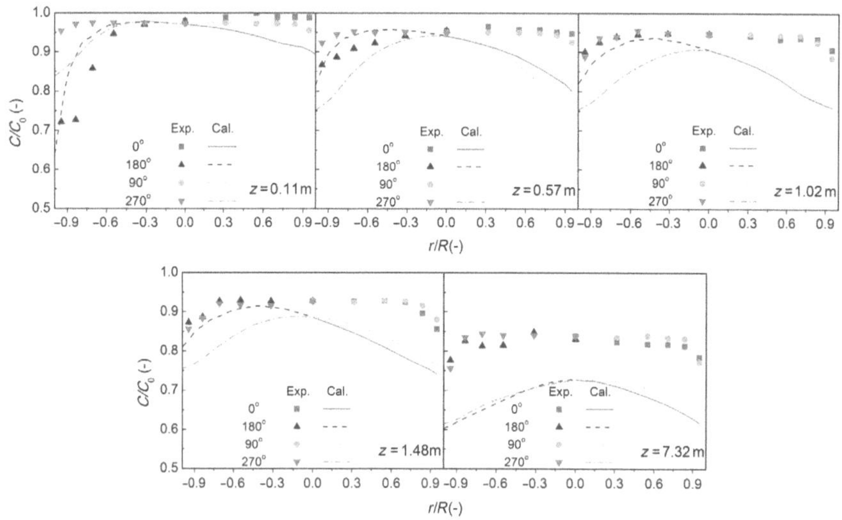

Figure 23 Time-averaged radial profiles of dimensionless ozone concentration at different heights in four directions ($U_g = 3.8$ m/s, $G_s = 106$ kg/m^2 s, $k_r = 3.88$ s^{-1}) (Liu et al, 2015).

without mesoscale modeling. However, as the reaction in industrial applications is rather complex (say, FCC or methanol to olefins), the reaction kinetics measured normally depends on empirical hydrodynamics (Ying et al, 2015). In such situations, the advantage of the structure-dependent mass transfer and reaction modeling is easy to be blurred by the overtuned parameters of reaction kinetics. In future, to make the reactive modeling more reliable, more efforts are needed first to obtain the reaction kinetics that is free of hydrodynamics.

6.4 Flow Regime Map with Mesoscale Modeling

As presented in the discussion for Fig. 2, a quantitatively reliable flow regime map always assumes the primary role for multiphase flow research. However, current knowledge limitation, in particular, on the understanding and quantification of mesoscale structures, keeps us far from achieving it. Therefore, here we have no intention to draw a unified regime map, but rather attempt to provide some hints on how the mesoscale modeling modifies or improves the widely applied regime maps (Ullah et al, 2013b).

6.4.1 Generalized Fluidization Diagram

By substituting the relative velocity, or slip velocity, between particles and fluid for the velocity of classical fluidization, Kwauk (1963, 1992) extended the steady-state fluidization diagram to describe the overall field of all possible combinations of particle-fluid motion, including concurrent-up, concurrent-down, countercurrent systems, and so on, which is called generalized fluidization diagram. The generalized fluidization diagram by Kwauk was drawn based on Richardson and Zaki (1954) correlation for homogeneous fluidization, that is,

$$\frac{U_{\text{slip}}}{u_{\text{t}}} = \varepsilon_{\text{g}}^{n}. \tag{131}$$

This can be rearranged as a force balance equation between drag force and effective gravity as follows

$$\frac{\beta_{\text{RZ}} u_{\text{slip}}}{\varepsilon_{\text{g}}} = \left(1-\varepsilon_{\text{g}}\right)\left(\rho_{\text{p}}-\rho_{\text{g}}\right)g, \tag{132}$$

where

$$\beta_{\text{RZ}} = \frac{\left(1-\varepsilon_{\text{g}}\right)\left(\rho_{\text{p}}-\rho_{\text{g}}\right)g}{u_{\text{t}}\varepsilon_{\text{g}}^{n-2}}. \tag{133}$$

For a given particle material, the exponent n is an empirical function of Re_{t} (Richardson and Zaki, 1954). Then a set of specified gas and solids velocities are fed into above equations to obtain the voidage expression. Figure 24 gives such a contour with exponent $n=4.4$.

On this diagram, one may locate, first, the points of minimum fluidization (point A: $U_{\text{g}}=U_{\text{mf}}$, $G_{\text{s}}=0$, $\varepsilon_{\text{g}}=\varepsilon_{\text{mf}}$) and free settling of single particles (point B: $U_{\text{g}}=u_{\text{t}}$, $\varepsilon_{\text{g}}=1$) in blue (gray in the print version) dots, then, the line of classical fluidization (solid blue (gray in the print version) line between A and B: $U_{\text{g}}>U_{\text{mf}}$, $G_{\text{s}}=0$), and finally, different flow regimes. The regimes to the right of the classical fluidization line are for concurrent-up flows, in which the regime to the left of the dash blue (gray in the print version) line ($U_{\text{g}}=u_{\text{t}}$) is the restrained concurrent-up flow that exists only with support of air distributor while to the right is the complete concurrent-up flow. To the left of the classical fluidization line, the solids are transported following the gravity, while the zero-gas-velocity line demarcates between the concurrent-down and countercurrent solids-down flows.

A significant character of the flooding line lies in that, in its range of gas velocity (between C and B) and for solids-downward flows ($G_{\text{s}}<0$), there may coexist two states at one specific gas velocity. That is, two ε coexist for a given set of U_{g} and G_{s}. In contrast, in the area of concurrent-up flow (to the right of the classical fluidization), there is only one state for any given set of

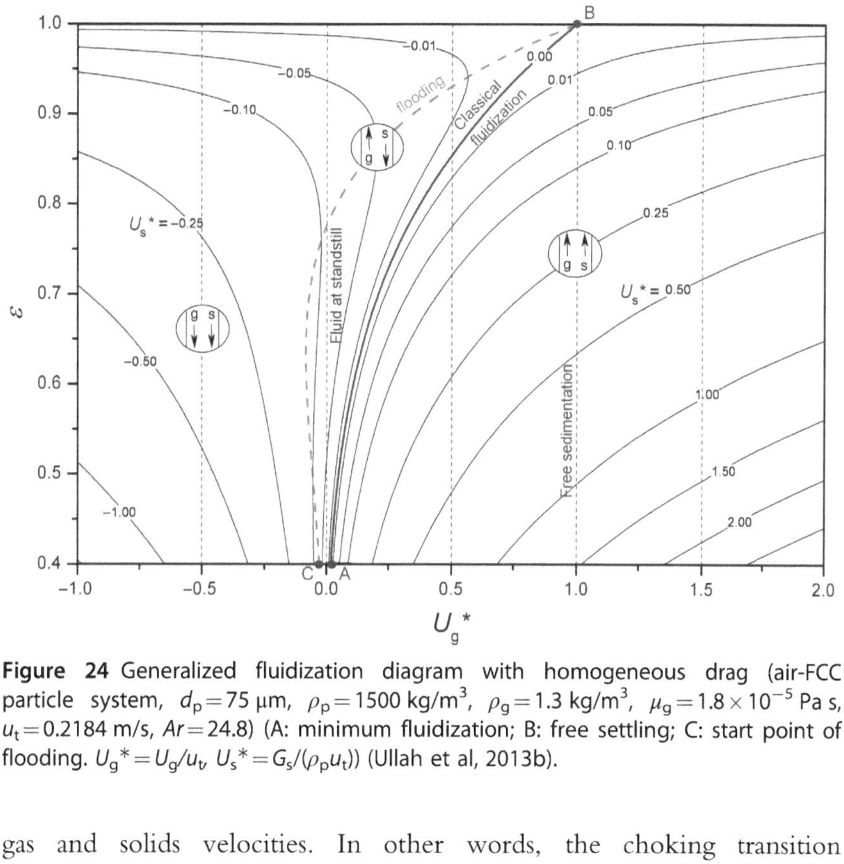

Figure 24 Generalized fluidization diagram with homogeneous drag (air-FCC particle system, $d_p = 75$ μm, $\rho_p = 1500$ kg/m^3, $\rho_g = 1.3$ kg/m^3, $\mu_g = 1.8 \times 10^{-5}$ Pa s, $u_t = 0.2184$ m/s, $Ar = 24.8$) (A: minimum fluidization; B: free settling; C: start point of flooding. $U_g* = U_g/u_t$, $U_s* = G_s/(\rho_p u_t)$) (Ullah et al, 2013b).

gas and solids velocities. In other words, the choking transition corresponding to the coexistence of the dilute top and the dense bottom cannot be predicted on this diagram. This is attributed to the use of homogeneous Richardson–Zaki correlation. To take into account the effect of mesoscale structure, we can rewrite Eq. (132) by incorporating an EMMS drag as follows,

$$\frac{\beta u_{\text{slip}}}{\varepsilon_g} = (1 - \varepsilon_g)(\rho_p - \rho_g)g, \tag{134}$$

where the structure-dependent drag coefficient can be given by

$$\beta = \beta_{\text{WY}} H_d = \frac{3\varepsilon_g(1 - \varepsilon_g)\rho_g C_{D0}|u_{\text{slip}}|}{4d_p} \varepsilon_g^{-2.7} H_d, \tag{135}$$

where the drag coefficient of Wen and Yu is used to scale the other drag coefficients. Redrawing the diagram with Eq. (134) yields Fig. 25. In contrast to Fig. 24, it looks twisted. And we can easily find that the maximum

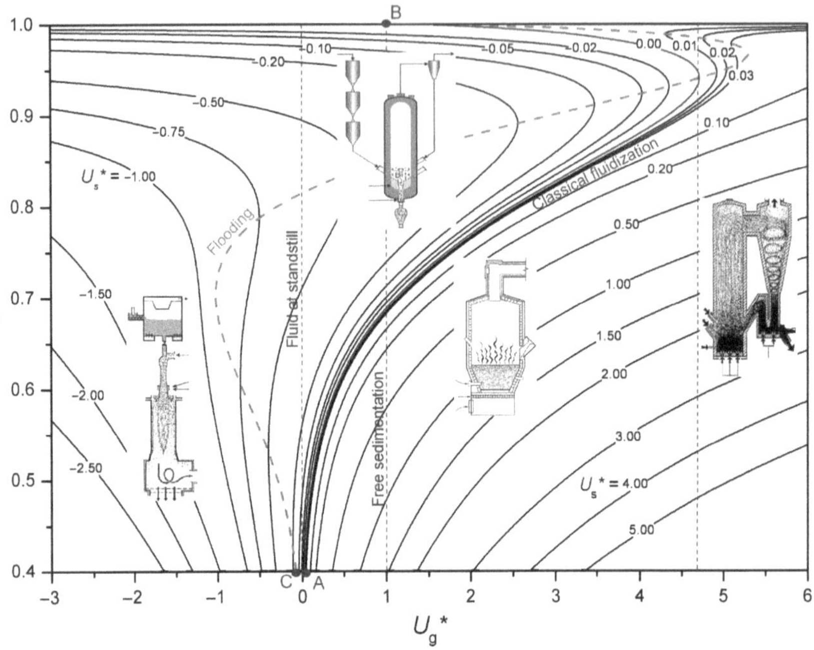

Figure 25 Generalized fluidization diagram with EMMS drag coefficient (air-FCC particle system, $d_p = 75\ \mu m$, $\rho_p = 1500\ kg/m^3$, $\rho_g = 1.3\ kg/m^3$, $\mu_g = 1.8 \times 10^{-5}$ Pa s, $u_t = 0.2184$ m/s) (Ullah et al, 2013b).

superficial gas velocity for the restrained concurrent-up flow is not equal to the terminal velocity. As indicated by the blue (gray in the print version) dotted line to the right side, it is about 4.7 times the terminal velocity for this case, which is comparable to the reported values of the incipient "transport velocity" (Grace et al, 2006). In addition, Fig. 25 predicts two or even three coexistent states in the concurrent-up, the countercurrent gas-upward, and the concurrent-down flows.

Figure 26 shows a close-up of the coexistent states in concurrent-up flows, which looks like the phase diagram predicted by the famous van der Waals equation. We can see that, for dimensionless solids velocity $U_s^* = 0.02$, though three states may coexist at around $U_g^* = 5.0$, the section between points "b" and "c" is actually unstable because the solids volume fraction $(1-\varepsilon)$ increases with gas velocity thereon. Therefore, only two states are stable within the concurrent-up flow area. Similar flow regime diagram has ever been predicted by using TFM CFD simulation with EMMS drag coefficient (Wang et al, 2007, 2008b).

For a pneumatic transport riser, if one keeps the solids flux constant (for this case, $U_s^* = 0.02$) and gradually decreases the superficial gas velocity, the

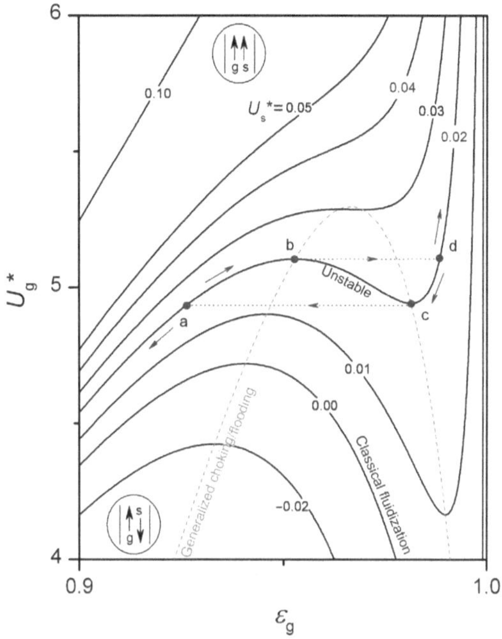

Figure 26 Close-up of the generalized fluidization diagram with EMMS drag (air-FCC particle system, $d_p = 75$ μm, $\rho_p = 1500$ kg/m^3, $\rho_g = 1.3$ kg/m^3, $\mu_g = 1.8 \times 10^{-5}$ Pa s, $u_t = 0.2184$ m/s) (Ullah et al, 2013b).

voidage will undergo a gradual decrease through point "d," to point "c" and then a jump decrease to point "a." Such a jump change explains the flow instability at the choking in pneumatic transport system, which can be attributed to the two coexisting, stable states as described by the generalized choking/flooding line. Further, we can also explain the flooding phenomenon in countercurrent flows, with respect to the maximum solids flux for a given superficial gas velocity, the detail of which is referred in Ullah et al (2013b).

6.4.2 Reh's Fluidization Diagram

In what follows, we return to the Reh's diagram (Reh, 1996). If we assume the fluidization is homogeneous, then Wen–Yu drag correlation can be used. Assuming the force balance between drag, buoyancy, and gravity, we can redraw the diagram as shown in Fig. 27, which is similar to Reh's result in Fig. 2. This figure presents 3D surface for Ar–Ω–ϕ (LHS) with projections onto Ar–Ω plane (RHS). Here, ϕ denotes the solids volume fraction. Lines for Fr^* and Re are also shown with direction of increasing magnitude. Clearly in the case that the homogeneous drag is used in force balance analysis, we can only predict the streaked area, which is surrounded

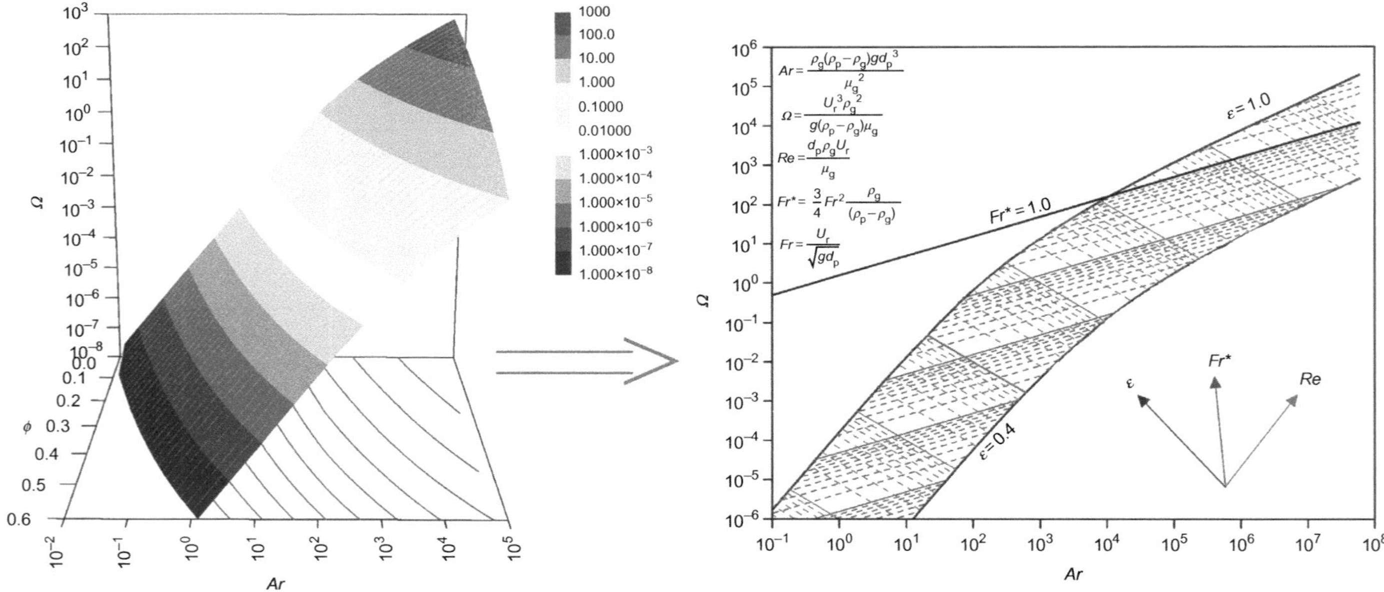

Figure 27 3D surface plot (left) and contour plot (right) for system where Wen–Yu drag is used (Ullah et al, 2013b).

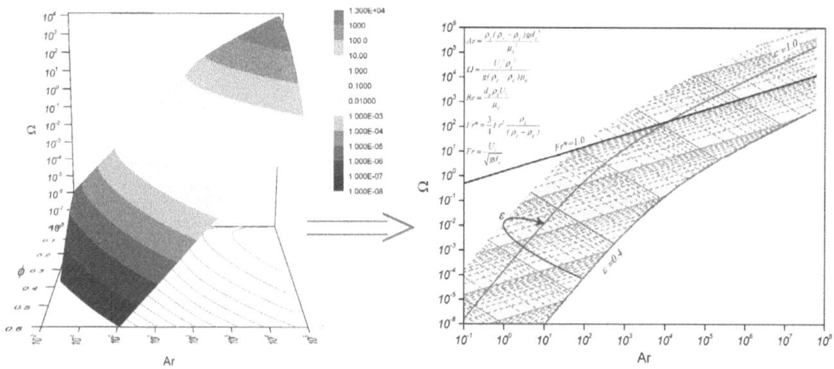

Figure 28 3D surface plot (left) and contour plot (right) for system where EMMS drag is used (Ullah et al, 2013b).

by the minimum fluidization line ($\varepsilon_g = 0.4$) and single-particle sedimentation line ($\varepsilon_g = 1$). However, large amount of experimental data show that CFB operates near the line defined by $Fr^* = 1$. Thus, there is a vacant area between the lines of $Fr^* = 1$ and $\varepsilon_g = 1$, the state of which cannot be predicted by using the homogeneous drag.

If we use the EMMS drag, however, in the force balance, we can have updated Reh's diagram as shown in Fig. 28. We can observe the turning point behavior on the contour plot on the right plane of Ar–Ω, which corresponds to the folded surface on the 3D plot of LHS. That implies bistable states may exist after the particles are fluidized beyond the minimum fluidization state. It should be noted that, by using the EMMS drag, one can predict the flow states between the lines of $Fr^* = 1$ and $\varepsilon_g = 1$, which do exist in CFBs but cannot be predicted by using the homogeneous drag that shows again the advantage of using the structure-dependent EMMS drag model.

It should be stressed that for simplicity and qualitative analysis, here we assume that the EMMS drag used in Figs. 21, 22, and 24 does not change with material properties. Indeed, we calculate the EMMS drag based on one specific air-FCC particle system and extend its use to the whole range of Archimedes number. In principle, such practice is not correct because the mesoscale structure and the heterogeneity index varies with the gas and particle properties. A strict flow regime map should be drawn with a huge set of different drag models which should be verified with the whole range of material properties. But that is definitely beyond our current capability. In fact, we are still short of a unified drag correlations that are valid for all particles, or the whole range of Ar. That is the reason why the quantitative disparity reflected by the vacant area between $Fr^* = 1$ and our prediction still exists. Indeed, a general-purpose diagram should be more complex since the domain-size or scale-up effect may also exist on the structure-dependent drag force and hence the force balance. We can expect lots of efforts are still needed to complete a unified flow regime map.

7. SUMMARY AND PROSPECTS

Mesoscale structure is the core of research for multiphase flows and, more generally, multiphase reaction engineering. Without mesoscale structure, there will be no essential difference between flow/reactions in a lab-scale flask and in a homogeneously distributed, industrial-scale multiphase flow reactor. In other words, there will be no scale-up effects for such multiphase systems. Recognizing the importance of mesoscale structure, chemical engineers have been focusing on it ever since the emergence of modern chemical engineering (fluidization research development is a good example for it, as discussed in Section 4.2), though at that time, there is no such terminology of "mesoscale." Until recently after several decades of efforts, chemical engineering researchers revealed that the same mechanism termed the principle of "compromise in competition" may be obeyed by various kinds of mesoscale structures (Li and Huang, 2014; Li et al, 2013). Though all these progress have been made, we are still far from quantitative understanding of the mesoscale structure due to its nonlinear nonequilibrium nature.

It has been said that "scientists tackle those problems which can be solved; engineers are faced with problems which must be solved"(Sherwood et al, 1975). Such a statement well describes the difference between chemical engineering and mechanics communities when encountering nonequilibrium multiphase flows. The past efforts on fluid mechanics modeling of fluidization largely stays on the hydrodynamic description of fluid-like granular flows, which is based on the local equilibrium assumption, seeming analogy between granular gas and molecular gas and successful application of Navier–Stokes equations for single-phase flows. Though the mesoscale structure (or termed as microstructure) has been recognized to greatly affect the flow state of rapid granular flow or granular gas (Campbell, 1990; Goldhirsch, 2003), the clear statement of breakdown of hydrodynamics for inelastic particles was made until Du et al (1995). Even now, the TFM still prevails in the simulation of fluidization.

Recent years have witnessed a clear tendency of fusion between the communities of granular matter physics, fluid mechanics, and fluidization engineering, due to more and more understanding of the nonequilibrium features inherent in granular flows and fluidization, e.g., non-Gaussian velocity distribution, energy nonequipartition, and correlated density fluctuations, and hence a blossom of mesoscale modeling research from various aspects of the problems and angles of view. In these efforts, the EMMS-based, SFM has found successful applications by greatly improving the prediction accuracy in terms of flow, mass transfer, and reactions as well as the understanding of flow regime transitions. In contrast, traditional TFM, which depends on local equilibrium and homogeneous distribution assumptions, fails to predict the dynamic, nonequilibrium phenomena in CFBs even with fine-grid resolution.

We can expect the disputes over the applicability of the local equilibrium assumption that underlies the TFM will remain for a longer time, as the secrets of scale-dependent, nonequilibrium multiphase flows are not fully understood. More in-depth investigation should be conducted with close coupling among experimental measurement, theoretical derivation, and numerical simulations:

- In experiments, microscopic, noninvasive observation of the velocity distribution, correlated density fluctuation, or clustering will be helpful to quantify the collective behavior of granular flows under different heating or energy driving conditions.
- With experimental quantification of mesoscale structures, statistical analysis based on nonequilibrium distribution may require novel mathematical skills to unravel the complex dependence of stress–strain relation on the mass exchange between phases. The structure-dependent energy analysis may help elucidate the dependence between energy dissipation and structural parameters, and how to relate such structural-dependent analysis and the nonlinear nonequilibrium thermodynamics remains a challenge, especially for the scale-dependent granular flow or fluidization systems.
- Direct numerical simulation (DNS) is a powerful tool to investigate the velocity distribution and density fluctuation of multiphase flow systems, thus facilitates revealing the mechanisms of nonequilibrium behavior. Due to its high demand in computing resources, current DNS is largely limited to simulations over static arrays of particles or small-sized, periodic flow domains, which are expected to be close to local equilibrium states. Thus, the nonequilibrium characteristics of multiphase flow are hard to be fully revealed. Recent release of hybrid computing hardware boosts the rapid development of DNS with respect to the scales of time and space

Goldhirsch I: Rapid granular flows, *Annu Rev Fluid Mech* 35:267–293, 2003.
Goldhirsch I: Introduction to granular temperature, *Powder Technol* 182:130–136, 2008.
Grace JR: Contacting modes and behavior classification of gas-solid and other 2-phase suspensions, *Can J Chem Eng* 64:353–363, 1986.
Grace JR, Leckner B, Zhu J, Cheng Y: Fluidized beds. In Crowe CT, editor: *Multiphase flow handbook*, Boca Raton, 2006, CRC Press.
Haff PK: Grain flow as a fluid-mechanical phenomenon, *J Fluid Mech* 134:401–430, 1983.
Halder P, Basu P: Mass transfer from a coarse particle to a fast bed of fine solids, *AIChE Symp Ser* 84:58–64, 1988.
Harris AT, Davidson JF, Thorpe RB: The prediction of particle cluster properties in the near wall region of a vertical riser (200157), *Powder Technol* 127:128–143, 2002.
Harth K, Kornek U, Trittel T, et al: Granular gases of rod-shaped grains in microgravity, *Phys Rev Lett* 110:5, 2013.
Helland E, Occelli R, Tadrist L: Computational study of fluctuating motions and cluster structures in gas-particle flows, *Int J Multiphase Flow* 28:199–223, 2002.
Herbst O, Muller P, Otto M, Zippelius A: Local equation of state and velocity distributions of a driven granular gas, *Phys Rev E* 70:051313, 2004.
Herbst O, Muller P, Zippelius A: Local heat flux and energy loss in a two-dimensional vibrated granular gas, *Phys Rev E* 72:041303, 2005.
Holloway W, Sundaresan S: Filtered models for reacting gas-particle flows, *Chem Eng Sci* 82:132–143, 2012.
Hong K, Wang W, Zhou Q, Wang J, Li J: An EMMS-based multi-fluid model (EFM) for heterogeneous gas-solid riser flows: part I. Formulation of structure-dependent conservation equations, *Chem Eng Sci* 75:376–389, 2012.
Hong K, Shi Z, Wang W, Li J: A structure-dependent multi-fluid model (SFM) for heterogeneous gas-solid flow, *Chem Eng Sci* 99:191–202, 2013.
Hong K, Chen S, Wang W, Li J: Fine-grid two-fluid modeling of fluidization of Geldart A particles, *Powder Technol* , 2015. http://dx.doi.org/10.1016/j.powtec.2015.07.003.
Hoomans BPB, Kuipers JAM, Briels WJ, van Swaaij WPM: Discrete particle simulation of bubble and slug formation in a two-dimensional gas-fluidised bed: a hard-sphere approach, *Chem Eng Sci* 51:99–118, 1996.
Hoomans BPB, Kuipers JAM, van Swaaij WPM: Granular dynamics simulation of segregation phenomena in bubbling gas-fluidised beds, *Powder Technol* 109:41–48, 2000.
Horio M, Morishita K, Tachibana O, Murata N: Solid distribution and movement in circulating fluidized beds. In *Circulating fluidized bed technology II*, New York, 1988, Pergamon.
Hou M, Liu R, Zhai G, et al: Velocity distribution of vibration-driven granular gas in Knudsen regime in microgravity, *Microgravity Sci Technol* 20:73–80, 2008.
Huan C, Yang X, Candela D, Mair RW, Walsworth RL: NMR experiments on a three-dimensional vibrofluidized granular medium, *Phys Rev E* 69:041302, 2004.
Igci Y, Sundaresan S: Constitutive models for filtered two-fluid models of fluidized gas-particle flows, *Ind Eng Chem Res* 50:13190–13201, 2011.
Igci Y, Andrews AT, Sundaresan S, Pannala S, O'Brien T: Filtered two-fluid models for fluidized gas-particle suspensions, *AIChE J* 54:1431–1448, 2008.
Igci Y, Pannala S, Benyahia S, Sundaresan S: Validation studies on filtered model equations for gas-particle flows in risers, *Ind Eng Chem Res* 51:2094–2103, 2012.
Ishii M, Zuber N: Drag coefficient and relative velocity in bubbly, droplet or particulate flows, *AIChE J* 25:843–855, 1979.
Jaynes ET: The minimum entropy production principle, *Annu Rev Phys Chem* 31:579–601, 1980.

Jenkins J, Mancini F: Balance laws and constitutive relations for plane flows of a dense, binary mixture of smooth, nearly elastic, circular disks, *J Appl Mech Trans ASME* 54:27–34, 1987.
Jenkins JT, Richman MW: Kinetic-theory for plane flows of a dense gas of identical, rough, inelastic, circular disks, *Phys Fluids* 28:3485–3494, 1985.
Johnson PC, Jackson R: Frictional-collisional constitutive relations for granular materials, with application to plane shearing, *J Fluid Mech* 176:67–93, 1987.
Jung J, Gidaspow D, Gamwo IK: Measurement of two kinds of granular temperatures, stresses, and dispersion in bubbling beds, *Ind Eng Chem Res* 44:1329–1341, 2005.
Koch DL, Sangani AS: Particle pressure and marginal stability limits for a homogeneous monodisperse gas fluidized bed: kinetic theory and numerical calculations, *J Fluid Mech* 400:229–263, 1999.
Kudrolli A, Henry J: Non-Gaussian velocity distributions in excited granular matter in the absence of clustering, *Phys Rev E* 62:R1489–R1492, 2000.
Kunii D, Levenspiel O: *Fluidization engineering,* Boston, 1991, Butterworth-Heinemann.
Kwauk M: Generalized fluidization: I. Steady state motion, *Sci Sin* 12:587–612, 1963.
Kwauk M: *Fluidization: idealized and bubbleless, with applications,* Beijing and New York, 1992, Science Press and Ellis Horwood.
Levenspiel O: *Chemical reaction engineering,* New York, 1999, John Wiley & Sons.
Li J: Approaching virtual process engineering with exploring mesoscience, *Chem Eng J* 278:541–555, 2015.
Li T, Benyahia S: Evaluation of wall boundary condition parameters for gas-solids fluidized bed simulations, *AIChE J* 59:3624–3632, 2013.
Li J, Huang W: *Towards mesoscience: the principle of compromise in competition,* Berlin Heidelberg, 2014, Springer.
Li Y, Kwauk M: The dynamics of fast fluidization. In Grace JR, Matsen JM, editors: *Fluidization,* New York, 1980, Pergamon Press, pp 537–544.
Li J, Kwauk M: *Particle-fluid two-phase flow: energy-minimization multi-scale method,* Beijing, 1994, Metallurgy Industry Press.
Li J, Kwauk M: Multiscale nature of complex fluid-particle systems, *Ind Eng Chem Res* 40:4227–4237, 2001.
Li J, Kwauk M: Exploring complex systems in chemical engineering—the multi-scale methodology, *Chem Eng Sci* 58:521–535, 2003.
Li J, Wen L, Ge W, Cui H, Ren J: Dissipative structure in concurrent-up gas-solid flow, *Chem Eng Sci* 53:3367–3379, 1998.
Li J, Zhang J, Ge W, Liu X: Multi-scale methodology for complex systems, *Chem Eng Sci* 59:1687–1700, 2004.
Li J, Ge W, Kwauk M: Meso-scale phenomena from compromise—a common challenge, not only for chemical engineering, *arXiv preprint* , 2009. arXiv:0912.5407.
Li T, Grace J, Bi X: Study of wall boundary condition in numerical simulations of bubbling fluidized beds, *Powder Technol* 203:447–457, 2010.
Li D, Ray AK, Ray MB, Zhu J: Rotational asymmetry of reactant concentration and its evolution in a circulating fluidized bed riser, *Particuology* 10:573–581, 2012a.
Li F, Song F, Benyahia S, Wang W, Li J: MP-PIC simulation of CFB riser with EMMS-based drag model, *Chem Eng Sci* 82:104–113, 2012b.
Li J, Ge W, Wang W, et al: *From multiscale modeling to meso-science,* Berlin Heidelberg, 2013, Springer.
Li J, Wei Y, Chen J, et al: Cavity controls the selectivity: insights of confinement effects on MTO reaction, *ACS Catal* 5:661–665, 2015.
Lin Q, Wei F, Jin Y: Transient density signal analysis and two-phase micro-structure flow in gas-solids fluidization, *Chem Eng Sci* 56:2179–2189, 2001.

Liu X, Jiang Y, Liu C, Wang W, Li J: Hydrodynamic modeling of gas–solid bubbling fluidization based on energy-minimization multiscale (EMMS) theory, *Ind Eng Chem Res* 53:2800–2810, 2014.

Liu C, Wang W, Zhang N, Li J: Structure-dependent multi-fluid model for mass transfer and reactions in gas-solid fluidized beds, *Chem Eng Sci* 122:114–129, 2015.

Losert W, Cooper DGW, Delour J, Kudrolli A, Gollub JP: Velocity statistics in excited granular media, *Chaos* 9:682–690, 1999.

Lu H, Gidaspow D, Manger E: Kinetic theory of fluidized binary granular mixtures, *Phys Rev E* 64:8, 2001.

Lu H, Wang S, He Y, Ding I, Liu G, Hao Z: Numerical simulation of flow behavior of particles and clusters in riser using two granular temperatures, *Powder Technol* 182:283–293, 2008.

Lu B, Wang W, Li J: Searching for a mesh-independent sub-grid model for CFD simulation of gas-solid riser flows, *Chem Eng Sci* 64:3437–3447, 2009.

Lu B, Wang W, Li J: Eulerian simulation of gas-solid flows with particles of Geldart groups A, B and D using EMMS-based meso-scale model, *Chem Eng Sci* 66:4624–4635, 2011.

Lun CKK, Savage SB, Jeffrey DJ, Chepurniy N: Kinetic theories for granular flow-inelastic particles in Couette-flow and slightly inelastic particles in a general flowfield, *J Fluid Mech* 140:223–256, 1984.

Lungu M, Wang J, Yang Y: Numerical simulations of a bubbling fluidized bed reactor with an energy minimization multiscale bubble based model: effect of the mesoscale, *Ind Eng Chem Res* 53:16204–16221, 2014.

Martyushev LM, Seleznev VD: Maximum entropy production principle in physics, chemistry and biology, *Phys Rep* 426:1–45, 2006.

McKeen T, Pugsley T: Simulation and experimental validation of a freely bubbling bed of FCC catalyst, *Powder Technol* 129:139–152, 2003.

Mclennan JA: *Introduction to nonequilibrium statistical mechanics,* Englewood Cliff, New Jersey, 1989, Prentice Hall.

Meer DVD, Reimann P: Temperature anisotropy in a driven granular gas, *EPL (Europhys Lett)* 74:384, 2006.

Morris JP, Fox PJ, Zhu Y: Modeling low Reynolds number incompressible flows using SPH, *J Comput Phys* 136:214–226, 1997.

Mudde RF: Gravity-driven bubbly flows, *Annu Rev Fluid Mech* 37:393–423, 2005.

Nieuwland JJ, Annaland MV, Kuipers JAM, van Swaaij WPM: Hydrodynamic modeling of gas/particle flows in riser reactors, *AICHE J* 42:1569–1582, 1996.

Nikolopoulos A, Papafotiou D, Nikolopoulos N, Grammelis P, Kakaras E: An advanced EMMS scheme for the prediction of drag coefficient under a 1.2 MWth CFBC isothermal flow-part I: numerical formulation, *Chem Eng Sci* 65:4080–4088, 2010.

Noije TPCV, Ernst MH: Velocity distributions in homogeneous granular fluids: the free and the heated case, *Granul Matter* 1:57–64, 1998.

Olafsen JS, Urbach JS: Velocity distributions and density fluctuations in a granular gas, *Phys Rev E* 60:R2468–R2471, 1999.

Ono IK, O'Hern CS, Durian DJ, Langer SA, Liu AJ, Nagel SR: Effective temperatures of a driven system near jamming, *Phys Rev Lett* 89:095703, 2002.

Parmentier J-F, Simonin O, Delsart O: A functional subgrid drift velocity model for filtered drag prediction in dense fluidized bed, *AICHE J* 58:1084–1098, 2012.

Prigogine I: *Introduction to thermodynamics of irreversible processes,* Charles C Thomas, 1955, Springfield.

Radl S, Sundaresan S: A drag model for filtered Euler-Lagrange simulations of clustered gas-particle suspensions, *Chem Eng Sci* 117:416–425, 2014.

Rahaman M, Naser J, Witt P: An unequal granular temperature kinetic theory: description of granular flow with multiple particle classes, *Powder Technol* 138:82–92, 2003.

Reh L: Fluidized bed processing, *Chem Eng Prog* 67:58–64, 1971.

Reh L: *Fluid dynamics of CFB combustors. Circulating fluidized bed technology V*, Beijing, 1996, Science Press, 1–15.

Richardson J, Zaki W: Fluidization and sedimentation—part I, *Trans Inst Chem Eng* 32:38–58, 1954.

Rouyer F, Menon N: Velocity fluctuations in a homogeneous 2D granular gas in steady state, *Phys Rev Lett* 85:3676–3679, 2000.

Sagaut P: *Large eddy simulation for incompressible flows*, ed 3, Berlin, 2001, Springer-Verlag.

Savage SB, Jeffrey DJ: The stress tensor in a granular flow at high shear rates, *J Fluid Mech* 110:255–272, 1981.

Schneiderbauer S, Pirker S: Filtered and heterogeneity-based subgrid modifications for gas-solid drag and solid stresses in bubbling fluidized beds, *AICHE J* 60:839–854, 2014.

Segre PN, Liu F, Umbanhowar P, Weitz DA: An effective gravitational temperature for sedimentation, *Nature* 409:594–597, 2001.

Serero D, Goldhirsch I, Noskowicz S, Tan M: Hydrodynamics of granular gases and granular gas mixtures, *J Fluid Mech* 554:237–258, 2006.

Sherwood TK, Pigford RL, Wilke CR: *Mass transfer*, New York, 1975, McGraw-Hill.

Shi Z, Wang W, Li J: A bubble-based EMMS model for gas-solid bubbling fluidization, *Chem Eng Sci* 66:5541–5555, 2011.

Shuai W, Huilin L, Guodong L, Zhiheng S, Pengfei X, Gidaspow D: Modeling of cluster structure-dependent drag with Eulerian approach for circulating fluidized beds, *Powder Technol* 208:98–110, 2011.

Shuai W, Guangbo Z, Guodong L, Huilin L, Fixiang Z, Tianyu Z: Hydrodynamics of gas-solids risers using cluster structure-dependent drag model, *Powder Technol* 254:214–227, 2014.

Sinclair JL, Jackson R: Gas-particle flow in a vertical pipe with particle-particle interactions, *AICHE J* 35:1473–1486, 1989.

Snider DM: An incompressible three-dimensional multiphase particle-in-cell model for dense particle flows, *J Comput Phys* 170:523–549, 2001.

Song F, Wang W, Hong K, Li J: Unification of EMMS and TFM: structure-dependent analysis of mass, momentum and energy conservation, *Chem Eng Sci* 120:112–116, 2014.

Song F, Li F, Wang W, Li J: Multiphase particle-in-cell simulation of fluidization with a sub-grid EMMS drag. 2015 (in preparation).

Squires AM, Kwauk M, Avidan AA: Fluid beds—at last, challenging two entrenched practices, *Science* 230:1329–1337, 1985.

Sundaresan S: Modeling the hydrodynamics of multiphase flow reactors: current status and challenges, *AICHE J* 46:1102–1105, 2000.

Syamlal M, Pannala S: Multiphase continuum formulation for gas-solids reacting flows. In Pannala S, Syamlal M, O'Brien TJ, editors: *Computational gas-solids flows and reacting systems: theory, methods and practice*, New York, 2011, IGI Global, pp 1–65.

Syamlal M, Rogers W, O'Brien TJ: *MFIX documentation: theory guide*, Technical note DOE/METC-95/1013 and NTIS/DE95000031. 1993, National Energy Technology Laboratory, Department of Energy.

Tan ML, Goldhirsch I: Rapid granular flows as mesoscopic systems, *Phys Rev Lett* 81:3022–3025, 1998.

Tebianian S, Motlagh HA, Vashisth S, et al: Extending the comparison of voidage measurement and modeling techniques in fluidized beds. In Proceedings of the 11th international conference on fluidized bed technology, Beijing, China. 2014.

Tsuji Y, Kawaguchi T, Tanaka T: Discrete particle simulation of 2-dimensional fluidized-bed, *Powder Technol* 77:79–87, 1993.

Turns SR: *An introduction to combustion,* New York, 2000, McGraw-Hill.
Ullah A, Wang W, Li J: Evaluation of drag models for cocurrent and countercurrent gas-solid flows, *Chem Eng Sci* 92:89–104, 2013a.
Ullah A, Wang W, Li J: "Generalized fluidization" revisited, *Ind Eng Chem Res* 52:11319–11332, 2013b.
Ullah A, Wang W, Li J: An EMMS based mixture model for turbulent fluidization. In 11th International conference on fluidized bed technology (CFB11), Science Press, Beijing, pp 267–272, 2014.
Van der Hoef M, Ye M, van Sint Annaland M, Andrews A, Sundaresan S, Kuipers J: Multiscale modeling of gas-fluidized beds, *Adv Chem Eng* 31:65–149, 2006.
Van der Hoef MA, Van Sint Annaland M, Deen NG, Kuipers JAM: Numerical simulation of dense gas-solid fluidized beds: a multiscale modeling strategy, *Annu Rev Fluid Mech* 40:47–70, 2008.
van der Meer D, Reimann P: Temperature anisotropy in a driven granular gas, *Europhys Lett* 74:384–390, 2006.
van Zon JS, MacKintosh FC: Velocity distributions in dissipative granular gases, *Phys Rev Lett* 93:038001, 2004.
van Zon JS, MacKintosh FC: Velocity distributions in dilute granular systems, *Phys Rev E* 72:051301, 2005.
Venderbosch RH: *The role of clusters in gas-solids reactors, an experimental study,* 1998, Universiteit Twente.
Wallis GB: *One-dimensional two-phase flow,* New York, 1969, McGraw-Hill.
Wang W, Li J: Simulation of gas-solid two-phase flow by a multi-scale CFD approach—extension of the EMMS model to the sub-grid level, *Chem Eng Sci* 62:208–231, 2007.
Wang H, Menon N: Heating mechanism affects equipartition in a binary granular system, *Phys Rev Lett* 100:158001, 2008.
Wang W, Lu B, Li J: Choking and flow regime transitions: simulation by a multi-scale CFD approach, *Chem Eng Sci* 62:814–819, 2007.
Wang J, Ge W, Li J: Eulerian simulation of heterogeneous gas-solid flows in CFB risers: EMMS-based sub-grid scale model with a revised cluster description, *Chem Eng Sci* 63:1553–1571, 2008a.
Wang W, Lu B, Dong W, Li J: Multi-scale CFD simulation of operating diagram for gas-solid risers, *Can J Chem Eng* 86:448–457, 2008b.
Wang J, van der Hoef MA, Kuipers JAM: Why the two-fluid model fails to predict the bed expansion characteristics of Geldart A particles in gas-fluidized beds: a tentative answer, *Chem Eng Sci* 64:622–625, 2009.
Wang W, Lu B, Zhang N, Shi Z, Li J: A review of multiscale CFD for gas-solid CFB modeling, *Int J Multiphase Flow* 36:109–118, 2010.
Wang W, Ge W, Yang N, Li J: Meso-scale modeling—the key to multi-scale CFD simulation, *Adv Chem Eng* 40:1–58, 2011.
Warr S, Huntley JM, Jacques GTH: Fluidization of a 2-dimensional granular system—experimental-study and scaling behavior, *Phys Rev E* 52:5583–5595, 1995.
Wen C, Yu Y: Mechanics of fluidization, *Chem Eng Symp Ser* 62:100–111, 1966.
Wilhelm RH, Kwauk M: Fluidization of solid particles, *Chem Eng Prog* 44:201–218, 1948.
Xie N, Battaglia F, Pannala S: Effects of using two- versus three-dimensional computational modeling of fluidized beds—part I, hydrodynamics, *Powder Technol* 182:1–13, 2008a.
Xie N, Battaglia F, Pannala S: Effects of using two- versus three-dimensional computational modeling of fluidized beds: part II, budget analysis, *Powder Technol* 182:14–24, 2008b.
Xiong Q, Deng L, Wang W, Ge W: SPH method for two-fluid modeling of particle-fluid fluidization, *Chem Eng Sci* 66:1859–1865, 2011.

Xu B, Yu A: Numerical simulation of the gas-solid flow in a fluidized bed by combining discrete particle method with computational fluid dynamics, *Chem Eng Sci* 52:2785–2809, 1997.
Xu B, Yu A, Chew S, Zulli P: Numerical simulation of the gas-solid flow in a bed with lateral gas blasting, *Powder Technol* 109:13–26, 2000.
Xu M, Ge W, Li J: A discrete particle model for particle-fluid flow with considerations of sub-grid structures, *Chem Eng Sci* 62:2302–2308, 2007.
Xue Q, Fox RO: Multi-fluid CFD modeling of biomass gasification in polydisperse fluidized-bed gasifiers, *Powder Technol* 254:187–198, 2014.
Yang W: "Choking" revisited, *Ind Eng Chem Res* 43:5496–5506, 2004.
Yang X, Huan C, Candela D, Mair RW, Walsworth RL: Measurements of grain motion in a dense, three-dimensional granular fluid, *Phys Rev Lett* 88:044301, 2002.
Yang N, Wang W, Ge W, Li J: CFD simulation of concurrent-up gas-solid flow in circulating fluidized beds with structure-dependent drag coefficient, *Chem Eng J* 96:71–80, 2003.
Yerushalmi J, Turner DH, Squires AM: The fast fluidized bed, *Ind Eng Chem Process Des Dev* 15:47–53, 1976.
Ying L, Yuan X, Ye M, Cheng Y, Li X, Liu Z: A seven lumped kinetic model for industrial catalyst in DMTO process, *Chem Eng Res Des* 100:179–191, 2015.
Zenz FA: Two-phase fluid-solid flow, *Ind Eng Chem* 41:2801–2806, 1949.
Zhu H, Zhou Z, Yang R, Yu A: Discrete particle simulation of particulate systems: theoretical developments, *Chem Eng Sci* 62:3378–3396, 2007.
Zhu H, Zhu J, Li G, Li F: Detailed measurements of flow structure inside a dense gas–solids fluidized bed, *Powder Technol* 180:339–349, 2008.

CHAPTER FIVE

MTO Processes Development: The Key of Mesoscale Studies

Mao Ye[1], Hua Li, Yinfeng Zhao, Tao Zhang, Zhongmin Liu
Dalian National Laboratory for Clean Energy and National Engineering Laboratory for MTO, Dalian Institute of Chemical Physics, Chinese Academy of Sciences, Dalian, PR China
[1]Corresponding author: e-mail address: maoye@dicp.ac.cn

Contents

Advances in Chemical Engineering, Volume 47
ISSN 0065-2377
http://dx.doi.org/10.1016/bs.ache.2015.10.008

Abstract

Methanol to olefins (MTO), which provides a new route to produce light olefins such as ethylene and propylene from abundant natural materials (e.g., coal, natural gas or biomass), has been recently industrialized by the Dalian Institute of Chemical Physics (DICP), Chinese Academy of Sciences. In this contribution, the process development of MTO is introduced, which emphasizes the importance of mesoscale studies and focuses on three aspects: a mesoscale modeling approach for MTO catalyst pellet, coke formation and control in MTO reactor, and scaling up of the microscale-MTO fluidized bed reactor to pilot-scale fluidized bed reactor. The challenges and future directions in MTO process development are also briefed.

1. INTRODUCTION

Light olefins such as ethylene and propylene are key components in the chemical industries. They are conventionally produced from petrochemical feedstock via naphtha thermal cracking and fluid catalytic cracking (FCC) processes (Corma, 2003; Primo and Garcia, 2014; Van Santen et al., 1999). The efforts to viable routes for producing light olefins from alternative resources other than oil have been continuously growing since the oil crisis in 1970s (Chang, 1984; Chang and Silvestri, 1977; Liang et al., 1990). Methanol, which can be readily produced from coal, natural gas, and biomass via synthesis gas ($CO + H_2$) by existing and proven technologies, offers an attractive choice (Chang, 1984; Chang and Silvestri, 1977; Keil, 1999; Tian et al., 2015). Early researches by scientists in Mobil discovered that a zeolite-based process could be used to convert methanol into gasoline. In this methanol to gasoline (MTG) process, a new class of synthetic shape-selective zeolites, namely ZSM-5, was used. Later on, Union Carbide reported the successful synthesis of a silicoaluminophosphate (SAPO) catalyst in 1980s. Scientists from the Dalian Institute of Chemical Physics (DICP), Chinese Academy of Sciences found that SAPO-34 catalyst could be used to convert the methanol to light olefins with high selectivity (Liang et al., 1990). Since then, methanol to olefins (MTO) process has become a subject of intense researches spanning catalyst synthesis, reaction mechanism, reaction kinetics, process development, and reactor scale-up. In August 2010, a commercial unit (600 kt/a of ethylene and propylene production) based on the DICP MTO process (DMTO) was successfully brought into stream in Shenhua's Baotou coal-to-olefins plant in north China (Liu et al., 2014). This is the world's first MTO commercial plant, and its success represents an important milestone and breakthrough of MTO process development. So far, several MTO processes have been commercialized, which include, in addition to the DMTO process, the MTO process by UOP (Zhang, 2013), the methanol to propylene (MTP) process by Lurgi (Nan et al., 2014), and the SINOPEC's MTO (SMTO) process by SINOPEC (Jiang et al., 2014).

The catalysts synthesis, reaction mechanism and kinetics, process development, and reactor design and operation for industrial MTO processes require the detailed researches from molecular to reactor level due to the multiphase and multiscale nature, which cover a considerably broad range of space and time scales. From practical point of view, the understanding of the MTO reaction mechanism and shape selectivity to target products (ethylene and propylene) over relevant zeolite catalyst, which is the basis of reactor selection, process optimization, and unit operation for MTO process development, is of critical importance. In the past decades, a huge amount of work has been published studying the MTO process at different space and time scales. At very fundamental scale, modeling approaches such as quantum mechanics (QM) and molecular dynamics (MD) (Van Speybroeck et al., 2014), as well as high temporal and spatial resolution measurement techniques such as high-energy Operando X-ray diffraction (HXRD) and nuclear magnetic resonance (NMR) (Buurmans and Weckhuysen, 2012; Hunger et al., 2001; Li et al., 2015b; Wragg et al., 2012) were used to study the MTO reaction mechanism. At reactor level, macroscale experiments under either cold flow or reaction conditions, as well as computational fluid dynamics study were carried out to derive the gross reaction kinetics and monitor the fluid flow. Except that in a few cases, the macroscale reaction data have been used to deduce the reaction network, it is generally difficult to get the information of element steps, and thus, reaction mechanism in the zeolites with sufficient accuracy based on the macroscale reaction data. Meanwhile, the mechanistic researches, although being applied to instruct the MTO reaction controlling, the quantitative translation of the knowledge obtained from the fundamental scale to macroscale remains a hard task. The systematic theories and methods linked the fundamental scale to macroscale are yet to be developed. Therefore, the MTO process development, at current stage, is still dependent on the experience built via experiments at different scales. Li et al. (2013) classified the methods to bridge such gap as the mesoscale methods. Apparently, the studies at mesoscales play an important part in establishing the theories or methods for reaction-transport process linking two adjunct scales.

In this contribution, the development of MTO processes will be introduced. We will emphasize the mesoscale challenges and the related studies in the DMTO process development, for which we focus on three aspects: a mesoscale modeling approach for MTO catalyst pellet, the coke formation and control in MTO reactor, and the scale-up of the microscale-MTO fluidized bed reactor to pilot-scale. A section followed is dedicated to the future directions in MTO process development in terms of mesoscale studies.

2. MTO PROCESS DEVELOPMENT

Methanol is an important C_1 compound, and is very active over a variety of acidic zeolites with different topologies (pores, channels, and cavities), compositions (acid sites), and morphologies (crystal size, micro-, and mesopores) to form different hydrocarbons (Olsbye et al., 2012). As discussed above, ZSM-5 and SAPO-34 are two zeolites of industrial interests in converting methanol to light olefins. SAPO-34 has pore size of 0.38 nm, which shows high selectivity to ethylene and propylene (about 80%) but could be deactivated in several hours. Thus, a continuous regeneration is required for MTO process based on SAPO-34 catalyst. ZSM-5 has 10-membered ring openings with 3D pore structure, and the pore size is 0.54×0.57 nm. The larger pore size of ZSM-5 leads to a much wider product distribution. Heavy gasoline compounds are formed over ZSM-5 as by-product. In DMTO process developed by DICP, the SAPO-34 catalyst is used to convert methanol in fluidized bed reactor. In the following, the history of MTO process development is introduced.

2.1 MTG Processes

The successful synthesis of ZSM-5 zeolites which have relatively larger pore size and higher Si-to-Al ratio in early 1970s stimulates the rapid development of MTG process (Chang, 1983; Keil, 1999). Mobil first built a fixed bed pilot plant to demonstrate the feasibility of MTG process. In 1985, the MTG process was implemented into a commercial plant in New Zealand with a gasoline capacity of 14,500 bpd. Almost at the same time, Mobil also developed fluidized bed MTG technology, which was demonstrated in a 4 bpd pilot plant in Paulsboro, New Jeasey and scaled up to a demonstration scale of 100 bpd during 1981–1984 in Wesseling, Germany. This fluidized bed MTG demonstration unit stopped operation at the end of 1985 due to the lower price of oil at that time (Keil, 1999). Apparently, the

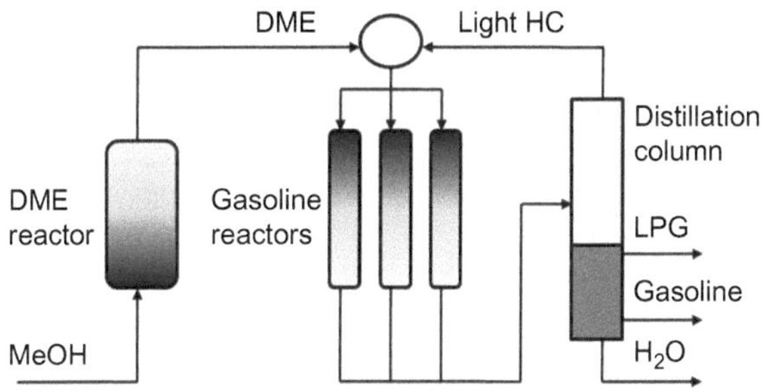

Figure 1 The schematic diagram of Mobil's MTG process.

early work by Mobil showed the confidence and feasibility of MTG process with ZSM-5 catalyst.

In Mobil's MTG process, the products spin from C_1 to C_{11} hydrocarbons, and among them the C_5^+ (benzene fraction) accounts to roughly 80%. Although ZSM-5 catalyst was shown feasible for MTG process, they would still deactivate slowly due to the deposition of coke. Therefore, several parallel reactors had been used in the MTG commercial unit in New Zealand, and the intermittent regeneration was used to maintain the continuous operation. Figure 1 shows the schematic diagram of Mobil's commercial MTG unit in New Zealand. In 1990s, ExxonMobil further optimized the MTG process and started to license the improved MTG process. In March 2010, the first MTG unit (100 kt/a gasoline product) in China was started up. This unit used the ExxonMobil MTG technology and was built in Jincheng, Shanxi. When oil price is low, the MTG process is economically less competitive.

2.2 DMTO Process

Light olefins are in fact intermediates in the MTG process. Careful controlling of the reaction conditions, e.g., temperature, pressure, and catalyst acidity can prompt the production of light olefins in the process of methanol conversion. Mobil's researchers demonstrated the MTO process based on ZSM-5 zeolites. In 1982, the scientists at DICP initiated a project on MTO research under the support by Chinese government and Chinese Academy of Sciences. A 300 t/a MTO-fixed bed pilot plant was constructed and operated in 1993 by DICP (Tian et al., 2015).

In 1986, Union Carbide reported the successful synthesis of a SAPO catalyst. Researchers from DICP found that SAPO-34 catalyst could be used to convert the methanol to light olefins with high selectivity (Liang et al., 1990). Except high selectivity to light olefins, SAPO-34 catalyst is readily deactivated due to coke deposition. In this sense, fluidized bed is the best choice for MTO process based on SAPO-34 catalyst as deactivated catalyst can be regenerated online by circulating catalyst between reactor and regenerator. Note that the synthesis of dimethyl ether is thermodynamically more favorable than that of methanol, researchers from DICP first developed a syngas/dimethyl ether to olefins (SDTO) method in early 1990s. A pilot fluidized bed reactor for SDTO process was constructed and successfully tested in 1995. However, the SDTO process was economically less competitive when the oil price goes lower. To this end, the researchers from DICP focused on zeolite synthesis and modifications in order to further improve the catalyst performance (Tian et al., 2015). Great success was achieved by the DICP team in synthesizing high efficient and economic MTO fluidized bed catalyst (Tian et al., 2015).

Meanwhile, DICP team also tried to scale-up the MTO process in the laboratory. In 2004, they finished the MTO pilot-scale experiments and decided to build a MTO demonstration unit (16 kt/a Methanol feed) and scale-up the MTO process to commercial scale. Two partners, Shaanxi Xinxin Coal Chemical Ltd. and Luoyang Petrochemical Engineering Corporation, joined in DICP team to construct the demonstration unit, which was completed in July 2005. In December 2005, the MTO demonstration unit was brought on stream. In June 2006, DICP announced the success of the operation of MTO demonstration unit and started to license the MTO technology, which is now called DMTO technology. Based on DMTO technology, the world's first MTO commercial unit (600 kt/a of ethylene and propylene production) was started up in August 2010 in Shenhua's Baotou coal-to-olefins plant in north China (Liu et al., 2014). Figure 2 is the schematic diagram of Shenhua's DMTO unit.

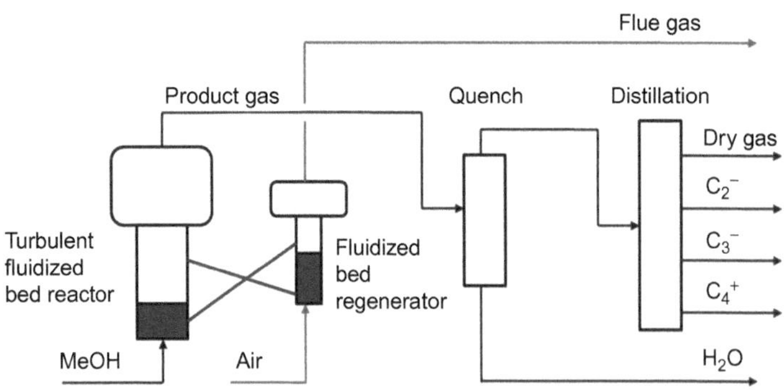

Figure 2 The schematic diagram of DMTO process.

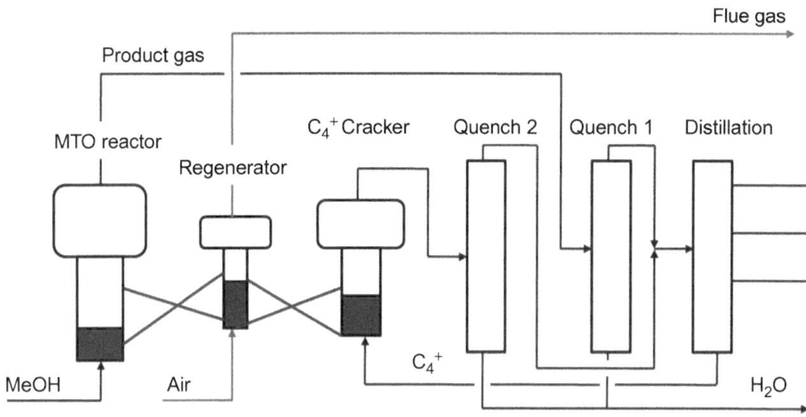

Figure 3 The schematic diagram of DMTO-II process.

In order to achieve a higher light olefins yield, DICP synthesize a new dual-functional catalyst, over which both the MTO reaction and C_4^+ hydrocarbon cracking reaction can be realized. Based on this catalyst, DICP developed the DMTO-II process. In the DMTO-II process, the C_4^+ compounds are recycled to the fluidized bed C_4^+ cracking reactor to increase the ethylene and propylene yield. As there is only one catalyst in this process, a single fluidized bed regenerator is possible. This can significantly simplify the process and improve the utility efficiency. In September 2009, DICP revamped the DMTO demonstration unit and upgraded it to a DMTO-II demonstration unit. In May 2010, the DICP announced that the experiments of DMTO-II process was successful and started to license the DMTO-II technology. At the end of 2014, the first DMTO-II unit (with 670 kt/a of ethylene and propylene production) was commissioned and started its commercial operation (the scheme of the DMTO-II process shown in Fig. 3). Until December 2014, there are six DMTO commercial units and one DMTO-II commercial unit in operation, with a total capacity of 417 kt/a of ethylene and propylene production.

2.3 MTO Process by UOP

UOP has long been working with the MTO catalyst and process development. In June 1995, UOP and Norsk Hydro (now INEOS) built a fluidized bed MTO pilot plant with a capacity of 0.75 t/d methanol feed. In this pilot plant, the reactor–regenerator and product separation system were included (Vora et al., 1997). To further improve the selectivity to ethylene and propylene, UOP combined the UOP/Hydro MTO process with TOTAL's

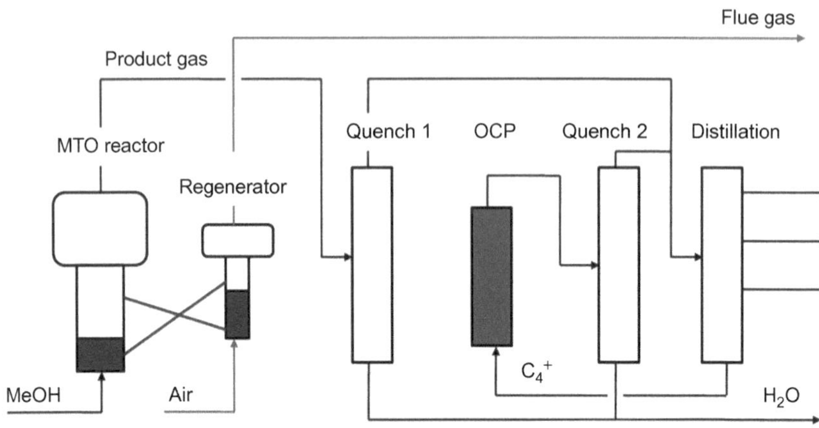

Figure 4 The schematic diagram of UOP's MTP/OCP process.

olefin cracking process (OCP), and the latter could crack the C_4^+ compounds into ethylene and propylene. Together with TOTAL, UOP constructed a demonstration unit in Feluy, Belgium, which, as schematically shown in Fig. 4, can process roughly 10 ton methanol feed per day. In 2009, this demonstration unit was started up. The combined MTO/OCP process has two separate reaction systems, i.e., a fluidized bed MTO system and a fixed bed C_4^+ cracking system. Compared to the DMTO-II process, two catalysts have to be used in UOP's MTO/OCP process. In 2013, the first commercial unit based on UOP's MTO process (the capacity is 295 kt/a of light olefins production) was commissioned in Nanjing, China (Zhang, 2013).

2.4 Other MTO Processes

Besides Mobil, DICP, and UOP, SINOPEC has been actively involved in the MTO process development since 2000 (Liu, 2015). In 2005, a pilot-scale MTO unit capable of processing 12 t/a methanol feed was built in Shanghai. The SMTO process follows that of DICP and UOP, in which SAPO-34 catalyst is used. The SMTO process (100 t/d of methanol feed) has been demonstrated in Yanshan in Beijing in 2007. In 2011, an industrial SMTO unit with a capacity of 200 kt/a ethylene and propylene was brought on stream in Henan, China (Jiang et al., 2014).

Lurgi follows the route as originally developed by Mobil, and the ZSM-5 zeolite catalyst is used to convert methanol. Unlike DICP and UOP, Lurgi particularly concentrates on the propylene yield, which leads to the development of a fixed bed MTP process. In order to maximize the propylene yield, the by-products are recycled to the fixed bed reactor for further conversion. In 2011, the first fixed bed MTP commercial unit was commissioned in Ningxia in China. The unit reached its full capacity close to 500 kt/a of propylene 1 year later (Nan et al., 2014).

In parallel to the Lurgi fixed bed MTP process, Tsinghua University proposed a fluidized bed MTP (FMTP) process based on the SAPO-18/34 zeolite catalyst. It was declared that this catalyst can limit the formation of the compounds of C_4 and beyond, and thus prompt the yield of ethylene and propylene. FMTP process has two fluidized bed reactors. The first one is used to convert methanol to light olefins, and the second one is mainly for further converting ethylene and butylenes to propylene. In 2008, Tsinghua, together with its partners, built a FMTP demonstration unit (capacity of 30 kt/a of methanol feed) in Anhui, China. In September 2009, the demonstration unit was started up. The experimental data shows that the methanol conversion is almost 100%, and the propylene yield is close to 67.3% (CH_2 basis). FMTP has not been industrialized so far.

3. MULTISCALE NATURE OF MTO PROCESS

Apparently, the development of MTO process has an inherent multiscale nature, as shown in Fig. 5. The optimal design of an industrial MTO fluidized bed reactor requires the detailed information from molecular to reactor scales, which covers the space scale of nm to m by an order of 9 to 10.

3.1 Reaction Mechanism at Molecular Scale

At molecular scale (~nm), the conversion of methanol molecules to either intermediates or final products over zeolite needs to be understood. Great efforts have been devoted to study the formation of initial C—C bonds as methanol is a typical C_1 compound. More than 20 possible mechanisms were proposed by various groups, and most of them, however, lack of direct experimental evidences. Song et al. (2002) and recently Qi et al. (2015) proposed that the impurities of feedstock from different sources might be the reason for the initial C—C bonds formation. Nevertheless, the consent

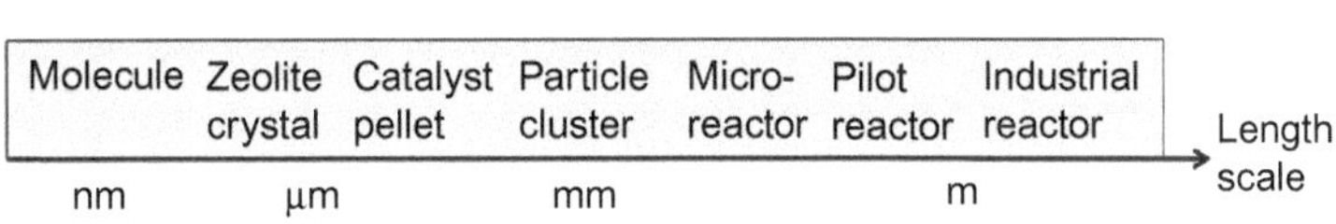

Figure 5 The multiscale nature of MTO process development.

on the mechanism of the initial C—C bonds formation is far from being reached. Of practical importance is the reaction path in MTO process. Hydrocarbon pool proposed by Dahl and Kolboe (1994) is now widely accepted as the main path for MTO reaction. The hydrocarbon pool refers to organic species, $(CH_2)_n$, confined in the zeolite cage or intersection of channels, which can further assemble olefins from methanol feed. Many researches concentrated on the mechanism underlying the hydrocarbon pool formation, the intermediate species, and the element steps in MTO reaction (Tian et al., 2015). The induction period is also critical for MTO reaction. The induction period is a stage during which the organic-free zeolite catalyst is transferred to a working catalyst. Basically, the methanol conversion in the induction period could be quite low, which is controlled by the complicated reaction kinetics of element steps (Qi et al., 2015). For DMTO catalyst, the induction period occurs at temperature below 350 °C (Yuan et al., 2012). Actually, the duration of induction period is a function of reaction temperature. This is important for instructing the operation of industrial DMTO reactors, where the reactor temperature has to be out of the range where the induction period occurs. In this regards, the quantitative incorporation of kinetic rates of the element steps into the reactor model is highly desired.

3.2 Reaction–Diffusion at Catalyst Scale

It has been well known that the topology (i.e., pore size, cavities, and pore network) and composition (i.e., acidity strength and acid sites distribution) of zeolite particles can affect the product distribution. One interesting finding is that SAPO-34 zeolite used in DMTO process shows a higher selectivity to ethylene with certain coke deposition. The mechanism underlying this finding is discussed in Section 5. Anyway, a simple explanation is that the pore blockage due to coke formation inhibits molecular diffusion in the pore channel (Guisnet, 2002), which in turn leads to an intense restriction for larger molecules to diffuse out of the cavities. However, it should be noticed that the coke deposition may cause some acid sites to be covered, and, in serious cases, make the zeolite particles deactivated rapidly. It is thus expected that the optimal coke deposition exists for SAPO-34 zeolite particles, by which a high selectivity to ethylene can be achieved while the catalyst is still not deactivated. The understanding of this shape selectivity in DMTO reaction requires a detailed study on the diffusion and reaction inside SAPO-34 zeolite particles. However, *in situ* measurements of the detailed diffusion and reaction inside SAPO-34 zeolite particles, if not impossible, would be extremely difficult. In this regards, the modeling approach seems to be much feasible.

An industrial DMTO fluidized bed catalyst pellet is basically composed of SAPO-34 zeolite particles and catalyst support (or matrix). The pores of zeolite particles and matrix are interconnected as a complex network. The pores inside zeolite particles are typically micropores (less than 2 nm) and the matrix normally has either mesopores (2–50 nm) or macropores (>50 nm), or both (Krishna and Wesselingh, 1997). The bulk diffusion coefficients in the meso- and macropores might be several orders of magnitude larger than surface diffusion coefficients in the micropores. Kortunov et al. (2005) found that the diffusion in macro- and mesopores also plays a crucial part in the transport in catalyst pellets. Therefore, other than a model for SAPO-34 zeolite particles, a modeling approach for diffusion and reaction in MTO catalyst pellets, which are composed of SAPO-34 zeolite particles and catalyst support, is needed.

3.3 Reaction and Solid–Gas Flow at Reactor Scale

The DMTO catalyst pellets are typically A type according to Geldart's classification in terms of fluidization characteristics (Tian et al., 2015). For this type of particles, the fluidization regime in the reactor can change from homogeneous fluidization to bubbling fluidization and turbulent fluidization, and to fast fluidization when the superficial gas velocity increases. Besides, the increase of reactor size also leads to significant variation of solid–gas two-phase flow patterns. The inelastic collisions between catalyst pellets lead to the formation of heterogeneous structure such as catalyst clusters. The intrinsic bubbling behavior in fluidized bed intensifies gas bypass and weakens the mass transfer in the reactor. The catalyst clusters and gas bubbles could further develop with an increasing reactor size or a higher gas velocity, which complicates the scaling up process of MTO fluidized bed reactor. The DMTO fluidized bed reactor was scaled up via various experiments at four different scales (Tian et al., 2015). The scale factor between two adjunct scales is roughly 100 in terms of methanol feed rate, and 10 in terms of reactor diameter. The microscale fluidized bed was operated at bubbling fluidization regime without catalyst circulation. In the pilot-scale experiments, the circulation between reactor and regenerator was established. In the demonstration and commercial scale, the turbulent fluidized bed reactor has been selected in order to achieve a high feed throughput.

As discussed above, the diversity of hydrodynamics in the reactors of different scale is one of the big challenges encountered in the fluidized bed scale-up, which has also been well addressed by a few researchers (Knowlton et al., 2005; Matsen, 1996; Rüdisüli et al., 2012). Another challenge that has been seldom discussed is about the translation of experimental results obtained from microscale DMTO fixed fluidized bed experiments to pilot-scale circulating fluidized bed reactor design and operation. In the microscale DMTO fluidized bed reactor, the catalyst is not circulated for regeneration. Thus, the coke content in catalyst increases with time on stream (TOS). But at any given time, the coke content can be considered as uniformly distributed in space because of the excellent mixing of solids in fluidized bed, which is shown in Fig. 6. However, in pilot-scale reactor, the catalyst is transported continuously to the regenerator to restore the activity by burning off the coke deposited on the catalyst. The regenerated catalyst with almost zero coke content is then transported back to the reactor. Figure 7 shows the typical distribution of coke content in catalyst in pilot-scale circulating fluidized bed reactor, which is not uniform in space due to the non-homogeneity of the residence time of catalyst in the fluidized bed. From Figs. 6 and 7, it can be seen that the coke distribution in the circulating fluidized bed differs significantly from that of fixed fluidized bed. Note that the coke content in catalyst is the key to achieve a high selectivity to ethylene, the translation of the microscale fluidized bed results to the circulating fluidized bed design actually represents a critical step in scaling up the DMTO reactor.

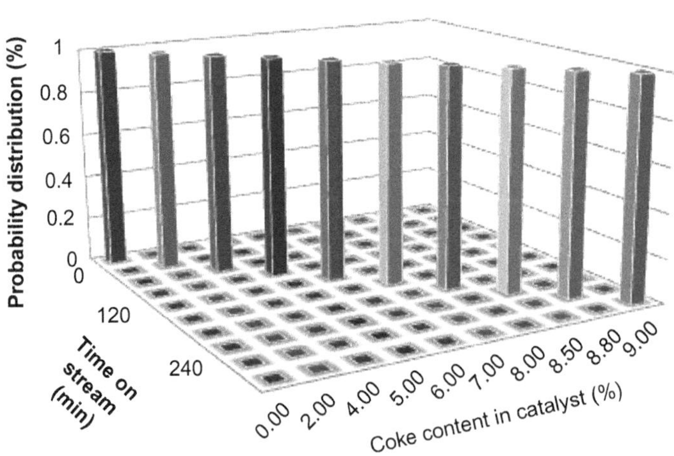

Figure 6 Typical distribution of coke content in catalyst in microscale DMTO fluidized bed reactor.

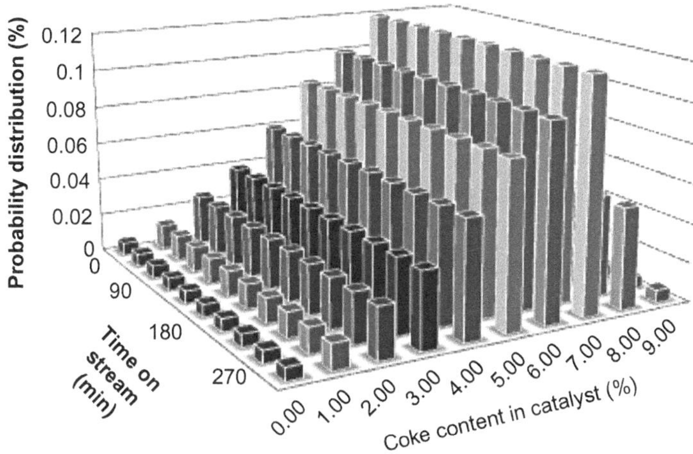

Figure 7 Typical distribution of coke content in catalyst in pilot-scale DMTO fluidized bed reactor.

3.4 Mesoscale Studies: The Key in MTO Process Development

In the DMTO process development, as discussed above, there were some specified issues at different scales. These issues are critical for instructing the scaling up, design, and operation of the fluidized bed reactor. The solutions to these issues, however, require knowledge and information obtained from a more fundamental scale. Quantitative translation of the knowledge and information from a lower (i.e., the characteristic size is smaller) scale to the issue of interest (normally at a higher scale) is actually the task of the mesoscale studies as proposed by Li et al. (2013). For example, in DMTO reaction mechanism study, the simulation results of QM can be used in MD only if the influences of topologies and morphologies of zeolites have been investigated and understood. In this chapter, the mesoscale challenges and the related studies in the DMTO process development are illustrated and discussed. Particularly, we focus on three aspects: a mesoscale modeling approach for MTO catalyst pellet, coke formation and control in MTO reactor, and scaling up of the microscale-MTO fluidized bed reactor to pilot-scale-fluidized bed reactor.

4. MESOSCALE MODEL FOR MTO CATALYST

4.1 Microscale Modeling for Reaction–Diffusion in Zeolites

The understanding of reaction–diffusion process in MTO catalyst pellet is of great importance in catalyst design optimization, and, as discussed above, understanding the coking behavior. A single DMTO catalyst pellet could be considered composed of microporous SAPO-34 zeolite particles and catalyst support (or matrix). The overall performance of a single catalyst pellet is dependent on both reaction–diffusion process of the zeolite crystal region and diffusion process of support region. Basically, the zeolites are active part of the pellet and the reactions mainly occur in this part.

As the online measurement of the reaction and diffusion inside a single zeolite particle is still not realistic, mathematical modeling offers an alternative way for describing the reaction–diffusion process in zeolite region. A reliable model should reveal the physical essence of the reaction–diffusion process. In principle, quantum chemical or *ab initio* dynamical simulation may provide the accurate prediction to the catalytic reaction process. But the quantum chemical approach is very time-consuming and mainly simulates the energy barriers of elementary reactions and vibration frequency spectrum of stationary geometries (Hansen and Keil, 2012; Keil, 2012), which prevents it from direct application in catalyst design and reactor optimization. A hierarchical multiscale approach by Keil (2012) and Hansen and Keil (2012) combined different approaches such as first principles, quantum chemistry, force field simulations and macroscopic differential equations to simulate the active centers of the catalyst, adsorption and diffusion of reactants and products, and reaction and diffusion in zeolite particles and fixed bed reactors, respectively. In this approach, the continuum model could be naturally constructed when the chemical kinetic and diffusion parameters are derived from lower scales.

This continuum model could be applied to describe reaction–diffusion process in zeolite region (or zeolite crystal particle). We call this model microscale model. In this model, the partial differential equations (PDEs) used to describe the change of loading with time of species i in zeolite region are expressed as (Li et al., 2015a)

$$\frac{\partial q_i}{\partial t} = -\nabla \cdot \vec{N_i} + r_i \quad (i = 1, 2, \ldots, n) \tag{1}$$

where q_i is the loading of species i, $\vec{N_i}$ the molecular flux of component i, and r_i the reaction rate of species i. And the molecular flux is calculated based on Maxwell–Stefen theory (Krishna and van Baten, 2009):

$$\begin{pmatrix} \vec{N_1} \\ \vec{N_2} \\ \vdots \\ \vec{N_n} \end{pmatrix} = -[B]^{-1}[\Gamma] \begin{pmatrix} \nabla q_1 \\ \nabla q_2 \\ \vdots \\ \nabla q_n \end{pmatrix}. \tag{2}$$

The detailed defines of matrix $[\Gamma]$ and matrix $[B]$ could be found in Li et al. (2015a). It is noted that the diffusion of each species, based on Eq. (2), is coupled by all other species.

This micromodel could be used to investigate the effect of zeolite particle size on its catalytic performance. The influence of crystal size actually represents the impact of species diffusion on the reaction, and could be quantified by the internal effective factors, which are defined as following

$$\eta_{Z,i} \equiv \frac{\int_{V_Z} r_i \mathrm{d}V + \frac{c_u^s}{c_u^v}\int_{S_Z} r_i \mathrm{d}S}{r_i^b V_Z} \tag{3}$$

where V_Z represents the total volume of zeolite region, and S_Z is the total area of the boundary faces of zeolite region, and r_i^b is the reaction rate of species i calculated at the region boundary condition. c_u^s means the number of unit cells per square meters, and c_u^v represents the number of unit cells per cubic meters.

Despite the development of microscale modeling for reaction–diffusion in zeolite, the complex of MTO reaction mechanism impedes the application of microscale modeling to MTO process. Up to now, the reliable reaction kinetics based on element reactions in MTO process is still under development (van Speybroeck et al., 2014). However, a reduced or simplified microscale model could be applied. Basically, the diffusion effect is negligible if the crystal radius is small enough. Then mass equation, i.e., Eq. (1), could be simplified by neglecting the species flux term. In this case, MTO processes over ZSM-5 and SAPO-34 catalyst could be simulated by use of the single-event kinetics by Alwahabi and Froment (2004a) as an input.

4.2 Macroscale Modeling for Reaction–Diffusion in MTO Reactor

Methanol transformation to olefins over SAPO-34 catalyst is featured by high exothermicity (Alwahabi and Froment, 2004b) and rapid deactivation (Chen et al., 1999; Park et al., 2008). Considering these, circulating fluidized bed is used for MTO process in industry. The model describing reaction–diffusion process in this macro reactor is called macroscale model. However, it should be pointed out that the focus of macroscale model on reaction and diffusion is far different from microscale model.

In microscale model, the reaction kinetics is constructed on the basis of element reaction system, and usually could be obtained by quantum chemistry calculation and transition theory. It means that the reaction kinetics in the microscale model is mainly determined by the structures of zeolite and reaction species, and is independent of other conditions, such as the size of catalyst particles. Such reaction kinetics has its corresponding appropriate computational unit. For example, the effective factor for alkylation of benzene over H-ZSM-5 crystal, as described above, is sensitive to computational unit in the size range of 0.1–1 μm. It is obvious that macroscale model of MTO reactor could not afford such a heavy calculation by using computational unit of this size. Therefore, the ensemble-average (or effective) reaction rates and simplified reaction network are normally used in the chemical kinetics in macroscale model. The parameters of chemical kinetics in macroscale model could be derived via two schemes: one is based on the simulation results from microscale models and another one is lumped from experimental data. As in the first scheme, the simulations can be considered as virtual numerical experiments. These two schemes, at first glance, show some similarity. However, the first scheme exhibits several advantages over the second scheme. The simulation methods used in the first scheme include exclusively quantum chemistry and MD, which have solid theoretical foundations. The second scheme is mainly dependent on the experimental conditions and, to certain extend, the researchers' experience. In addition to that, in numerical experiments, the effect of fluid flow and/or diffusion on reaction kinetics in the reactor could be readily reduced by incorporating well-designed simulations conditions. Such ideal conditions, in general, could not be achieved in realistic experiments. Thus the first scheme, in many cases, is much more efficient. However, the second scheme is still widely employed by researchers due to the complexity of MTO reaction process.

Since in the macroscale model, the reaction rate and diffusion coefficient are effective ones that are obtained on an ensemble-averaged basis, the internal diffusion will not appear in the controlling equations explicitly. The effective reaction rate already includes the influence of internal diffusion inside catalyst pellets. The external mass transfer term, which mainly accounts for the species transport outside catalyst pellets, is used in the controlling equations in macroscale models. So, the diffusion mentioned in macroscale model normally represents species diffusion outside catalyst pellets. In fluidized bed, species diffusion is closely related to the flow regime in the reactor (Abba et al., 2003). Abba et al. (2003) summarized the formulae for calculating diffusion coefficients in different flow regimes in fluidized bed.

If the effective reaction rates and diffusion coefficients are known, the mass equation of reaction species i in fluidized bed could be written as following:

$$\frac{\partial\left(\varepsilon_g\rho_g m_{g,i}\right)}{\partial t}+\nabla\cdot\left(\varepsilon_g\rho_g\vec{v}_g m_{g,i}\right)=\nabla\cdot\left(\varepsilon_g\rho_g D_{g,i}\nabla m_{g,i}\right)+R_i. \quad (4)$$

When Eq. (4) is coupled with controlling equations of mass and momentum for gas phase and solid phase, the detailed flow-reaction–diffusion process in a MTO fluidized bed reactor can be simulated (Zhao et al., 2013). In practical applications, simplified models for two-phase hydrodynamics are also proposed (Abba et al., 2003; Bos et al., 1995; Zhang et al., 2012), in which the detailed flow patterns cannot be calculated but it is very efficient in the overall reactor performance evaluation.

Despite the controlling equations used in macroscale models for MTO process, another important issue needs to be considered is the reaction term. In MTO process, the coke deposits increasingly on catalyst particles with TOS, and the reaction rate changes correspondingly. When the coke accumulates to a certain amount, the activity of catalyst will become lower. In real industrial MTO unit, the spent catalyst with certain coke deposition in the reactor is transported to regenerator via standpipes. And a large portion of coke on the catalyst is burned off in the regenerator, and these regenerated catalyst particles, which have sufficient reaction activity, then will be returned to the MTO reactor. In fact, the catalyst particles in MTO reactor stay in a dynamic balance due to the continuous outflow of spent catalyst and inflow of regenerated catalyst. This indicates that the residence time of catalyst in the reactor shows a certain distribution. Note that the coke content on catalyst varies proportionally with residence time of the catalyst, the coke content on catalyst also demonstrates a distribution in MTO reactor, which means that different catalyst particle in the reactor may have different coke content. Therefore, the correct formula of reaction term in mass equation should be expressed as $\int_{C_{C,\text{ini}}}^{C_{C,\max}} p\left(C_C, C_{C,\text{ini}}\right)\cdot r_i(C_C)\mathrm{d}C_C$, where $C_{C,\text{ini}}$ is the initial coke content on catalyst, $C_{C,\max}$ is the maximum coke content on catalyst, and $p(C_C, C_{C,\text{ini}})$ is the probability density function of coke content on catalyst. The variable C_C is the coke content on catalyst particle, and function $r_i(C_C)$ is the reaction rate of species i. By assuming that $p(C_C, C_{C,\text{ini}})$ is independent of space coordinates, we can derive the following expression based on the population balance theory:

$$p(C_C, C_{C,\text{ini}}) = \frac{1}{\tau \cdot r_{\text{coke}}(C_C)} \exp\left(-\int_{C_{C,\text{ini}}}^{C_C} \frac{1}{\tau \cdot r_{\text{coke}}(C_C)} \mathrm{d}C_C\right) \quad (5)$$

where τ indicates residence time of catalyst particles in MTO reactor.

Another approach accounting for the influence of coke content distribution is to use the age distribution of catalyst particles (Bos et al., 1995; Zhang et al., 2012). Suppose that the catalyst particles are perfectly mixed, the age distribution of catalyst particles is

$$E(t) = \frac{1}{\tau} \exp\left(-\frac{t}{\tau}\right) \quad (6)$$

where t is time. Then average rate constant and average coke content can be described as in the following forms:

$$\bar{k}_i = \int_0^{\infty} k_i(C_C(t))E(t)dt \quad (7)$$

$$\bar{C}_C = \int_0^{\infty} C_C(t)E(t)dt \quad (8)$$

Both approaches on reaction terms discussed above could give the same results. However, the coke distribution approach based on the population balance theory shows the physical picture in a much clear way. By introducing coke content distribution or age distribution of catalyst particles in the controlling equations, the flow-reaction–diffusion process in MTO reactor could be simulated more realistically.

4.3 Mesoscale Modeling for Reaction–Diffusion in Catalyst Pellet

Apparently, there lacks an explicit link between the microscale and macroscale models discussed above. In this section, a mesoscale model is introduced to describe the reaction–diffusion in a single catalyst pellet. The significance of this model can be embodied at least in two aspects: a necessary link between the microscale model and macroscale model and the theoretical basis for MTO catalyst design optimization.

Before discussing the reaction–diffusion process over a single catalyst pellet, we should focus on the structure of the catalyst pellet first, as it is critical for the internal reaction–diffusion process. As mentioned above, an MTO catalyst pellet is composed of zeolite regions and catalyst support (or matrix) regions. The support regions have meso-/macropores. And the zeolite regions are composed of small microporous crystal particles, which normally have micropores and are distributed discretely in the support regions. Figure 8 shows the sketch of the structure of an MTO catalyst pellet. Reactions occur mainly in the zeolite regions, and species transport occurs in both zeolite and catalyst support regions. However, the diffusion mechanisms are different due to the difference of pore sizes. For microporous crystal particles, surface diffusion of adsorbed molecular species along the pore wall surface is dominant. For meso-/macropores, the bulk or molecular diffusion and Knudsen diffusion are critical if no strong adsorption exists. When there is a net change in the number of moles inside the porous catalyst pellet, the internal pressure gradient is always not negligible and this pressure gradient leads to viscous or Darcy flow (Krishna and Wesselingh, 1997). Figure 9 shows the schematic diagram.

The coefficients of bulk diffusion combined with Knudson diffusion in meso-/macropores is about 4 to 6 orders of magnitude higher than that of the surface diffusion in micropores, which means a large difference between the diffusion resistances of these two regions. It also implies that the difference of characteristic time of these two regions could be several orders of magnitude. The convenient approach to solve such a problem is to describe the reaction–diffusion processes in these two regions by two separate set of PDEs, which are coupling by the continuity of mass and heat flux for the same species at the interfaces.

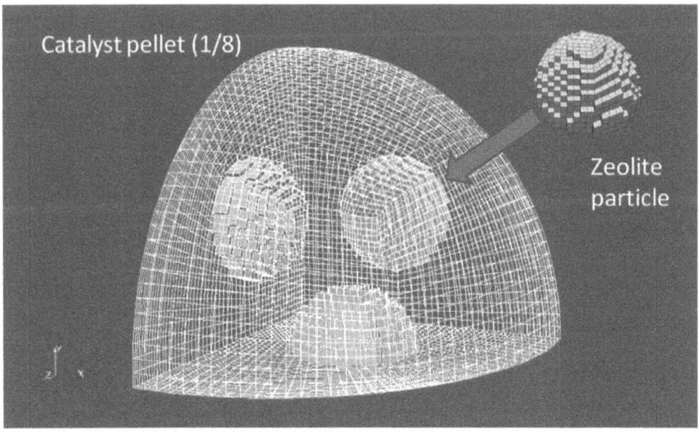

Figure 8 A catalyst pellet (1/8) formed with microporous zeolite particles and support regions.

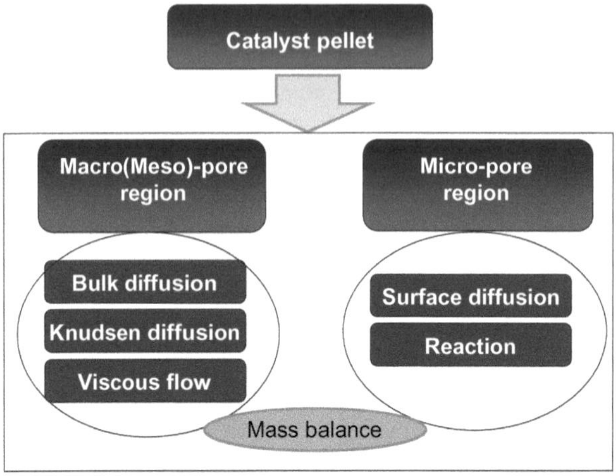

Figure 9 The schematic diagram of the mesoscale model for a single catalyst pellet formed with microporous zeolite particles and support regions (macro-/mesopore region).

For the zeolite regions, the reaction–diffusion process could be described by microscale model, as introduced above, since in these regions are mostly zeolite crystal particles. And for the support regions, the mass equation could be written as:

$$\varepsilon \frac{\partial c_i}{\partial t} = -\nabla \cdot \overrightarrow{\mathbb{N}_i} + \mathbb{r}_i \qquad (i = 1,2, \ldots, n) \tag{9}$$

where ε (in dimensionless) denotes porosity of catalyst support, c_i is the concentration of species i, $\overrightarrow{\mathbb{N}}_i$ is the molar flux of species i, and $\mathbb{r}_i$ is the reaction rate of species i. The value of $\mathbb{r}_i$ in the whole support region should be set to zero if the surface reactions are ignored. When incorporating the surface reactions, $\mathbb{r}_i$ should be set as zero in most cells. And only in the cells which are adjacent to the microporous crystal particles $\mathbb{r}_i$ needs to be calculated. The fluxes, $\overrightarrow{\mathbb{N}}_i$, could be given by (Li et al., 2015a):

$$\begin{pmatrix} \overrightarrow{\mathbb{N}}_1 \\ \overrightarrow{\mathbb{N}}_2 \\ \vdots \\ \overrightarrow{\mathbb{N}}_n \end{pmatrix} = -[\mathbb{B}^e]^{-1} \begin{pmatrix} \frac{1}{\mathrm{RT}} \nabla p_1 \\ \frac{1}{\mathrm{RT}} \nabla p_2 \\ \vdots \\ \frac{1}{\mathrm{RT}} \nabla p_n \end{pmatrix}. \tag{10}$$

The numerical methods have been developed for solving two sets of PDEs for two regions. This mesoscale model for a single catalyst pellet is called *multiregion model.*

The mesoscale model is of significant importance in catalyst design, since it could be used to investigate the effect of size, distribution, and amount of microporous crystal particles in the catalyst pellet on the overall catalytic performance. For convenience, as Eq. (3) in microscale model, another internal effective factor is defined to quantify catalytic performance of the pellet:

$$\eta_{\text{pellet},i} \equiv \frac{\int_{V_Z} r_i \mathrm{d}V + \frac{c_u^s}{c_u^v} \int_{S_Z} r_i \mathrm{d}S}{r_i^b V_{\text{pellet}}}. \tag{11}$$

It is meaningful to examine the relation between microscale model, mesoscale model, and micromodel. For reaction kinetics, microscale and mesoscale models adopt the same kinetics that based on element reaction system. For diffusion, mesoscale model embodies two diffusion mechanisms (one for micropores and another for mesopores and macropores), and microscale model considers one diffusion mechanism since it only has micropores. No diffusion was considered within the macropores. It is obvious that the mesoscale model possesses the same theoretical foundation as the microscale model, but its application scope has been enlarged compared to the microscale model. Therefore, it could be reliably used as a tool to derive some parameters, such as effective chemical kinetics and effective diffusion parameters, for macroscale model. In the section following, we discuss the method on how to link the microscale kinetics to the lumped macroscale kinetics via the mesoscale modeling approach.

4.4 Mesoscale Modeling: Linking the Microscale Kinetics and Macroscale Lumped Kinetics

4.4.1 Microscale Kinetics for MTO Reaction

In microscale model, the reactions generally refer to elementary reaction steps. The reaction network is closely related to the reaction mechanism and could be well obtained by quantum chemistry or *ab initio* calculations. The corresponding parameters, such as pre-exponential factors and activation energies, could be predicted based on transition state theory (TST) or variational transition state theory (VTST).

Although the MTO mechanism has been a topic of intensive research, the kinetics for elementary reaction steps is still a big challenge. The reaction

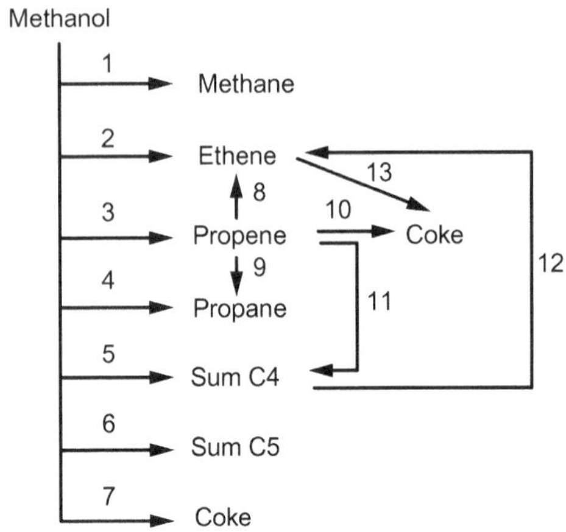

Figure 11 Kinetic scheme for MTO process over SAPO-34 catalyst by Bos et al. (1995).

of 12 reactions involving six species lumps plus coke (see Fig. 11), and the effect of coke on both the activity and selectivity is considered by using an empirical correlation for rate constants:

$$k_i(C_C) = k_i^0 e^{-\alpha_i C_C}, \tag{12}$$

where C_C is the coke content on catalyst. The effect of coke on the selectivity was predicted by taking different values for the empirical constant α_i. Another kinetic model for MTO process over SAPO-34 catalyst was proposed by Gayubo et al. (2000), in which the effect of water on activity and selectivity was included. Meanwhile, the model is further simplified by considering four individual steps for the production of ethylene, propylene, butylenes, and remaining hydrocarbons. The effect of water is described by multiplying a fraction term on rate constants:

$$\theta_W = \frac{1}{1 + K_W X_W} \tag{13}$$

where K_W indicates the resistance to species formation due to the presence of water and X_W represents the weight fraction of water. Recently, Ying et al. (2015) developed seven-lumped kinetic model for industrial catalyst in DMTO process. They proposed a new function to quantify the effect of coke on the activity and selectivity of DMTO catalyst

$$\phi_i = \frac{1}{1 + A\exp(B \times (C_C - D))} \exp(-\alpha_i C_C). \tag{14}$$

Here, A, B, D, and α_i are empirical constants.

In actual applications, a simpler kinetic model is more favored for large-scale reactor simulation, suppose that it can adequately describe the reaction and transport process in the reactor. The simplification is normally based on the experimental findings. For example, Gayubo et al. (2000) reduced the number of reactions from eight to four by eliminating some reaction steps with slow reaction rates. Therefore, the application scope of the macroscale reaction kinetics is highly dependent on the experimental conditions studied.

4.4.3 Mesoscale Modeling: Linking the Microscale Kinetics and Lumped Kinetics

As mentioned above, in the macroscale kinetics, the reaction rate constants are ensemble averaged, and, in most of cases, representing gross effect of reaction rates for elementary steps and the corresponding mass transfer. This can be well explained by Eqs. (3) and (11). If the surface reactions are negligible, $\eta_{Z,i}$ and $\eta_{\text{pellet},i}$ could be viewed as the ratios of volume-averaged reaction rate to elementary step rate for zeolite crystal particles and catalyst pellet, respectively. The $\eta_{Z,i}$ and $\eta_{\text{pellet},i}$ become smaller if the diffusion is of increasingly importance. So the microscale kinetics in most situations may not be directly compared to the macroscale experimental results, if the internal and external diffusion cannot be accounted for. Using catalyst of smaller size might reduce the diffusion resistance to certain extend in kinetic study.

The mesoscale multiregion model discussed above may open a way to link the microscale kinetics to the macroscale kinetics. The macroscale kinetics derived from microscale kinetics at least ensures that the reaction mechanism at the microscale can be correctly reflected. As mentioned by Campbell (1994), knowing a mechanism can give an intelligent way to extrapolate kinetics to unknown conditions. As far as we know, there is no MTO macroscale kinetics at present derived directly from microscale kinetics.

It is also possible to carry out detailed study on the reaction and diffusion process at catalyst pellet scale. Different diffusion–reaction mechanisms, as well as the microscale kinetics can be implemented in the zeolites and support regions through the multiregion model, as described above. By considering the realistic size and distribution of zeolite crystal particles, we can simulate the reaction–diffusion inside a single catalyst pellet. A direct numerical simulation (DNS) approach is also under development in the authors' group to study the catalyst–gas interaction. The mesoscale multiregion modeling approach, if coupled with the DNS method (Van der Hoef et al., 2006), may eventually compute the lumped kinetic parameters. We would stress, however, the application of the mesoscale model approach in MTO process is still far from being reached because of the complexity of MTO reaction.

5. COKE FORMATION AND CONTROL FOR MTO PROCESS

MTO reaction mechanisms over SAPO-34 catalyst have been investigated by many researchers (Olsbye et al., 2012). The formation of coke species can be readily occurring in MTO process over SAPO-34 catalyst. Here, the species of coke refers to the carbon depositions that can plug the pore and cover the active centers. However, in acidic molecular sieve-catalyzed reactions including methanol transformation reaction, the molecules cannot diffuse out of the channels and stay in the channels or cages if the molecular size is too large or the molecules have strong proton affinity. On one hand, this prompts the selectivity to smaller molecules that can diffuse out of the channels, for example ethylene. On the other hand, the accumulation of large molecules components would limit the mass transfer of reactant and accelerate the coverage of acidic centers, causing side reactions and catalyst deactivation. Apparently, coke formation is not only the major cause of deactivation in MTO reaction but also decisive to the light olefins selectivity. The understanding of coke formation and its control in MTO process is of practical importance. But the coke formation over zeolite catalyst is essentially a multiscale phenomenon. Froment (1997) proposed that the coke formation and catalyst deactivation could be studied at three scales: microscale for activity center such as acid center inside channel, pore or cage of zeolites; mesolevel for topology of zeolite such as pore, channel, or cage; and macrolevel for reactor.

5.1 Coke Formation at Microscale: Effect of Acidity of Catalyst

The acidity of the catalyst has a significant effect on the deactivation of the catalyst. Researches by Wilson and Barger (1999), Mores et al. (2011), and other research groups (Dahl et al., 1999a,b; Haw, 2002; Stocker, 1999; Yuen et al., 1994; Zhu et al., 2008) discovered that the catalyst with a strong acidity and higher acid center density has higher deactivation rate in MTO reaction. Guisnet et al. (2009) related the catalyst acidity to chemical reaction rate, as shown in Fig. 12. The stronger the acidity of catalyst, the higher the reaction rate in MTO process. In this case, the coke precursor formation, and thus, the coke deposition accelerates. This in turn prompts catalyst deactivation. In addition to the acidity, the density of acid sites on

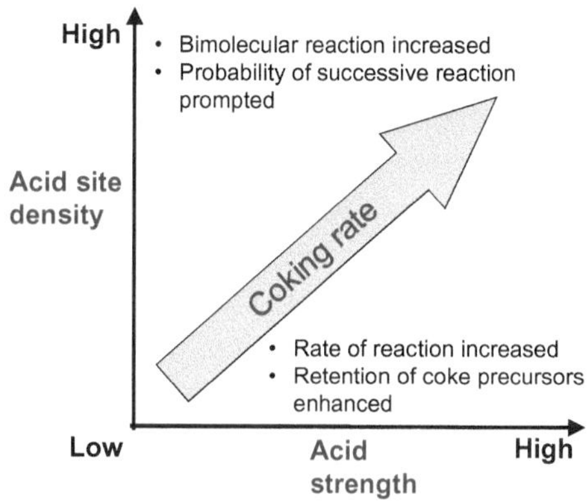

Figure 12 Influence of the acidity and acid site density in catalyst on the rate of coking.

the catalyst also plays a crucial role in catalyst deactivation. A higher density of acid sites normally means shorter distance between two adjacent acid centers. In this case, the reactant molecules under diffusion in channels and cages may have higher probability to be absorbed by acid centers, react, and generate coke species. Therefore, the catalyst deactivation rate is also enhanced with a higher density of acid sites.

Yuen et al. (1994) have compared the methanol conversion over SAPO-34 and H-SSZ-13 catalyst, which have same CHA topology, density of acid sites, and grain size. The results show that H-SSZ-13 has higher deactivation rate owning to its higher acid strength. Under the same temperature, the coke species on SAPO-34 catalyst are mainly methyl benzene and methyl naphthalene, while on H-SSZ-13, the monocyclic, bicyclic, and tricyclic aromatic hydrocarbons and their derivatives are observed. It is suggested that the polycyclic aromatic hydrocarbons can be generated more easily in methanol conversion over H-SSZ-13 than that over SAPO-34 due to the difference of acid strength of zeolites.

5.2 Coke Formation at Mesoscale: Effect of Topological Structure of Zeolites

In methanol conversion over molecular sieves having three-dimensional pore and cage structures, methyl benzene is recognized as a reactive intermediate that can enhance the reaction rate, and the bicyclic aromatic hydrocarbons show very low activity in prompting methanol conversion. Actually, both the bicyclic and polycyclic aromatic hydrocarbons can be coke species during catalyst deactivation in MTO process over molecular sieves (Wei et al., 2012a). The formation of polycyclic aromatic hydrocarbons requires a certain spatial structure inside the molecular sieves. Therefore, the formation of main components of coke should be closely related to the topology of molecular sieves in MTO process.

The different topology structures of ZSM-5 and SAPO-34 lead to different deactivation modes during the conversion of MTO. ZSM-5 molecular sieve has smaller 10-ring cross channel, which cannot provide sufficient space to allow the formation of macromolecular bicyclic aromatic hydrocarbons and polycyclic aromatic hydrocarbons. The channels can only allow generating relatively small number of methyl benzene. And these species are able to diffuse from 10-ring pore channel to the gas phase. The channel structure of ZSM-5 decides the coke species generation (Bjorgen et al., 2007). The deactivation of ZSM-5 in MTO reaction is mainly caused by coke deposition on outer surface (Fig. 13A; Guisnet et al., 2009).

SAPO-34 has small 8-ring channels and big cage structure. Methyl benzene, the intermediates in MTO process, can be converted to polycyclic aromatic hydrocarbons in the cage. As the polycyclic aromatic hydrocarbons have large volume and occupy most of the space in the cage, which hinders the contact between methanol molecules with the active sites and reduces the mass transfer rate substantially (as shown in Fig. 13B), the catalyst will lose the activity rapidly (Haw et al., 2003). The study of Hereijgers et al. (2009)

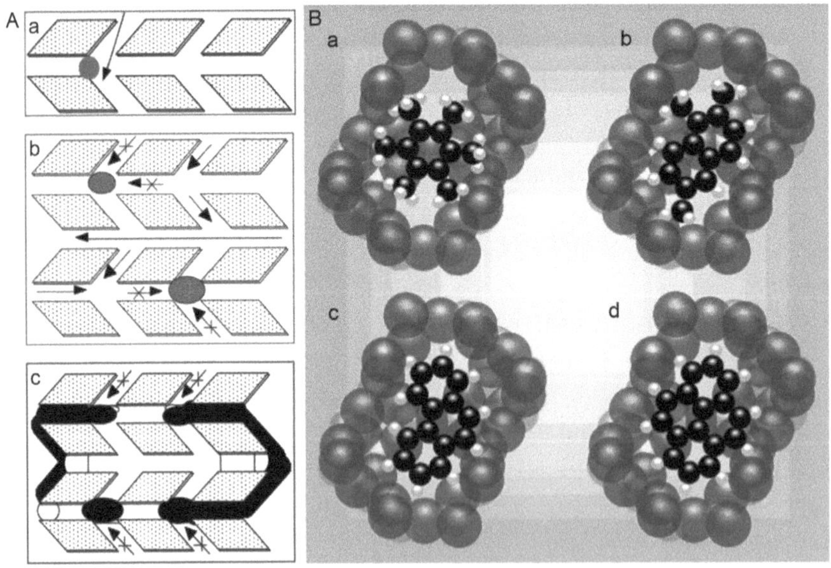

Figure 13 Deactivation process in MTO reaction over ZSM-5 (A) (Guisnet et al., 2009) and SAPO-34 (B) (Haw et al., 2003) zeolites.

also supports that low diffusion rate can cause the deactivation of SAPO-34 catalyst. Due to the restriction of the opening of 8-ring window, it is difficult for large molecules to move out of the channels in SAPO-34 zeolites, and thus, the heavy compounds of coke can be readily accumulated in MTO reaction. Mores et al. (2008) have studied the coke formation in SAPO-34 and ZSM-5 zeolites in methanol conversion process by use of *in situ* measurement techniques and confirmed the difference between these two catalysts.

Bleken et al. (2011) have compared the methanol conversion of four kinds of catalysts with 10-membered ring three-dimensional pore structure (i.e., IMF, TUN, MEL, and MFI). They found that although all catalysts have 10-ring cross channel, but there are differences between them in terms of life time and coke compositions. Since the cross channels in IMF and TUN structure have wide space near the cross, which allows the formation of heavy coke species, the zeolites (IM-5 and TUN-9) having IMF and TUN structure appear rapid deactivation in methanol conversion reaction. On the contrary, zeolites with MEL and MFI structure (such as ZSM-11 and ZSM-5), in which the space near the cross is relatively narrow and limits the formation of coke compositions, have a long life time in the methanol conversion reaction. In this case, there are no heavy coke compositions in the channels, and the deactivation is mainly caused by the coke formation on external surface of zeolites.

5.3 Coke Formation at Mesoscale: Effect of Reaction Temperature

Schulz (2010) found that in MTO process over H-ZSM-5 catalyst, the operation temperature affects catalyst life time and deactivation mechanism significantly. At temperature of 543–573 K, large volume methyl benzene molecules (three methyl isopropyl ethyl benzene and two methyl benzene) may form and occupy the pores in H-ZSM-5 zeolites. While at a higher temperature of 625 K, these large volume methyl benzenes would be cracked into olefins and small volume benzene, and methyl benzenes, which may cause catalyst deactivation, do not exist in the channels of H-ZSM-5. When the temperature becomes higher than 625 K, coke generated at the external surface of H-ZSM-5 catalyst will be the dominant reason for catalyst deactivation. Bleken et al. (2009) have studied methanol conversion reaction and coke generation on two CHA cage structure zeolites, SAPO-34 and H-SSZ-13. They found that, by altering reaction temperature, the life time, coke content, and coke species show similar trend for these two zeolites, although the acidic strength is different for Si–Al and Si–P–Al zeolites. If the reaction is kept at 573 K for 25 min,, the coke content of SAPO-34 zeolites can grow to 16%. When the reaction temperature is further increased to 673 K, the coke content drops to 6% in SAPO-34 zeolites. Under the same reaction conditions, the coke content in H-SSZ-13 zeolites first increases to 20% at 573 K, and then goes down to 9% at 673 K. Apparently, there is an optimized operation window for reaction temperature, which is from 573 to 698 K for both SAPO-34 and H-SSZ-13 catalysts. In this window, both catalysts have a relatively long life time. At an operation temperature departed from this range, catalysts may deactivate more quickly.

Yuan et al. (2012) investigated methanol conversion reaction and coke deposition over SAPO-34 catalyst in a microscale fluidized bed reactor, which presented some interesting results in their temperature-programmed experiments (Yuan et al., 2012). As shown in Fig. 14, methanol was fed to the reactor at 250 °C, but the hydrocarbon products generated in the temperature range of 250–300 °C is negligible. The conversion of methanol increased from temperature of 300 °C, and reached a peak conversion at 325 °C and then dropped until 350 °C. When the temperature further rose from 350 °C, the conversion of methanol increased continuously. In order

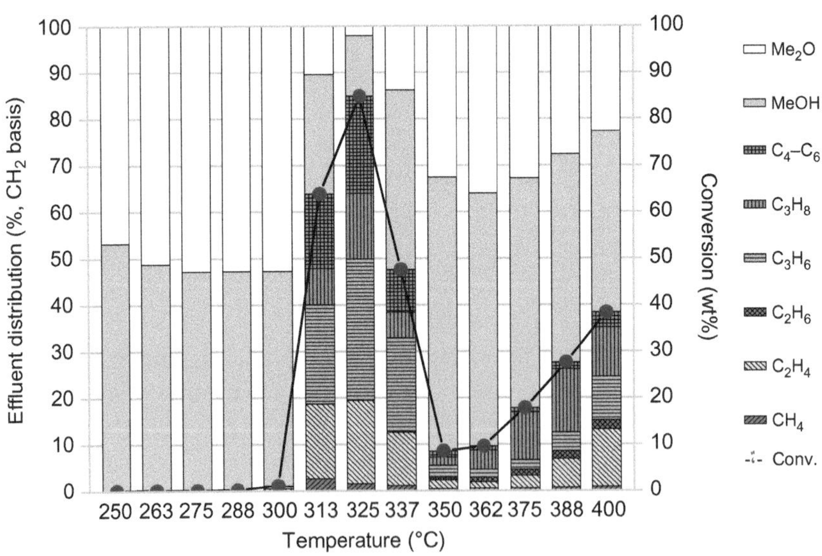

Figure 14 Effluent distribution of MTO process in a microscale fluidized bed reactor with temperature programmed (Yuan et al., 2012).

to understand this phenomenon, Wei et al. (2012a,b) studied the MTO reaction at constant temperatures, with a focus on the coke deposition and catalyst deactivation. Their results showed that methanol conversion reaction had distinguished characteristics at high and low temperature over SAPO-34 catalyst. At low temperature (<350 °C), the reaction is featured by an induction period, which may cause a fast deactivation of the SAPO-34 catalyst. The duration of the induction period is dependent on the reaction temperature. The higher the reaction temperature, the shorter the induction period. At high reaction temperature (>350 °C), catalyst deactivation is caused by rapid coke deposition. But in the range of 400–450 °C, the SAPO-34 catalyst has low coke formation rate and long life time, as shown in Fig. 15 (Wei et al., 2012b). The organic species of coke in the cage of SAPO-34 zeolites at high temperature are mainly fusing ring aromatic hydrocarbons, which attributes to the rapid deactivation of catalyst at high temperature but does not present at low temperature. Further researches show that at low temperature saturated alkane products such as adamantane compounds appears as coke species, as shown in Fig. 16 (Wei et al., 2012a). These adamantane compounds are different from methyl benzene and other aromatic coke species formed at high temperature. The latter can act as the hydrocarbon pool with functions of assembling C_1 compounds to higher hydrocarbons. These saturated naphthenic hydrocarbons occupy the cage and limit mass transfer of methanol molecules. Thereby, it suppresses the continuous formation of hydrocarbon pool species and results in rapid deactivation at low temperature.

Based on the results at constant temperature, it is possible to explain the peak conversion at 325 °C in the temperature-programmed experiments, as

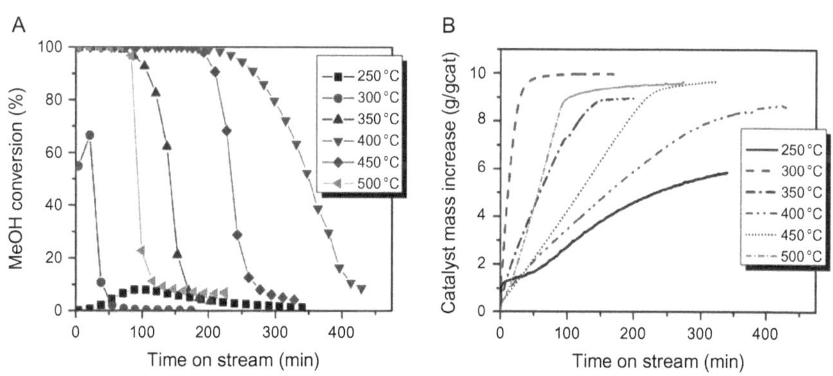

Figure 15 (A) Methanol conversion (■ 250, • 300, ▲ 350, ▼ 400, ◆ 450, and ◀ 500 °C) and (B) real-time catalyst mass increase observation (Wei et al., 2012a).

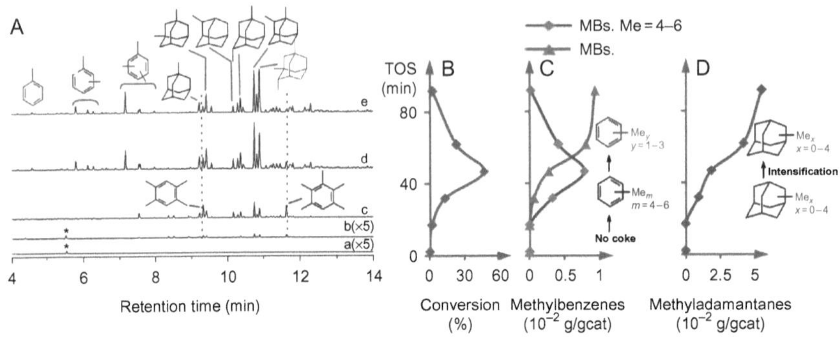

Figure 16 GC–MS analyses (A, left) of confined organics after methanol conversion at 300 °C for 17 (a), 32 (b), 47 (c), 62 (d), and 92 min (e). Methanol conversion with time on stream (B and C, middle) and confined methylbenzenes and methyladamantanes variation with time on stream (D, right) (Wei et al., 2012b).

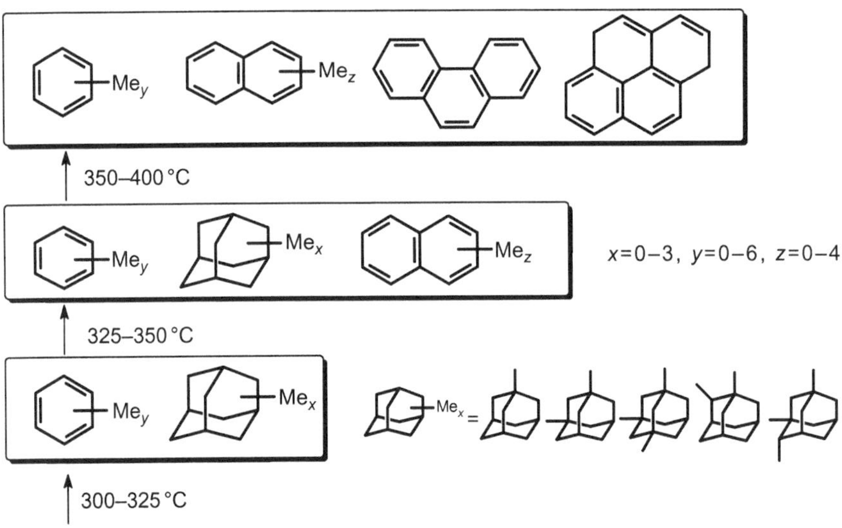

Figure 17 Coke species evolution in the temperature-programmed methanol conversion over SAPO-34 (Yuan et al., 2012).

shown in Fig. 14. At low temperature, the coke species is mainly adamantane species, and the formation and accumulation of which can lead to a rapid deactivation of catalyst. This species cannot act as hydrocarbon pool. When gradually increasing the reaction temperature, the adamantane compounds generated at low temperature are converted to naphthalene derivatives, and eventually form fused ring aromatic hydrocarbons phenanthrene and pyrene (Fig. 17). The coke species inside SAPO-34 zeolites changes with temperature (Yuan et al., 2012), which complicates our understanding on coke formation in MTO process.

5.4 Coke Formation at Macroscale: Effect of Selectivity to Light Olefins

As discussed above, coke formation affects the selectivity to light olefins in MTO process over SAPO-34 catalyst. It has been found that at a given temperature, the ethylene-to-propylene ratio in MTO reaction is increased when coke content in catalyst increases (Barger, 2002; Song et al., 2001). Figure 18 shows the typical results in a microscale fluidized bed reactor at temperature of 450 °C and weight hourly space velocity (WHSV) of 1.5 h^{-1} without catalyst regeneration. As can be seen, when the coke content on catalyst increases from 2% to roughly 8%, the selectivity to ethylene increases from 37% to 48% while the conversion is still sufficiently high (>98.5%). The selectivity to propylene keeps almost unchanged. Thus, a gain of selectivity to light olefins can be achieved, with certain coke content on catalyst.

Two hypotheses have been proposed to explain this behavior: the first is that the accumulation of coke suppresses the free space in the cavities of zeolites thus limits the formation of methyl benzenes with 5–6 methyl groups, which thereby favors the ethylene formation. The second is that the diffusion of large product molecules from the cavities is hindered by partial blockage of pores and opening windows access to the cavities due to coke formation. Only smaller molecules such as ethylene can freely pass through

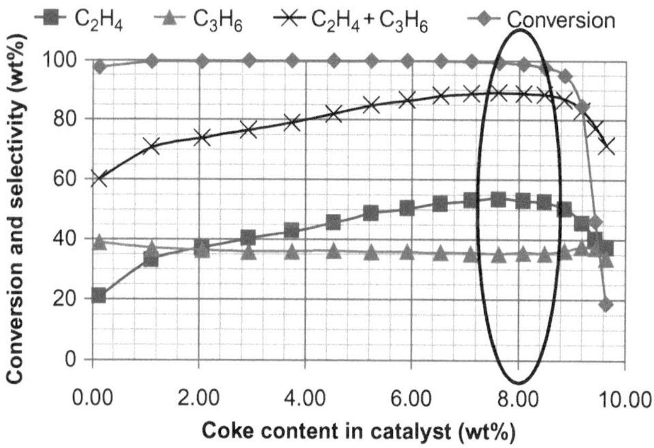

Figure 18 The methanol conversion and light olefins selectivity under different coke content on catalyst. Reaction temperature 450 °C, WHSV = 1.5 h^{-1}.

the partially blocked pores in SAPO-34 zeolites. Chen et al. (2007) studied the influence of coke content on the selectivity to different hydrocarbons. Their results clearly showed that the selectivity to ethylene increases with increasing coke content, and the selectivity to the following products drops to different extend with an increasing coke content: $C_6 > C_5 > C_4$. Dahl et al. (1999a,b) studied the product shape selectivity to olefins on different-sized crystals by use of ethanol and 2-propanol as probe molecules. It has been demonstrated that the diffusion of molecules is slow if the size of molecules is comparable to the channel size, which is named configuration diffusion. Dahl et al. (1999a,b) also found that ethanol conversion was not limited by the ethanol diffusion while 2-propanol conversion was controlled by 2-propanol diffusion. Barger (2002) has also proposed product shape selectivity by comparing the measured ethylene-to-propylene ratio with thermodynamically predicted ratio in the gas phase at different temperatures. It was found that the measured ethylene-to-propylene ratio was much higher than the thermodynamically equilibrium ratio.

5.5 Coke Control at Macroscale: Optimize the DMTO Fluidized Bed Reactor Design and Operation

In pilot-scale DMTO fluidized bed reactor, due to circulation of catalyst between reactor and regenerator, there exists a certain distribution of residence time of catalyst. Thus, the coke content on catalyst also shows a certain distribution, as shown in Fig. 7. But from Fig. 18, it is participated that, for a given temperature, an optimal value of coke content can be identified by which the selectivity to light olefins is maximized. Therefore, catalyst particles with coke content either higher or lower than the optimal value might lead to a lower selectivity to light olefins; the overall selectivity is then affected by the coke distribution in DMTO circulating fluidized bed reactor.

5.5.1 Counter-Current Fluidized Bed Configuration

In pilot-scale DMTO fluidized bed reactor, the regenerated catalyst normally has very low coke content. As discussed above, such catalyst may not favor the selectivity to light olefins. Therefore, a counter-current fluidized bed configuration is adopted. In this configuration, the regenerated catalyst is injected into the reactor via catalyst distributor from the top of the dense bed, and the coked catalyst is taken from the draw-off bin beneath the gas distributor at the bottom. Thus, the methanol feed from the gas distributor first contacts the coked catalyst, by which a higher selectivity to light olefins can be achieved. Our experimental results confirmed that the yield of light olefins can increase by 5% when a counter-current configuration is used.

5.5.2 Minimize the Induction Period

At low temperature (<350 °C), the DMTO reaction is featured by an induction period, which may cause a fast deactivation of the SAPO-34 catalyst. Thus in the real operation, the induction period has to be avoided. Normally, the higher the reaction temperature, the shorter the induction period. Basically above 350 °C, the induction period can be minimized. Therefore, in the start-up of the DMTO reactor, the catalyst should be heated to above 350 °C before feeding methanol. Note that the organic species of coke in the cage of SAPO-34 zeolites at high temperature are mainly aromatics, which however are not found at low temperature; it is expected that the introduction of aromatics can shorten the induction period at lower temperature. Qi et al. (2015) showed that the induction period could be remarkably shortened by adding only 4 ppm of aromatics. This is very meaningful for DMTO reactor operation.

The coke formation is critical for MTO reaction over SAPO-34 catalyst. The influence of coke formation is twofold: a certain amount of coke deposition can prompt the selectivity to light olefins, while it also makes the catalyst deactivate rapidly. Thus, the understanding of the coke formation at microscale is extremely important for controlling coke distribution in the reactor. The influences of zeolite structure and reaction temperature on coke formation have been discussed to illustrate the essence of the mesoscale researches. However, there is still a lot of work to be explored at mesoscale concerning the coke formation. These results are eventually expected to benefit the reactor design and operation.

6. DMTO FLUIDIZED BED REACTOR SCALE-UP

Scale-up of fluidized bed has long been considered as a big challenge in catalytic reactor development (Knowlton et al., 2005; Matsen, 1996; Rüdisüli et al., 2012). On one hand, good understanding of the chemistry must be obtained. This includes the reaction mechanism, reaction kinetics, and catalysis behavior. On the other hand, the influence of hydrodynamics on the reactor performance plays another critical role (Knowlton et al., 2005). The hydrodynamics in fluidized bed has inherent multiscale nature. The fluidization behavior of catalyst can vary significantly with the change of the physical properties (i.e., size and density) of catalyst particles and fluidizing gas (Geldart, 1973). The mild distribution of catalyst particles at microscale may lead to dynamic mesoscale heterogeneous structures such as bubbles and clusters, which are closely related to the fluidized bed size as well as the operation conditions. These dynamic mesoscale heterogeneous structures can further affect macroscale mixing and mass transfer in fluidized bed reactor (Sundaresan, 2013). In view of these complexities, scale-up of a new fluidized bed process, at current stage, is still an engineering practice rather than an exact science (Knowlton et al., 2005; Matsen, 1996; Rüdisüli et al., 2012).

DMTO fluidized bed reactor has been scaled up via experiments at four different scales (Tian et al., 2015). The scale factor between two adjunct scales is roughly 100 in terms of methanol feed rate, and 10 in terms of reactor diameter. The microscale fluidized bed was operated at bubbling fluidization regime without catalyst circulation. In the pilot-scale experiments, the circulation between reactor and regenerator was established. In the demonstration and commercial scale, the turbulent fluidized bed reactor has been selected in order to achieve a high feed throughput. One thing determined at the early stage of DMTO process development is that the DMTO catalyst particles should have similar physical properties as FCC catalyst particles. In this way, both DMTO catalyst and FCC catalyst are typical Geldart type A particles, which could maintain good fluidity in the fluidized bed operation. In previous reviews, such as Matsen (1996), Knowlton et al. (2005), and Rüdisüli et al. (2012), the challenges and general methods for scaling up fluidized bed reactor were summarized. Especially, these reviews specially focused on the influence of hydrodynamics in fluidized bed reactor scale-up. In this section, we intend to share our method on the analysis of the results at microscale and how to relate the microscale results to the design of pilot-scale fluidized bed reactor.

6.1 Microscale MTO Fluidized Bed Reactor

In the MTO process development, the purpose of microscale-MTO reactor experiments is threefold: (1) evaluation of the catalyst performance, (2) study of the influence of reaction conditions, and (3) exploration of the optimal reaction conditions that are used for pilot-scale reactor design. Due to the rapid deactivation of SAPO-34 catalyst and high exothermicity in MTO reaction, the fluidized bed reactor has been considered as the most suitable MTO reactor. In the laboratory, microscale-MTO fluidized bed reactor was

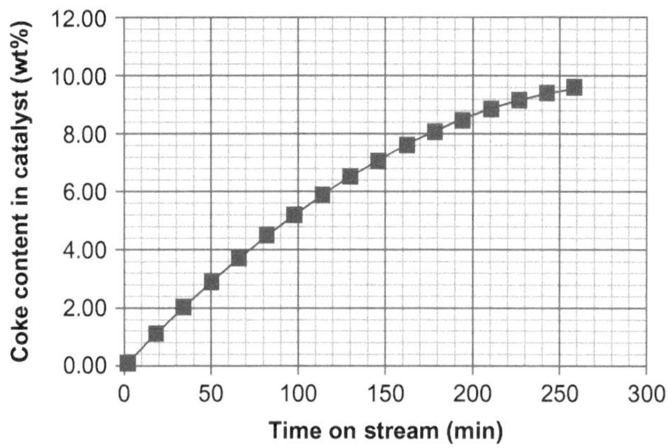

Figure 19 The coke content in catalyst as function of time on stream in microsale fluidized bed reactor. Reaction temperature 450 °C, WHSV = 1.5 h^{-1}.

constructed to study the reaction and kinetics. The notable difference between microscale fixed bed reactor and fluidized bed reactor is that the catalyst particles can be fluidized and well mixed in the latter. In the absence of circulation, the catalyst in microscale fluidized bed shows a uniform residence time distribution. And thus, the spatial distribution of coke content in catalyst, at any given time, can be considered as uniform. Figure 19 shows the typical results of coke content in catalyst (defined as the percentage of the mass of coke to the mass of catalyst) as a function of TOS. The corresponding methanol conversion and selectivity to ethylene and propylene are depicted in Fig. 18. Since there was no online regeneration, the coke content in catalyst experienced a continuous increase with the TOS in the microscale fluidized bed reactor until a considerably low conversion of 18.82% is achieved. A rapidly drop of methanol conversion was found at 210 min with a coke content in catalyst of 8.87%, which indicates that the activity of catalyst started to decline. In order to connect the results from microscale fluidized bed reactor to the pilot-scale fluidized bed reactor, important parameters, i.e., coke content in catalyst, coke formation rate, and catalyst-to-methanol ratio, have been studied.

6.1.1 Coke Content in Catalyst

In pilot-scale MTO experiments, the regeneration of spent catalyst deactivated in the MTO reactor was conducted in a continuous way by circulating catalyst from reactor to regenerator and vice versa. The activity of spent catalyst was then restored in the regenerator. When the steady circulation is established, and which is most likely the case in pilot experiments, the average coke content in catalyst becomes time independent. The optimal average coke content in catalyst has to be known *in prior* to maximize the selectivity to ethylene and propylene while maintaining high methanol conversion. Typical results were shown in Fig. 18. As can be seen, under the conditions studied, the average coke content in catalyst of around 7.6–8.5% is favorable in terms of selectivity to ethylene and propylene (ca. 88–89%) and methanol conversion (>98.5%).

6.1.2 Catalyst Residence Time in Reactor

As mentioned above, the coke content in catalyst around 7.6–8.5% is favorable in terms of selectivity to ethylene and propylene (ca. 88–89%) as well as methanol conversion (>98.5%). Further check with Fig. 19, we can find that this is corresponding to the TOS of 160–195 min. Under the operation conditions specified in Fig. 19; therefore, a catalyst residence time of 160–195 min is optimal for light olefins selectivity in pilot-scale experiments. It should be noted that the WHSV influence the catalyst residence time significantly. An estimation of the catalyst residence time is based on $RT_1 = \beta \cdot RT_2 \cdot \mathrm{WHSV}_2/\mathrm{WHSV}_1$, where RT_i is the optimal catalyst residence time under WHSV_i, and β is the coefficient obtained from experiments. A simple estimation can be made by assuming β as 1.

6.1.3 Coke Formation Rate

Note that coke deposition in the MTO catalyst can lead to the coverage of part of the active sites and reduce the catalyst activity. When the coke content is sufficiently high, the methanol conversion shows a rapid decrease and most of the active sites have been covered and the catalyst becomes deactivated. From Fig. 18, we can find that in a wide range of coke content (0–8.87 wt%) the catalyst in the microreactor can maintain a high methanol conversion (>98.5%). This implies that the complete conversion of methanol can be realized with a small mass of active catalyst (and thus, a small amount of active sites) in MTO reaction. Therefore, it is important to know the coke formation rate in MTO process. Based on the data reported in Fig. 18, we can estimate the coke formation rate:

$$\dot{c}(t) = \frac{\mathrm{d}}{\mathrm{d}t}(W_{\mathrm{RX}} \cdot C_{\mathrm{cat}}(t)) = W_{\mathrm{RX}} \frac{\mathrm{d}}{\mathrm{d}t} C_{\mathrm{cat}}(t) \tag{15}$$

where W_{RX} is the catalyst loading in the microscale reactor and $C_{\mathrm{cat}}(t)$ is coke content. The coke formation rate, normalized by the catalyst loading, is shown in Fig. 20. As can be seen, the coke formation rate declines with an

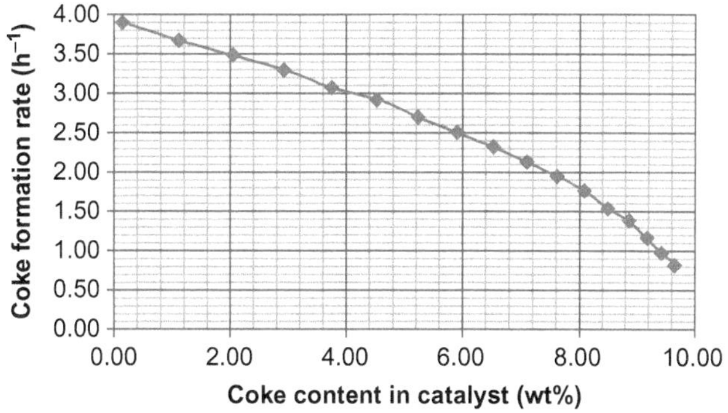

Figure 20 The coke formation rate as a function of the coke content. The coke formation rate is normalized by the catalyst loading, and the experimental conditions are the same as in Fig. 19.

increasing coke content in catalyst. This is not surprised since high coke content means more coverage of the active sites in catalyst, and thus a lower conversion of methanol. Lower conversion of methanol leads to a limited coke production.

6.1.4 Catalyst-to-Methanol Ratio

The third parameter of critical importance is the catalyst-to-methanol ratio. This parameter is the key to control the circulation rate of catalyst between reactor and regenerator in pilot-scale setup. Normally it is hard to derive the relation between the catalyst-to-methanol ratio and reaction results via direct measurement in the microscale experiments as there is no circulation. However, by analyzing the coke formation in the MTO reaction, we can predict the optimal catalyst-to-methanol ratio. From Eq. (15), we can obtain the coke formation rate. We assume that the coke formation rate can be directly used in the pilot-scale experiments. Thus, the mass flow rate of catalyst required to transport this amount of coke is estimated as following:

$$\dot{m}(t) = \frac{\dot{c}(t)}{C_{\text{cat}}(t)}. \tag{16}$$

And the cat-to-methanol ratio can be obtained as:

$$\text{CTM} = \frac{\dot{m}(t)}{\dot{Q}_m(t)} = \frac{\dot{c}(t)}{\dot{Q}_m C_{\text{cat}}(t)} \tag{17}$$

with $\dot{Q}_m(t)$ is the methanol feed rate. Figure 21 shows the methanol conversion and ethylene and propylene selectivity as a function of

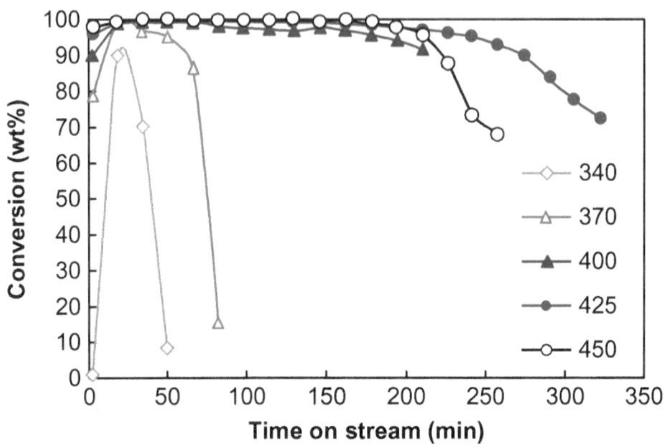

Figure 23 The methanol conversion as function of time in stream for different reaction temperature in the microscale fluidized bed reactor with a WHSV of 1.5 h^{-1}.

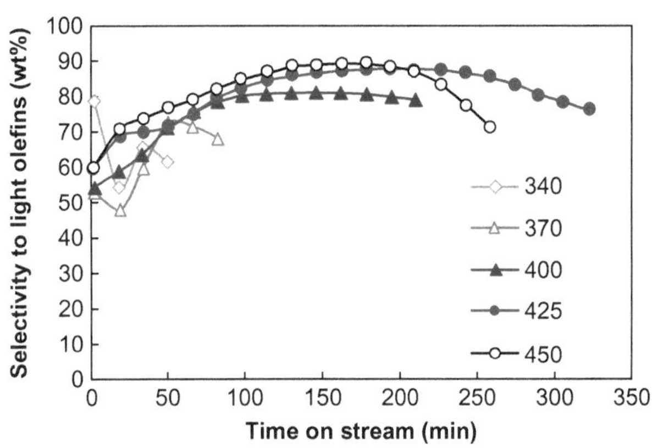

Figure 24 The selectivity to light olefins as function of time on stream for different reaction temperature in the microscale fluidized bed reactor with a WHSV of 1.5 h^{-1}.

370 °C, almost no steady period can be achieved in terms of methanol conversion. This can be explained as the longer induction period at a lower reaction temperature. Actually, the methanol conversion is only 1% at the initial reaction stage under reaction temperature of 340 °C. When the reaction temperature is above 400 °C, the duration of the steady state for methanol conversion increases to 200 min. Such a long steady period is beneficial for the operation in the pilot-scale reactor. Note that the product distribution may change when reaction temperature increases. The selectivity to light olefins (ethylene + propylene) can be maximized above 425 °C.

6.1.5.2 Gas–Solid Contact Time/Space Velocity

The catalyst–gas contact time plays an important part in designing the pilot-scale MTO reactor. Our experimental study confirms that complete methanol conversion can be reached even reducing the reaction contact time to 0.04 s. The contact time has little effect on the conversion is sufficiently high such that the induced period can be avoided. Basically, the shorter the contact time, the higher the selectivity to light olefins. Longer contact time will prompt the side reaction and formation ofby-products, which thus decreases the selectivity to light olefins and enhances the coke formation rate. Table 1 lists the methanol conversion and light olefins selectivity for different WHSV. As can be seen, when WHSV increases from 2 to 10 h^{-1}, the MeOH conversion only shows a negligible decline, from 100% to 99.3%. The selectivity to light olefins, on the other hand, increases 81.8–87.5%. This increase is mainly contributed by the ethylene. The selectivity to propylene keeps almost unchanged. Apparently, high WHSV is favorable for reducing the size of industrial reactor while maintaining the methanol feed rate. But in pilot-scale setup, it is generally difficult to design fluidized bed reactor with high WHSV. Table 1 suggests that the results under small WHSV can be extended to large WHSV confidentially. This certainly simplifies the scale-up process of MTO fluidized bed reactor.

6.1.5.3 Side Reactions

In MTO reactor, some side reactions are closely related to the reaction mechanism as well as reaction conditions. As discussed above, the contact time influences the reaction significantly. Ideally, it is expected to control the gas–solid contact time accurately. In laboratory scale microscale fluidized

Table 1 Influence of the Contact Time/Space Velocity: Time on Stream = 5 min, T = 450 °C

Contact time (s)	3.05	1.53	1.02	0.76	0.61
WHSV (h^{-1})	2	4	6	8	10
MeOH conversion (%)	100	99.9	99.7	99.6	99.3
Selectivity (wt%)					
C_2H_4	31.2	36.0	37.4	37.3	37.8
C_3H_6	50.6	48.8	49.8	50.1	49.7
$C_2H_4 + C_3H_6$	81.8	84.8	87.2	87.4	87.5

bed reactor, porous filters can be effectively used to separate the catalyst from product gas. In pilot-scale or industrial-scale fluidized bed reactor, the use of filter is not feasible. For example, three stage cyclones have been a common practice for removing the catalyst dust from product gas in FCC processes. Depending on the gas velocity, the entrainment of catalyst to the freeboard can be severe. The amount of catalyst entrained to the freeboard can be much as 20% of the catalyst in the dense bed. Meanwhile, in order to improve the separate efficiency, the gas velocity in the disengaging section can be lower than that in the dense bed. Thus, the side reactions in the freeboard cannot be avoided as the catalyst–gas contact time is prolonged. When scaling up the MTO fluidized bed reactor, it is necessary to realize such difference between the microscale fluidized bed reactor in the laboratory and the large device used in the pilot-scale and/or industrial-scale.

In order to assess the severity of the side reaction, we connect two microscale fluidized bed reactors in series. The first reactor simulates the main MTO reactor and the second simulates the disengaging section in pilot-scale and/or industrial-scale fluidized bed. Methanol is fed to the first reactor, and product gas stream is introduced directly into the distributor of the second reactor. Both reactors are operated at temperature of 450 °C. Catalyst loading in the second reactor is diluted by 33 times the weight of inert particles. The WHSV is the first reactor is 2 h^{-1}. Product gas streams from these two reactors are analyzed with two online GCs. The gas stream from the first reactor is sampled and analyzed every 10 min. The sampling and analysis time of the gas stream from the second reactor is slightly lagging behind. The catalyst loading in the second reactor is ca. 20% of that in the first reactor. Table 2 presents the reaction results. As can be seen, at the beginning of the reaction, the percentages of both ethylene and propylene in the gas product decrease after the product gas passing through the second reactor. The total selectivity to ethylene and propylene drops significantly from 85.83% to 79.74%. Meanwhile, the selectivity to C_4 plus C_5^+ rises from 11% to 17%. This indicates that when the product gas from the first reactor contacts with (relatively) fresh catalyst, the small molecular olefins will convert to high-molecular olefins via polymerization reaction. With the prolonging reaction time (after 10 min), the difference of the selectivity to light olefins between these two product gas streams decreases somehow. Particular interest is the large variation of the selectivity to propylene and C_4 after the second reactor. Clearly, the side reactions will certainly influence the final selectivity to light olefins. In DMTO process, we define our operation window to the regime of high coke content in catalyst, which can prevent the methanol from

Table 2 The Experimental Results for Two Microfluidized Bed Reactors in Series

TOS (min)	2	10	20	30	40	50	60	5	10	20	30	40	50	60
Product Gas (wt%)	**First Reactor: MTO Reaction**							**Second Reactor: Side Reactions**						
CH_4	1.4	1.83	2.24	1.83	1.83	2.11	2.09	1.65	1.95	1.87	1.84	2.03	2.12	2.19
C_2H_4	44.72	41.42	42.91	42.63	42.82	43.91	43.62	42.07	42.67	42.99	43.77	43.87	44.42	44.45
C_2H_6	0.3	0.3	0.48	0.3	0.3	0.33	0.32	0.3	0.33	0.33	0.33	0.33	0.35	0.35
C_3H_6	41.11	41.15	38.58	40.85	41.28	40.61	40.69	37.68	35.63	37.11	38.42	39.08	36.79	36.83
C_3H_8	0.97	1.32	1.85	1.23	1.18	1.18	1.18	1.25	1.53	1.43	1.36	1.25	1.33	1.35
DME	0	0	0	0	0	0	0.16	0	0	0	0	0	0	0
MeOH	0	0	0	0	0	0	0.06	0	0	0	0	0	0	0
C_4	8.89	10.58	10.72	10.26	10.1	9.55	9.55	12.57	13.6	12.73	11.58	10.57	11.73	11.63
C_5^+	2.61	3.4	3.22	2.9	2.49	2.31	2.33	4.48	4.29	3.54	2.7	2.87	3.26	3.2
Total	100	100	100	100	100	100	100	100	100.00	100	100	100	100	100
$C_2^- + C_3^-$	85.83	82.56	81.49	83.49	84.09	84.52	84.5	79.74	78.38	80.1	82.19	82.95	81.2	81.28

complete conversion and maintain a relatively low activity of the catalyst. This can inhibit the side reactions to a certain extent.

6.2 Pilot-Scale MTO Fluidized Bed Reactor

We designed and built a pilot-scale circulating fluidized reactor apparatus, which includes a bubbling fluidized bed reactor and a bubbling fluidized bed regenerator. The overall catalyst loading is 5 kg. The size of the MTO fluidized bed reactor is about 10 cm, and the catalyst inventory is roughly 1 kg. The coke combustion capacity of the regenerator is over-designed in order to ensure the complete regeneration of the catalyst under various operation conditions. Two plug valves are installed to control the circulation rate. Pressure controllers are placed at the product gas outlet of the reactor and the flue gas outlet of the regeneration. The product gas is sampled and analyzed via an online GC. The heat balances has not been considered in our pilot-scale system. Heat required for heating the system and maintaining the reaction and regeneration temperature is supplied by electric heater. Thus, the temperature of the reactor and regenerator can be independently controlled. This is important for flexible operation of the pilot-scale unit.

6.2.1 Fluidized Bed Reaction Without Regeneration

It is necessary to compare the results of the microscale fluidized bed reactor and the pilot-scale fluidized bed. For this purpose, we carried out experiments in the pilot-scale setup under the same operation conditions as in the microscale reactors. Here, the catalyst will not circulate to regenerator. Figures 25 and 26 show the methanol conversion and selectivity to ethylene and propylene, respectively, as a function of TOS in the reactor. Apparently, the results from the pilot-scale reactor as shown in Figs. 25 and 26 fit very well with the results in the microscale fluidized bed reactor (as shown in Fig. 18). A more direct comparison is made for the catalyst-to-methanol ratio shown in Figs. 27 and 28 with Fig. 21, which suggest that the results from microscale reactor and pilot-scale reactor are comparable. Especially, both experiments can reflect the declination of the selectivity to ethylene and propylene with an increasing catalyst-to-methanol ratio. However, the catalyst-to-methanol ratio corresponding to the highest selectivity to ethylene plus propylene is slightly higher in the pilot-scale reactor. This minor difference might be due to the different fluidization behavior in these two reactors. The superficial gas velocity in the microscale fluidized bed reactor is roughly 1/10 of that in the pilot-scale fluidized bed. Thus, the

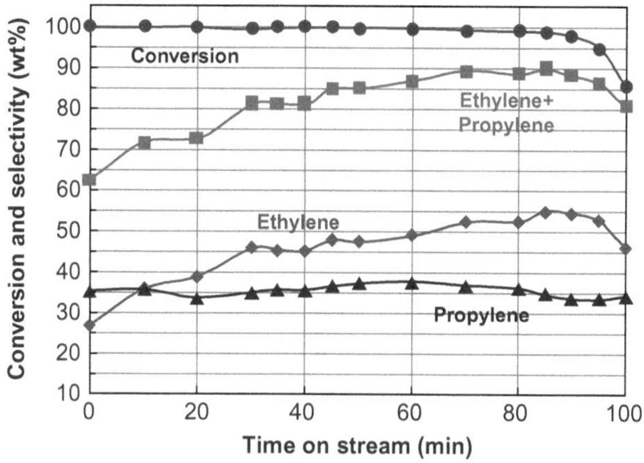

Figure 25 Fluidized bed reaction without regeneration in pilot-scale experiments. Reaction conditions: T=460–470 °C, catalyst inventory=1 kg, WHSV=2 h^{-1}, water: MeOH=20:80, and superficial gas velocity=25 cm/s.

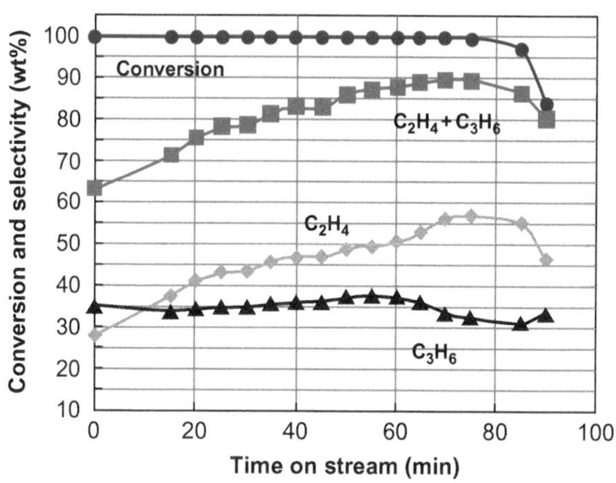

Figure 26 Fluidized bed reaction without regeneration in pilot-scale experiments. Reaction conditions: as the same as in Fig. 25 except T=470–480 °C.

bubble size in the pilot-scale fluidized bed is larger and a worse mass transfer might be expected. Overall, the influence of the hydrodynamics on MTO reaction is not essential, which certainly leads the scaling up of the MTO fluidized bed reactor easier. The agreement between the results from micro-scale fluidized bed reactor and pilot-scale fluidized bed reactor, on the other hand, indicates that the scaling up is successful. The methodologies used in the scaling up indeed provide us right direction.

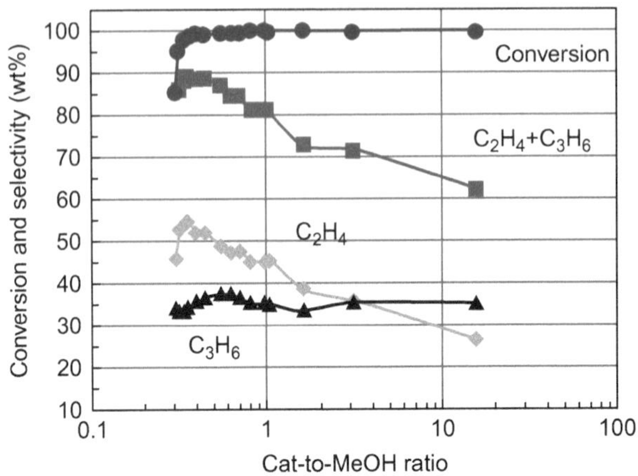

Figure 27 Fluidized bed reaction without regeneration in pilot-scale experiments, as function as catalyst-to-methanol ratio. Reaction conditions: as the same as in Fig. 25.

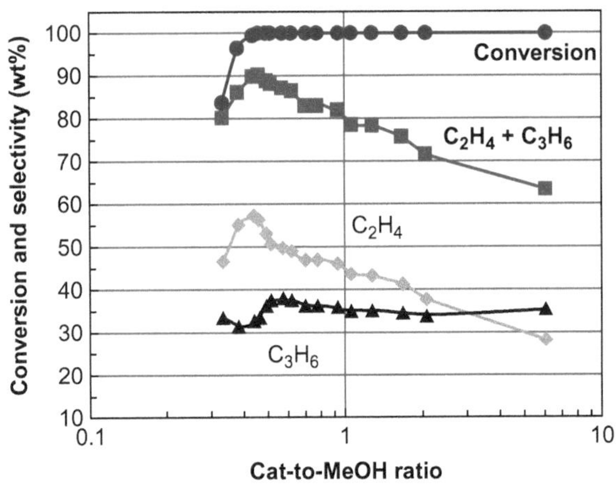

Figure 28 Fluidized bed reaction without regeneration in pilot-scale experiments, as function as catalyst-to-methanol ratio. Reaction conditions: as the same as in Fig. 26.

6.2.2 Fluidized Bed Reaction with Continuous Regeneration

In the pilot-scale experiments, the continuous reaction–regeneration is also investigated. The details of the results will not be discussed here. However, we will focus on the influence of the average residence time and catalyst to methanol on the MTO reaction in pilot-scale fluidized bed reactor.

Figure 29 shows the average residence time of catalyst in the pilot-scale fluidized bed reactor by adjusting the catalyst circulation rate while keeping other conditions such as reaction temperature, inventory, feed rate, and WHSV unchanged. Unlike that in the fluidized bed reactor without catalyst

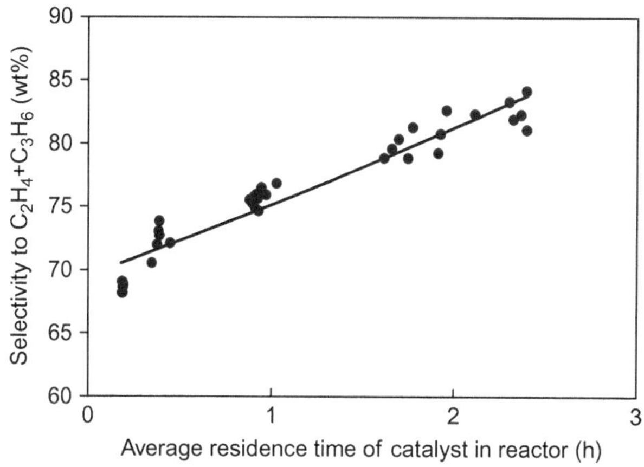

Figure 29 The influence of average residence time of catalyst in reactor on the selectivity of ethylene and propylene. Reaction conditions: $T=500$ °C, catalyst inventory = 1 kg, WHSV = 2 h^{-1}, water:MeOH = 20:80, and gas–solid contact time = 1.3 s.

circulation, the catalyst circulation will lead to residence time distribution of catalyst in the bed. This means that at any given time, catalyst having different residence time in the reactor may coexist. Some might stay in the reactor for much longer time and meanwhile some might be transported to the reactor just for a short while. Recall that the coke content in catalyst is a function of the residence time as shown in Fig. 19, we can simply assume that a longer residence time of a catalyst might cause higher coke content in it. Note that the optimal coke content is around 7.6–8.5%, it is therefore important to extend the reaction time to prompt the coke content. Therefore, in Fig. 29, the selectivity to light olefins is increased with an increasing average residence time. However, careful check has to be performed to ensure the high conversion of methanol feed.

Figure 30 shows the relationship of the catalyst-to-methanol ratio with the selectivity to light olefins. Compared with Fig. 21, it is interesting to note the qualitative agreement between the pilot-scale results and microscale results. Figure 30 can be used to optimize and control the catalyst circulation rate during the reaction, which on other side suggests that microscale fluidized bed reactor can be effectively used in scaling up the fluidized bed reactor with suitable methodology for analysis.

6.2.3 Continuous Operation of Pilot-Scale Setup

A continuous operation of the pilot-scale MTO fluidized bed reactor has been carried out. Table 3 depicts the mass balance calculation for typical

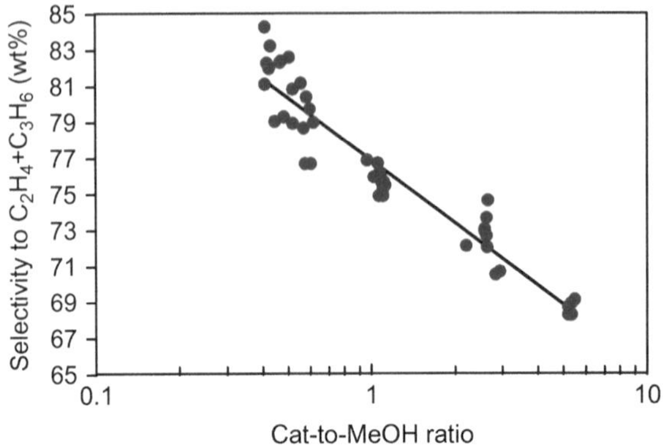

Figure 30 The influence of catalyst-to-methanol ratio in reactor on the selectivity of ethylene and propylene. Reaction conditions: T=500 °C, catalyst inventory=1 kg, WHSV=2 h^{-1}, water:MeOH=20:80, and gas–solid contact time=1.3 s.

Table 3 The Typical Mass Balance from Continuous Operation of the Pilot-Scale MTO Fluidized Bed Reactor

Elements			**C**	**H**	**O**
Feed rate (g/h)	CH_3OH	2048.00	768.00	256.00	1024.00
	H_2O	512.00		56.89	455.11
	Sum	2560.00	768	312.89	1479.11
Product gas (g/h)	H_2	0.28		0.28	
	CO	5.00	2.14		2.86
	CO_2	10.00	2.73		7.27
	H_2O	1612.70	0.00	179.19	1433.51
	CH_4	16.40	12.30	4.10	
	C_2H_4	404.71	346.89	57.82	
	C_2H_6	11.14	8.91	2.23	
	C_3H_6	289.12	247.81	41.30	
	C_3H_8	21.84	17.87	3.97	
	CH_3OH	13.90	5.21	1.74	6.95
	Me_2O	3.05	1.59	0.40	1.06
	C_4	89.91	77.07	12.84	
	C_5	34.87	29.89	4.98	
	C_6	8.80	7.54	1.26	
	Coke	20.48	19.25	1.23	
	Sum	2542.20	779.21	311.33	1451.65
Balance		0.99	1.01	1.00	0.98

Reaction conditions: T=500 °C, Inventory=1 kg, water:MeOH=20:40, and WHSV=2 h^{-1}.
Regeneration condition: T=600 °C.

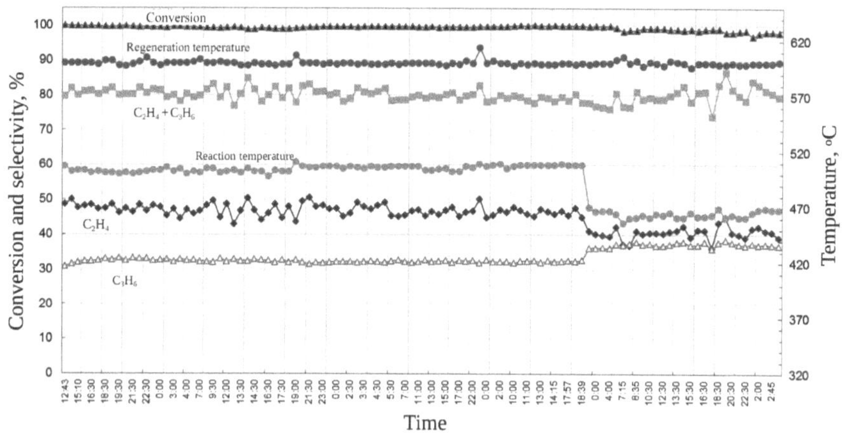

Figure 31 The typical results in the continuous operation of the pilot-scale MTO fluidized bed reactor.

operation conditions. Figure 31 shows the data measured during the operation. As can be seen, methanol conversion is nearly 100% at the reaction temperature of 500–510 °C. The average selectivity to ethylene is 48 wt% and to propylene is 32 wt%. When the reaction temperature is reduce to 460 °C, the average selectivity to ethylene drops to 42 wt% while to propylene increased to 38 wt%. Apparently, the ethylene/propylene ratio can be adjusted by altering reaction temperature. Lower reaction temperature favors propylene production. Based on the mass balance, it is easy to predict that the methanol consumption for producing 1 ton ethylene plus propylene is 2.925 in the continuous operation in the pilot-scale MTO fluidized bed reactor.

7. CHALLENGES AND FUTURE DIRECTIONS

MTO reaction received considerable interests from the viewpoint of either fundamental research or industrial applications. It can be regarded as one of the best examples that can link the work of chemists with chemical engineers. The understanding of the chemistry underlying the methanol conversion over SAPO or ZSM zeolites has been a subject of intense research. The reaction mechanism, i.e., the element reaction steps as well as the intermediate products, is very involved. The reaction itself is not a simple rate-controlling process, and the shape selectivity of the light olefins product can also be related to the molecular diffusion inside the cages and channels of the zeolite crystals. Nevertheless, great advancement has been achieved in the past decades concerning the MTO reaction mechanism. We expect breakthroughs in the understanding of the first C—C bond formation and reaction path for light olefins generation in the future. From the chemical engineering point of view, there also remain several challenges that need to be addressed in the coming years.

The establishment of mesoscale methods to connect the understanding at molecular (or zeolite crystal scale) to the process controlling at catalyst particle scale (or reactor scale) is highly desired. For example, the mechanism studies seem to support that the hydrocarbon pool mechanism is generally effective for MTO reaction; however, the intermediate carbenium ions vary with the cavity size, and thus, the reaction path may change by altering the crystal structures. Quantitative methods must be developed in order to transfer such understandings into our practice of catalyst design and synthesis. At the reactor scale, the reaction kinetic models are mostly obtained for lumped components via ensemble-averaged experiments. The validity of these models is normally limited to the operation conditions that the experiments have covered. Microscale kinetic models have been researched recently by several groups. A mesoscale approach that can link the microkinetics with the lumped kinetics would be a big challenge.

Another important aspect is how to control coke content distribution in the fluidized bed reactor with catalyst circulation. As we know, there exists optimal coke content for catalyst particle which can maximize the selectivity to light olefins. If there is circulation of catalyst particles, the coke content in catalyst shows a certain distribution. Ideally, if the coke content distribution is uniform such as that encountered in the fluidized bed reactor without circulation, the selectivity to light olefins can reach 90% over SAPO molecular sieves. Therefore, how to optimize the coke content distribution in MTO fluidized bed reactor to improve the selectivity to light olefins represents another future direction.

8. CONCLUSIONS

In this contribution, the process development of MTO process has been introduced. We emphasize the importance of the mesoscale studies in the MTO process development. Particularly, we focus on three aspects: a mesoscale modeling approach for MTO catalyst pellet, coke formation and control in MTO reactor, and scaling up of the microscale-MTO fluidized bed reactor to pilot-scale fluidized bed reactor. The applications of results obtained from these mesoscale studies have been outlined and demonstrated.

As a typical multiphase and multiscale process, the research of MTO process spanning molecules, zeolites, catalyst particles, microscale reactors, and pilot-scale reactors to industrial equipments, cross a wide time and length scales. The development of efficient mesoscale methods are expected for further optimizing the DMTO process and improving fluidized bed reactor design and operation.

ACKNOWLEDGMENTS

The DMTO process development is supported by Chinese Academy of Sciences (CAS), Chinese National Development and Reforming Committee (NDRC), Chinese Ministry of Science and Technology, National Natural Science Foundation of China (NSFC), Government of Shanxi Province, and China Petroleum and Chemical Industry Federation. We are grateful to SINOPEC Luoyang Petrochemical Engineering Co. and SYN Energy Technique Co., and all colleagues involved in the DMTO process development. The authors are supported by the National Natural Science Foundation of China (Grant no. 91334205) and the Strategic Priority Research Program of the Chinese Academy of Sciences (Grant no. XDA07070100).

REFERENCES

Abba IA, Grace JR, Bi HT, Thompson ML: Spanning the flow regimes: generic fluidized-bed reactor model, *AIChE J* 49:1838–1848, 2003.

Alwahabi SM, Froment GF: Single event kinetic modeling of the methanol-to-olefins process on SAPO-34, *Ind Eng Chem Res* 43:5098–5111, 2004a.

Alwahabi SM, Froment GF: Conceptual reactor design for the methanol-to-olefins process on SAPO-34, *Ind Eng Chem Res* 43:5112–5122, 2004b.

Barger P: *Zeolites for cleaner technologies,* London, 2002, Imperial College Press.

Bjorgen M, Svelle S, Joensen F, et al: Conversion of methanol to hydrocarbons over zeolite H-ZSM-5: on the origin of the olefinic species, *J Catal* 249:195–207, 2007.

Bleken F, Bjørgen M, Palumbo L, et al: The effect of acid strength on the conversion of methanol to olefins over acidic microporous catalysts with the CHA topology, *Top Catal* 52:218–228, 2009.

Bleken F, Skistad W, Barbera K, et al: Conversion of methanol over 10-ring zeolites with differing volumes at channel intersections: comparison of TNU-9, IM-5, ZSM-11 and ZSM-5, *Phys Chem Chem Phys* 13:2539–2549, 2011.

Bos ANR, Tromp PJJ, Akse HN: Conversion of methanol to lower olefins. Kinetic modeling, reactor simulation, and selection, *Ind Eng Chem Res* 34:3808–3816, 1995.

Buurmans ILC, Weckhuysen BM: Heterogeneities of individual catalyst particles in space and time as monitored by spectroscopy, *Nat Chem* 4:873–886, 2012.

Campbell CT: Micro- and macro-kinetics: their relationship in heterogeneous catalysis, *Top Catal* 1:353–366, 1994.

Chang CD: A kinetic model for methanol conversion to hydrocarbons, *Chem Eng Sci* 35:619–622, 1980.

Chang CD: Hydrocarbons from methanol, *Catal Rev Sci Eng* 25:1–118, 1983.

Chang CD: Methanol conversion to light olefins, *Catal Rev Sci Eng* 26:323–345, 1984.

Chang CD, Silvestri AJ: The conversion of methanol and other O-compounds to hydrocarbons over zeolite catalysts, *J Catal* 47:249–259, 1977.

Chen NY, Reagan WJ: Evidence of autocatalysis in methanol to hydrocarbon reactions over zeolite catalysts, *J Catal* 59:123–129, 1979.

Chen D, Moljord K, Fuglerud T, Holmen A: The effect of crystal size of SAPO-34 on the selectivity and deactivation of the MTO reaction, *Microporous Mesoporous Mater* 29:191–203, 1999.

Chen D, Grlnvold A, Moljord K, Holmen A: Methanol conversion to light olefins over SAPO-34: reaction network and deactivation kinetics, *Ind Eng Chem Res* 46:4116–4123, 2007.

Chen D, Moljord K, Holmen A: A methanol to olefins review: diffusion, coke formation and deactivation on SAPO type catalysts, *Microporous Mesoporous Mater* 164:239–250, 2012.

Corma A: State of the art and future challenges of zeolites as catalysts, *J Catal* 216:298–312, 2003.

Dahl IM, Kolboe S: On the reaction-mechanism for propene formation in the MTO reaction over SAPO-34, *Catal Lett* 20:329–336, 1993.

Dahl IM, Kolboe S: On the reaction-mechanism for hydrocarbon formation from methanol over SAPO-34 1. Isotopic labeling studies of the co-reaction of ethylene and methanol, *J Catal* 149:458–464, 1994.

Dahl IM, Wendelbo R, Andersen A, Akporiaye D, Mostad H, Fuglerud T: The effect of crystallite size on the activity and selectivity of the reaction of ethanol and 2-propanol over SAPO-34, *Microporous Mesoporous Mater* 29:159–171, 1999a.

Dahl IM, Mostad H, Akporiaye D, Wendelbo R: Structural and chemical influences on the MTO reaction: a comparison of chabazite and SAPO-34 as MTO catalysts, *Microporous Mesoporous Mater* 29:185–190, 1999b.

Froment GF: Coke formation in catalytic processes: kinetics and catalyst deactivation, *Stud Surf Sci Catal* 111:53–68, 1997.

Gayubo AG, Aguayo AT, del Campo AES, Tarrio AM, Bilbao J: Kinetic modeling of methanol transformation into olefins on SAPO-34 catalyst, *Ind Eng Chem Res* 39:292–300, 2000.

Geldart D: Types of gas fluidization, *Powder Technol* 7:285–292, 1973.

Guisnet M: "Coke" molecules trapped in the micropores of zeolites as active species in hydrocarbon transformations, *J Mol Catal A Chem* 182–183:367–382, 2002.

Guisnet M, Costa L, Ribeiro FR: Prevention of zeolite deactivation by coking, *J Mol Catal A Chem* 305:69–83, 2009.

Hansen N, Keil FJ: Multiscale modeling of reaction and diffusion in zeolites: from the molecular level to the reactor, *Soft Matter* 10(1–3):179–201, 2012.

Haw JF: Zeolite acid strength and reaction mechanisms in catalysis, *Phys Chem Chem Phys* 4:5431–5441, 2002.

Haw JF, Song WG, Marcus DM, Nicholas JB: The mechanism of methanol to hydrocarbon catalysis, *Acc Chem Res* 36:317–326, 2003.

Hemelsoet K, Nollet A, Speybroeck VV, Waroquier M: Theoretical simulations elucidate the role of naphthalenic species during methanol conversion within H-SAPO-34, *Chem Eur J* 17:9083–9093, 2011.

Hereijgers BPC, Bleken F, Nilsen MH, et al: Product shape selectivity dominates the methanol-to-olefins (MTO) reaction over H-SAPO-34 catalysts, *J Catal* 264:77–87, 2009.

Hunger M, Seiler M, Buchholz A: In situ MAS NMR spectroscopic investigation of the conversion of methanol to olefins on silicoaluminophosphates SAPO-34 and SAPO-18 under continuous flow conditions, *Catal Lett* 74:61–68, 2001.

Ilia S, Bhan A: Mechanism of the catalytic conversion of methanol to hydrocarbons, *ACS Catal* 3:18–31, 2013.

Jiang R, Wang J, Sun P: Optimization of reactor operating factors a 600,000 TPY MTO plant, *Pet Refin Eng* 44:7–10, 2014 (in Chinese).

Keil FJ: Methanol-to-hydrocarbons: process technology, *Microporous Mesoporous Mater* 29:49–66, 1999.

Keil FJ: Multiscale modeling in computational heterogeneous catalysis, *Top Curr Chem* 307:69–108, 2012.

Knowlton TM, Karri SBR, Issangya A: Scale-up of fluidized-bed hydrodynamics, *Powder Technol* 150:72–77, 2005.

Kortunov P, Vasenkov S, Karger J, et al: Investigations of molecular diffusion in FCC catalysts, *Diffus Fundam* 2:97.1–97.2, 2005.

Krishna R, van Baten JM: Unified Maxwell-Stefan description of binary mixture diffusion in micro- and meso-porous materials, *Chem Eng Sci* 64:3159–3178, 2009.

Krishna R, Wesselingh JA: The Maxwell-Stefan approach to mass transfer, *Chem Eng Sci* 52:861–911, 1997.

Kumar P, Thybaut JW, Svelle S, Olsbye U, Marin GB: Single-event microkinetics for methanol to olefins on H-ZSM-5, *Ind Eng Chem Res* 522:1491–1507, 2013.

Lesthaeghe D, Horre A, Waroquier M, Marin GB, Spey broeck VV: Theoretical insights on methylbenzene side-chain growth in ZSM-5 zeolites for methanol-to-olefin conversion, *Chem Eur J* 15:10803–10808, 2009.

Li J, Ge W, Wang W: *From multiscale modeling to meso-science: a chemical engineering perspective,* Heidelberg, 2013, Springer.

Li H, Ye M, Liu Z: A multi-region model for reaction–diffusion process within a porous catalyst pellet, *Chem Eng Sci* 2015a (submitted).

Li JZ, Wei Y, Chen J, et al: Cavity controls and selectivity: insights of confinement effects on MTO reaction, *ACS Catal* 5:661–665, 2015b.

Liang J, Li H, Zhao S, Guo W, Wang R, Yang M: Characteristics and performance of SAPO-34 catalyst for methanol-to-olefin conversion, *Appl Catal* 64:31–40, 1990.

Liu Z: *Methanol to olefins,* Beijing, 2015, Science Press (in Chinese).

Liu Z, Liu Y, Ye M, Qiao L, Shi L, Ma X: Process technology for DMTO unit with a capacity of 1.8 MM TPY methanol feed and unit features, *Pet Refin Eng* 44:1–6, 2014 (in Chinese).

Matsen JM: Scale-up of fluidized bed processes: principle and practice, *Powder Technol* 88:237–244, 1996.

Mores D, Stavitski E, Kox MHF, Kornatowski J, Olsbye U, Weckhuysen BM: Space- and time-resolved in-situ spectroscopy on the coke formation in molecular sieves: methanol-to-olefin conversion over H-ZSM-5 and H-SAPO-34, *Chem Eur J* 14:11320–11327, 2008.

Mores D, Kornatowski J, Olsbye U, Weckhuysen BM: Coke formation during the methanol-to-olefin conversion: in situ microspectroscopy on individual H-ZSM-5 crystals with different bronsted acidity, *Chem Eur J* 17:2874–2884, 2011.

Nan H, Wen Y, Wu X, Xu C, Guan F, Gong L: Recent development of methanol to olefins technology, *Mod Chem Ind* 34:41–46, 2014.

Olsbye U, Svelle S, Bjorgen M, et al: Conversion of methanol to hydrocarbons: how zeolite cavity and pore size controls product selectivity, *Angew Chem Int Ed* 51:5810–5831, 2012.

Park TY, Froment GF: Kinetic modeling of the methanol to olefins process. 1. Model formulation, *Ind Eng Chem Res* 40:4172–4186, 2001.

Park JW, Lee JY, Kim KS, Hong SB, Seo G: Effects of cage shape and size of 8-membered ring molecular sieves on their deactivation in methanol-to-olefin (MTO) reactions, *Appl Catal A Gen* 339:36–44, 2008.

Primo A, Garcia H: Zeolites as catalysts in oil refining, *Chem Soc Rev* 43:7548–7561, 2014.

Qi L, Wei Y, Xu L, Liu Z: Reaction behaviors and kinetics during induction period of methanol conversion on HZSM-5 zeolite, *ACS Catal* 5:3973–3982, 2015.

Rüdisüli M, Schildhauer TJ, Biollaz SMA, van Ommen JR: Scale-up of bubbling fluidized bed reactors—a review, *Powder Technol* 217:21–38, 2012.

Schoenfelder H, Hinderer J, Werther J, Keil FJ: Methanol to olefins-prediction of the performance of a circulating fluidized-bed reactor on the basis of kinetic experiments in a fixed-bed reactor, *Chem Eng Sci* 49:5377–5390, 1994.

Schulz H: Coking of zeolites during methanol conversion: basic reactions of the MTO-, MTP- and MTG processes, *Catal Today* 154:183–194, 2010.

Song WG, Fu H, Haws JF: Supramolecular origins of product selectivity for methanol-to-olefin catalysis on HSAPO-34, *J Am Chem Soc* 123:4749–4754, 2001.

Song WG, Marcus DM, Fu H, Ehresmann JO, Haw JF: An oft-studied reaction that may never have been: direct catalytic conversion of methanol or dimethyl ether to hydrocarbons on the solid acids HZSM-5 or HSAPO-34, *J Am Chem Soc* 124:3844–3845, 2002.

Stocker M: Methanol-to-hydrocarbons: catalytic materials and their behavior, *Microporous Mesoporous Mater* 29:3–48, 1999.

Sundaresan S: Role of hydrodynamics on chemical reactor performance, *Curr Opin Chem Eng* 2:325–330, 2013.

Svelle S, Ronning PO, Olsbye U, Kolboe S: Kinetic studies of zeolite-catalyzed methylation reactions. Part 2. Co-reaction of [12C]propene or [12C]n-butene and [13C]methanol, *J Catal* 234:385–400, 2005.

Tian P, Wei Y, Ye M, Liu Z: Methanol to olefins (MTO): from fundamentals to commercialization, *ACS Catal* 5:1922–1938, 2015.

Van der Hoef MA, Ye M, van Sint Annaland M, Andrews AT IV, Sundaresan S, Kuipers JAM: Multi-scale modeling of gas-fluidized beds, *Adv Chem Eng* 31:65–149, 2006.

Van Santen RA, van Leeuwen PWNW, Moulijn JA, Averill BA: *Catalysis: an integrated approach,* Amsterdam, 1999, Elsevier.

Van Speybroeck V, van der Mynsbrugge J, Vandichel M, et al: First principle kinetic studies of zeolite-catalyzed methylation reactions, *J Am Chem Soc* 133:888–899, 2011.

Van Speybroeck V, De Wispelaere K, van der Mynsbrugge J, Vandichel M, Hemelsoet K, Waroquier M: First principle chemical kinetics in zeolites: the methanol-to-olefin process as a case study, *Chem Soc Rev* 43:7326–7357, 2014.

Vora BV, Marker TL, Barger PT, Nilsen HR, Kvisle S, Fuglerud T: Economic route for natural gas conversion to ethylene and propylene, *Stud Surf Sci Catal* 107:87–98, 1997.

Wang CM, Wang YD, Liu HX, Xie ZK, Liu ZP: Catalytic activity and selectivity of methylbenzenes in HSAPO-34 catalyst for the methanol-to-olefins conversion from first principles, *J Catal* 271:386–391, 2010.

Wei YX, Li JZ, Yuan CY, et al: Generation of diamondoid hydrocarbons as confined compounds in SAPO-34 catalyst in the conversion of methanol, *Chem Commun* 48:3082–3084, 2012a.

Wei YX, Yuan CY, Li JZ, et al: Coke formation and carbon atom economy of methanol-to-olefins reaction, *ChemSusChem* 5:906–912, 2012b.

Wilson S, Barger P: The characteristics of SAPO-34 which influence the conversion of methanol to light olefins, *Microporous Mesoporous Mater* 29:117–126, 1999.

Wragg DS, O'Brien MG, Bleken FL, Di Michiel M, Olsbye U, Fjellvag H: Watching the methanol-to-olefin process with time- and space-resolved high-energy operando X-ray diffraction, *Angew Chem Int Ed* 51:7956–7959, 2012.

Xu S, Zheng A, Wei Y, et al: Direct observation of cyclic carbenium ions and their role in the catalytic cycle of the methanol-to-olefin reaction over chabazite zeolites, *Angew Chem Int Ed* 52:11564–11568, 2013.

Ying L, Yuan X, Ye M, Cheng Y, Li X, Liu Z: A seven lumped kinetic model for industrial catalyst in DMTO process, *Chem Eng Res Des* 100:179–191, 2015.

Yuan C, Wei Y, Li J, et al: Temperature-programmed methanol conversion and coke deposition on fluidized-bed catalyst of SAPO-34, *Chin J Catal* 33:367–374, 2012.
Yuen LT, Zones SI, Harris TV, Gallegos EJ, Auroux A: Product selectivity in methanol to hydrocarbon conversion for isostructural compositions of AFI and CHA molecular-sieves, *Microporous Mesoporous Mater* 2:105–117, 1994.
Zhang L: Wison engineering successfully commercializes MTO project, *Mod Chem Ind* 33:107, 2013 (in Chinese).
Zhang K, Cheng Y, Li X: Simulation of fluidized bed reactor form methanol to olefins (MTO) process, *J Chem Eng Chin Univ* 26:69–76, 2012.
Zhao Y, Li H, Ye M, Liu Z: 3D numerical simulation of a large scale MTO fluidized bed reactor, *Ind Eng Chem Res* 52:11354–11364, 2013.
Zhu QJ, Kondo JN, Ohnuma R, Kubota Y, Yamaguchi M, Tatsumi T: The study of methanol-to-olefin over proton type aluminosilicate CHA zeolites, *Microporous Mesoporous Mater* 112:153–161, 2008.

CHAPTER SIX

Mesoscale Effects on Product Distribution of Fischer–Tropsch Synthesis

Mingquan Shao*,†, Youwei Li*,†, Jianfeng Chen*,†, Yi Zhang*,†,[1]

*State Key Laboratory of Organic-Inorganic Composites, Beijing University of Chemical Technology, Beijing, China

†Research Centre of the Ministry of Education for High Gravity Engineering and Technology, Beijing University of Chemical Technology, Beijing, China

[1]Corresponding author: e-mail address: yizhang@mail.buct.edu.cn

Contents

Abstract

Fischer–Tropsch synthesis (FTS), directly converting a mixture of carbon monoxide and hydrogen (syngas) into sulfur-free hydrocarbons, has attracted much attention from academic and industrial community. However, the development of FTS mainly depends on experience, resulting in the inefficient development of catalysts and industrialization design. Recently, a new analysis method, mesoscale analysis, has attracted more attention due to researching on between different scales or crossing several scales, which would contribute to efficient R&D process of FTS. This chapter will summarize the multiscale effects on FTS products distribution such as ASF distribution, kinetic model,

Advances in Chemical Engineering, Volume 47
ISSN 0065-2377
http://dx.doi.org/10.1016/bs.ache.2015.10.002

reactor flow model and discuss mesoscale effects on FTS product distribution, such as modified ASF product distribution, olefin readsorption mechanism model and diffusion-reaction model of FTS process.

1. INTRODUCTION

1.1 The World Energy Situation

Nowadays, the need of energy has been increasing everyday with the population growth and the advancements of technology. Hence, the research of energy resources becomes a very important subject for the entire world (Bongiorno and Thiringer, 2013). As well known, energy is a primary element of economics and social development in the world.

Due to the shortage of crude oil reserves and environmental constraints, the increased demand of alternative energy resources or clean utilization of coal significantly contributes to resolving both energy and environmental problem. Fischer–Tropsch synthesis (FTS), directly converting a mixture of carbon monoxide and hydrogen (so-called synthesis gas or syngas) into sulfur-free hydrocarbons, has attracted much attention from academic and industrial community, because hydrocarbons can be obtained from coal, natural gas, and biomass instead of crude petroleum (Liu et al., 2010), especially the selective production of chemical feedstock such as ethylene, propylene, and butylenes ($C_2^{=}$–$C_4^{=}$) (Colley et al., 1989).

1.2 History of FTS Research and Industry

1.2.1 History of FTS Progress

In 1922, Hans Fischer and Franz Tropsch proposed the Synthol process, which gave, under high pressure (>100 bar), a mixture of aliphatic oxygenated compounds via reaction of carbon monoxide with hydrogen over alkalized iron chips at 673 K. This product was transformed after heating under pressure into "Synthine," a mixture of hydrocarbons (Khodakov et al., 2007).

Important progress in the development of FTS was made in 1923. It was found that more and more heavy hydrocarbons could be produced. When the Synthol process was conducted at lower pressure, heavy hydrocarbons were the main products of carbon monoxide hydrogenation on Fe/ZnO and Co/Cr_2O_3 catalysts. In 1926, Hans Fischer and Franz Tropsch published their first report (Fischer and Tropsch, 1926a, 1926b) about hydrocarbon synthesis. However, because of the problems of reactor design, the first large-scale FTS plant was built 10 years later. After World War II, ARGE

(Arbeitsge-meinschaft Ruhrchemie and Lurgi) developed a large-scale process with a fixed-bed FT reactor. At the same period, Kellog proposed a technology based on circulating catalyst bed. Both the ARGE and Kellog processes were realized by Sasol in South Africa.

Sasol in South Africa kept on their research on FTS and the Sasol One plant was built in Sasolburg in 1955. In 1969, the Natref crude oil refinery was commissioned. In 1980 and 1982, Sasol Two and Sasol Three, respectively, began production in Secunda (Dyk et al., 2006). Major accomplishments of Sasol in the design of catalysts, reactors, and processes for FTS have been summarized in recently published monograph (Aasberg, 2004).

In the 1980s, investments in the FT research and development programs were picked up again by major petroleum companies. The global resurgence of interest in FTS has been primarily driven by the problems of utilization of stranded gas, diversification of sources of fossil fuels, and environmental concerns. Synthetic liquid fuels generally have a very low content of sulfur and aromatic compounds compared to gasoline or diesel from crude oil.

FT synthesis has been considered as a part of gas-to-liquids (GTL) technology, which converts natural and associated gases to more valuable middle distillates and lubricants. The abundant reserves of natural gas in many parts of the world have made it attractive to commission new plants based on FT technology. From 1993, the Shell Bintulu 12,500 barrels per day (bpd) plant came into operation. And the Sasol Oryx 34,000 bpd plant was inaugurated. Shell and Exxon signed the agreement on building 140,000 and 150,000 bpd GTL-FT plants in Qatar. Thus, after several decades of research and development, FTS technology has finally come to the stage of full-scale industry and worldwide commercialization (Heidemann, 2004).

In recent years, the development of FTS in China gained extensive attention. In 1997, Institute of Coal Chemistry developed two kinds of catalysts which based on precipitation of iron: ICCI and ICCII. In 2006, the low-temperature iron series catalyst and slurry bed FTS technology were developed by the Institute of Coal Chemistry and Synfuels China. And then, three synthetic oil pilot plants were started to be established, which are the scale of 160–180 thousand tons of oil/a. These plants have been started to operate in March 2010.

1.2.2 Catalysts of FTS

To convert the syngas to hydrocarbons, all group VIII metals have taken into experiment to exam the activity. Ruthenium followed by iron, nickel, and cobalt are the most active metals for the hydrogenation of carbon monoxide. Vannice and coworkers (Vannice, 1975) showed that the average molecular weight of hydrocarbons produced by FT synthesis decreased in the following sequence: Ru > Fe > Co > Rh > Ni > Ir > Pt > Pd. Thus, only Ru, Fe, Co, and Ni have catalytic characteristics which allows considering them for commercial production. Ni catalysts under practical conditions produce too much methane. Ru is too expensive; moreover, its worldwide reserves are insufficient for large-scale industry. Cobalt and iron were proposed by Fischer and Tropsch as the first applied catalyst for syngas conversion (Dry, 1990; Steynberg et al., 1999; Jager, 1997; Jager and Espinoza, 1995; Liu et al., 2013). Both cobalt and iron catalysts have been widely used in the industry for FTS (Khodakov et al., 2007). A brief comparison of cobalt and iron catalysts is given in Table 1 (Khodakov et al., 2007).

Table 1 Comparison of Cobalt and Iron FT Catalysts (Khodakov et al., 2007)

Parameter	Cobalt Catalysts	Iron Catalyst
Cost	More expensive	Less expensive
Lifetime	Resistant to deactivation	Less resistant to deactivation (coking, carbon deposit, iron carbide)
Activity at low conversion	Comparable	
Productivity at high conversion	Higher; less significant effect of water on the rate of carbon monoxide conversion	Lower; strong negative effect of water on the rate of carbon monoxide conversion
Maximal chain-growth probability	0.94	0.95
Water–gas shift reaction	Not very significant; more noticeable at high conversions	Significant
$CO + H_2O \rightarrow CO_2 + H_2$	At high conversions	
Maximal sulfur content	<0.1 ppm	<0.2 ppm
Flexibility (temperature and pressure)	Less flexible; significant influence of temperature and pressure on hydrocarbon selectivity	Flexible; methane selectivity is relatively low even at 613 K
H_2/CO	~2	0.5–2.5
Attrition resistance	Good	Not very resistant

1.3 New Challenges for Further Development of FTS Industry

There are many challenges for further development of FTS industry. On the one hand, the FTS is a complex reactions process with a series of reactions, and the mechanism of FTS is still not clear. Therefore, most of the catalyst designs were based on the experience of research. Due to the limitation of the FTS mechanism and kinetics, the design of the reactor needs a lot of experience and long-period small plant test before the industrialization.

On the other hand, the key problem in FTS is to control its product distribution. From the Anderson–Schulz–Flory (ASF) distribution model, we can theoretically see that only methane and FT wax have the very high selectivity. However, high-value product like light olefins and liquid fuel have a low selectivity which was under 50%. Several research groups have suggested that the texture properties of the support had a significant influence on the FTS activity and selectivity. Generally, small pore size favored the formation of small supported metal particles (Zhang et al., 2002). However, the small pore size of supports results in poor diffusion efficiency of reactants and products, which is a disadvantage to catalytic performance of catalysts. In contrast, the support with a larger pore size improves the reducibility, favors the diffusion of reactants and products, suppresses the readsorption of 1-olefins, and leads to high heavy hydrocarbon content in the product (Fan et al., 2004; Tsubaki et al., 2001; Xu et al., 2005). However, the increased chain-growth probability occurring on the catalyst of larger pore size might be attributed to the combined effects of larger metal particle size and larger pore size. It was found that both the dispersion of the supported metal and the pore size of the catalyst show their tight relationship with the propagation of the carbon chain.

Although a few studies were carried out on the influence of many factors on the FTS reaction activity and selectivity, such as the metal particle size, the pore size of support or pellet size of catalysts, controversy persists, because these observations are the result of complex interplay among many factors. These factors include not only the readsorption probability of α-olefins in the confined space and the diffusion situation but also the effect of changes in the site density, such as the changes in loading, reducibility, and the particle size of active metal (Zhang et al., 2011). On the other hand, mass transfer is also considered to be a very important factor that affects both activity and selectivity (Iglesia et al., 1993). Even though in the gas-phase FTS reaction, the catalyst pores may be filled with liquid products. The mass transfer in the liquid phase is typically 4–5 orders of magnitude slower than in the gas phase. Some reactions are even perceived slow due to slow and limited diffusion in the liquid phase (Hilmen et al., 2001). However, applying monolith catalysts (Liu et al., 2009) or supercritical fluids (Liu et al., 2005) in FTS reaction have significantly improved the mass transfer of products resulting in a narrow products distribution. Liu (2002) reported that using appropriate material structures and flow conditions enable manipulation of external and internal mass transfer steps to enhance the reaction activity and selectivity.

Therefore, in FTS reaction system, there are a lot of factors, crossing several scales, including molecular reaction (element scale), internal and external diffusion (system scale), as well as type of reactor (system scale). And the combined effects of those factors significantly influence the product distribution. It is considered that the mesoscale study of FTS, which cross two or several scales, is the key point of controlling the product distribution in FTS reaction.

1.4 Mesoscale Viewpoint and Its Purpose to the FTS Process

1.4.1 Mesoscale and Multiscale Viewpoint

Traditional R&D modes in chemical and process engineering, featuring longer period and higher risks and cost, have become a bottleneck for the fast development of chemical and process industries (NSFC, 2013). Breaking the bottleneck should be based on the mesoscience and virtual process engineering. Firstly, the scientific challenge roots in the multiscale and multilevel structures in which the mesoscales, as the key, are far from being well understood. Understanding the mesoscale complexity plays an important role in the study (Doering and Cremer, 1995) of structure–function relationships of materials, mixing and transfer in reactors as well as the system integration (Li and Huang, 2014; Li et al., 2013, 2014). As shown in Fig. 1, in chemical engineering process, only via coupling the flow, mass, and heat transfer as well as reactions on the mesoscales, the multiscale structure in multiphase reactors could be accurately predicted (Ge et al., 2010).

The mesoscale phenomenon could be on element (small) scale, system (large) scale, or more likely that is, somewhere in between different scales or crossing several scales (Ge et al., 2010). However, it still remains a challenge to correlate the knowledge on element scales with that on system scales, regardless of the level (material, reactor, or system) involved. It is also uncertain how to integrate different disciplines (for instance, the three levels in Fig. 1) to optimize a whole chemical process from the molecular scale to the environmental scale. In fact, even when the phenomenon being studied is just at the element or system scale, it is difficult for us to gain a

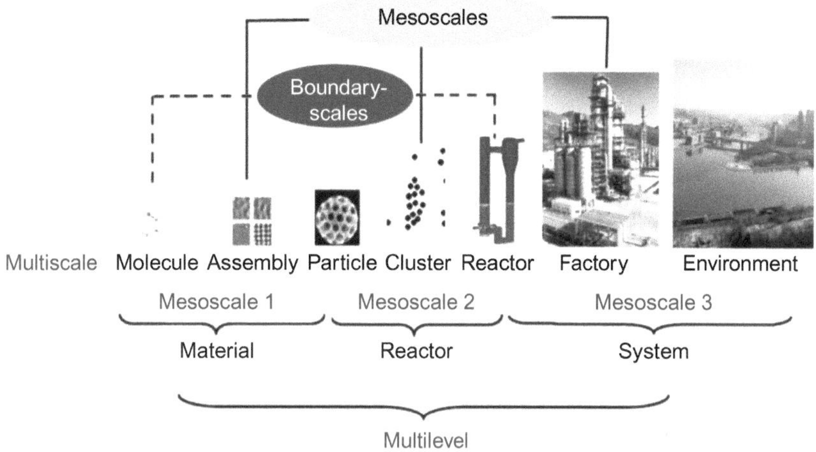

Figure 1 Three mesoscales in chemical engineering (Ge et al., 2010).

comprehensive understanding without considering the effects beyond its specific scale.

No matter in the area of chemical engineering or the others, the research of multiscale is becoming more and more popular. Three main types (Kwauk, 2003) of methods have been proposed to solve multiscaled problems: the describing method, which means using different scale to describe different time and area in the same system. The association method is describing the upper-level scale by basing on the lower scale. The extremum method is describing the whole system by relating different scale model through stability conditions. For example, combining kinetics simulation with Finite Element Analysis to describe the crack propagation and microflow process inside the materials is the so-called describing method (Cao and Chen, 2006; Ogata et al., 2001; Rudd and Broughton, 1998; Smirnova et al., 1999). Measuring the physical properties of flowing matter through kinetics analog method, and then inputting Navier–Stokes equation value to solve the permeability of the porous media is the association method (Hoover et al., 1980; Sirignano and Reviewer, 2000; Travis and Evans, 1997). Another example, EMMS was proposed to deal with the multiphase flow system, belonging to the extremum method (Li et al., 1988). The key point of these different methods is the coupling of multiscales. In the described method, it was reflected in the transforming of the multiscales model variable, such as the position, velocity (molecular discrete attributes), temperature, and pressure (the statistical properties of continuous medium) in the boundary area. In the association method, it was reflected in the extracting and transferring parameter of multiscale (determination of physical parameters and state equation). However, in the extremum method, it was reflected by proposing stability condition to solve the problem.

In fact, multiscales analysis is the combination of various scales problems, which does not become a whole chemical process from the molecular scale to the environmental scale, regardless of the level (material, reactor, or system) involved. Without the coupling of different scales, it would not take advantages of multiscales methods. It is easy to find that the coupling of different scales could clearly describe the phenomena and factors which could be on element (small) scale, system (large) scale, somewhere in between different scales, then to clarify the mechanism of its formation, which is the so-called mesoscale analysis.

1.4.2 Mesoscale Phenomena and Problems of FTS Process

Until now, scientists have paid attention to element and system scales, but have neglected the governing principle at mesoscales, although average and statistical approaches have been proposed to describe the complex phenomena at mesoscales.

For FTS process, two main mesoscale phenomena can be summarized: (1) the interaction between internal diffusion and external diffusion, which is determined by the texture properties of catalyst and flow type of reactor. Generally, the texture properties, such as the size and morphology of catalyst pellet, always influence the flow type of reactor. Therefore, those should be combined to discuss their effects on FTS reaction (Kim et al., 1989; Lee et al., 2004; Mikkola et al., 2007; Schneider and Mitschka, 1966). (2) The relationship between internal diffusion and the intrinsic properties of active site, which is influenced by texture properties and active phase of catalyst. As well known, texture properties and active phase of catalyst always influence each other, such as the catalyst with low-surface area is always difficult to form highly dispersed active metal particle. The above mesoscale phenomenon significantly affects on the deviation from the ASF distribution of FTS products.

Lots of theoretical model was proposed to clarify the deviation and the most acceptable model was olefin readsorption model. The basic viewpoint of the theory was that the primary products of FTS would be readsorbed on the active sites and the secondary reaction occurs resulting in the deviation from the ASF distribution. Meanwhile, mass transfer behavior of primarily formed olefins has a significant influence on olefins readsorption.

However, it was found that a solid–solution compound had formed when promoters were added to the iron-based catalysts, resulting in the promoter-induced textural structure variation of iron catalyst, which would change the particle size and reducibility of iron and the diffusion rate of reactants and products. Meanwhile, the electronic state of active phase would be influenced via the electronic transfer between iron and support. Generally, these factors always interact with each other to influence the activity and selectivity of obtained catalysts in the FTS reaction, resulting in different product selectivities and confusing the effects of promoter. It is considered that in FTS reaction system, the effects of element scales on selectivity of products, such as the properties of active site, are difficult to solely clarifying, due to complex influence of several scales. Therefore, in FTS, coupling the element (small) scale, system (large) scale, somewhere in between, is the key to clarify.

On the other hand, identifying levels should be the first step when considering mesoscale issues. Distinction of levels and scales is also important in traditional disciplines, because any individual boundary is in the domains of two disciplines, one as the element scale, and the other as the system. Complete understanding of this scale requires integration of knowledge from both sides, to which sufficient attention has not been paid in most cases. Therefore, in addition to understanding mesoscale issues, integration of the knowledge from different disciplines at boundary scales is also important.

2. MULTISCALE ISSUES IN THE PROCESS OF FTS

2.1 Introduction of Multiscales Analysis Method of FTS

As well known, several factors influence the product distribution of FTS. Mechanisms and kinetics are the general reasons which decide the product distribution, and the most important product distribution law: model of ASF was summarized from polymerization mechanism. Catalysts also influence the product distribution through their different active site, particle size, pore size, and acid intensity. For example, Co/SiO_2 with different SiO_2 pore size has different product distribution (Xu et al., 2005). Another important part of FTS is the reactor type and the reaction conditions, because the effects of system scale, such as flow model, temperature, and pressure, would significantly influence the product distribution (Moutsoglou and Sunkara, 2011). These factors can be divided into three main scales: molecular scale (internal diffusion and active site), catalyst pellet scale (internal and external diffusion), and reactor scale (external diffusion).

We will discuss multiscale issues in the FTS process by analyzing these three scales particularly and find how the mesoscale problem affects FTS reaction on the product distribution by coupling multiscale issues.

2.2 FTS Mechanism and Molecular Reaction

2.2.1 Surface Carbide Mechanism

The surface carbide mechanism was first proposed by Fischer and Tropsch (1926a, 1926b). They considered that Fe, Ni, and Co would generate their own carbide, when they reacted with CO, and then the carbide would be transformed into hydrocarbon (Henrici-Olivé and Olivé, 1976). They proposed that when CO and H_2 contacted the catalyst at the same time, CO would dissociate on the surface of the catalyst and then metallic carbide would be formed, the latter would form methylene intermediate, then olefins and paraffins would be formed by polymerization. The surface carbide mechanism is shown in Fig. 2. This is also the earliest FTS mechanism which could properly describe the formation of the hydrocarbon products; however, the generation of oxygenated chemicals was not explained in a proper way. Moreover, they could not explain why the generation rate of surface carbide significantly was lower than the generation rate of liquid hydrocarbon. Besides, although Ru does not form stable carbides, it is very effective in FTS.

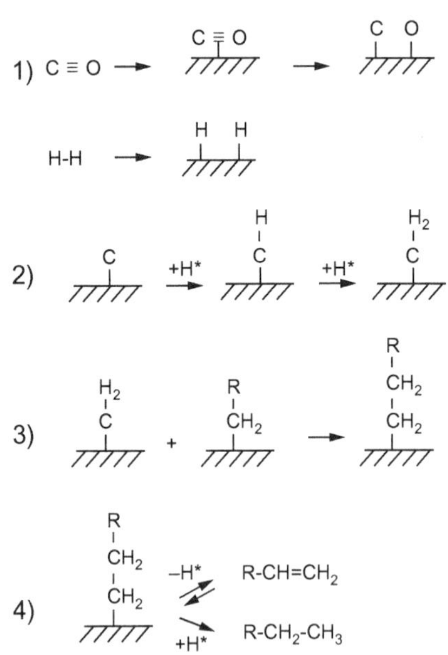

Figure 2 Diagram of FTS surface carbide mechanism (Sarup and Wojciechowski, 1989).

2.2.2 Surface Enol Mechanism

Due to the limitation of surface carbide mechanism, Anderson and his coworkers proposed (Storch et al., 1951) a more detailed mechanism to explain FTS product distribution. The theory regards that the chemical adsorption of H_2 and CO would generate surface enol complex (HCOH), then it would generate methylene after dehydration and then the hydrocarbon would be formed. In surface enol mechanism, CO has been hydrogenated without dissociation in advance, as shown in Fig. 3, which is well explained on the energy level.

2.2.3 Mechanism of Carbonyl Interposition

Pichler and coworkers (Pichler and Schluz, 1970) proposed another theory which regards that the chain initiation starts from the insertion of CO into metal–hydrogen bond. The characteristic of the mechanism supposed that after CO forming formyl group by hydrogenation, the formyl group would form bridge methylene oxide content by further hydrogenation, then the bridge methylene oxide species would be further hydrogenated to the methyl and carbine. Subsequently, the insertion of CO into metal–hydrogen bond would repeat, and the growth of the carbon chain can be completed (Tam et al., 1979), as shown in Fig. 4. The mechanism is more reasonably to explain the formation process of straight-chain product. However, for

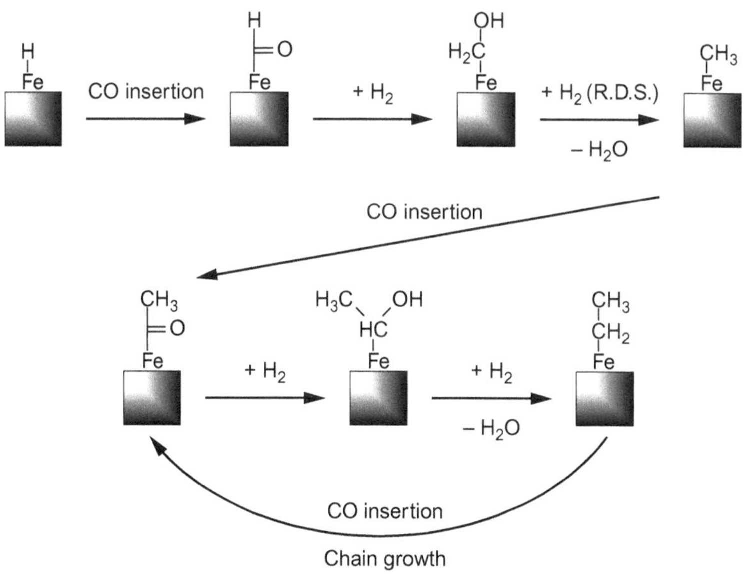

Figure 3 Diagram of FTS surface enol mechanism (de Smit and Weckhuysen, 2008).

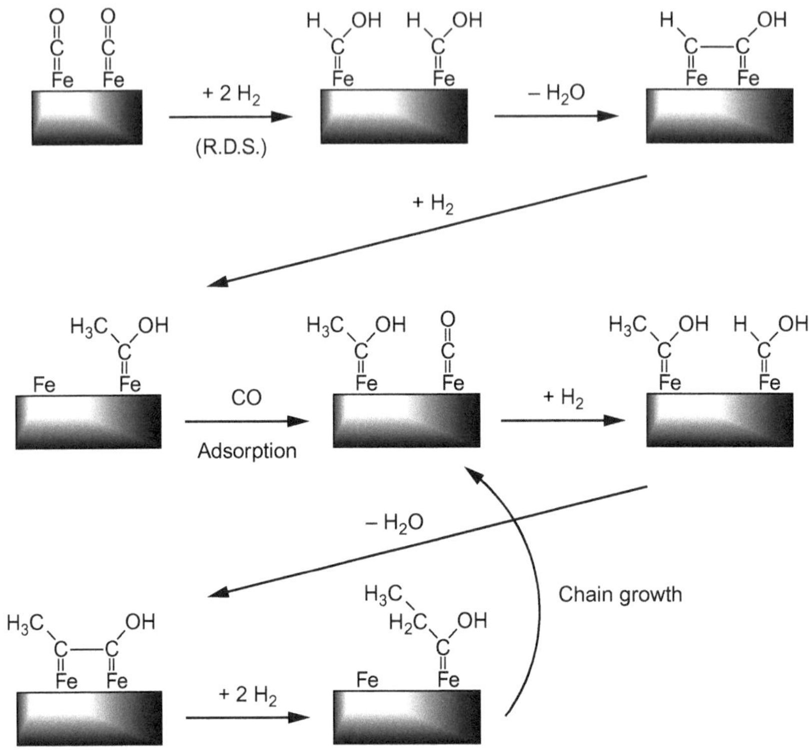

Figure 4 Diagram of FTS carbonyl interposition mechanism (de Smit and Weckhuysen, 2008).

explaining why most of the product is *n*-paraffin, it proposes that the formation rate of paraffins is much faster than that of the olefins.

2.2.4 Polymerization Mechanism and ASF Product Distribution

Many researchers proposed that FTS follows the polymerization mechanism (Friedel and Anderson, 1950; Rofer-Depoorter, 2002). It is proposed that CO undergoes dissociative or hydrogen-assisted dissociative chemisorption on the surface of Ru, Co, or Fe metal, or metal carbide nanoparticles, forming CH_x ($x = 0 - 3$) intermediates as the monomers for polymerization. The coupling between CH_x monomers leads to chain growth, providing C_nH_m intermediates. The C_nH_m intermediates with different carbon numbers can then undergo hydrogenation or dehydrogenation to afford paraffins or olefins as the final products (Madon, 1979). The products of FT synthesis follow the ASF distribution because of the polymerization mechanism. According to ASF distribution, ideally, the molar fraction (M_n) of the hydrocarbon product with a carbon number of *n* is only dependent on the chain-growth probability

(α), which is a function of the rates of chain growth and chain termination, by the following Eq. (1):

$$M_n = (1-\alpha)\alpha^{n-1} \tag{1}$$

In other words, the product selectivity is determined by α value in an ideal case. A smaller α value means more lighter (C_1–C_4) hydrocarbons in products, while a larger α value means formation of more heavier (C_{21+}) hydrocarbons. However, the products of FTS with ASF distribution are unselective for the middle-distillate products, which are usually the target products. For examples, the maximum selectivity to gasoline-range (C_5–C_{11}) hydrocarbons is less than 45% and that to diesel-range (C_{10}–C_{20}) hydrocarbons is less than 35%, as shown in Fig. 5. The development of novel FT catalysts, which can tune the selectivities to desired products, is one of the most challenging targets in FT synthesis.

Because of the wide product distribution, FT syn-crudes obtained by conventional FT catalysts must be subjected to further refining to produce high-quality liquid fuels such as gasoline, jet fuel, and diesel fuel (Klerk, 2008, 2011). For this purpose, it is necessary to increase the selectivity to C_{5+} hydrocarbons and to decrease those to CH_4 and C_2–C_4 paraffins. Most of the publications related to FTS have aimed at developing efficient FT catalysts with high C_{5+} selectivity as well as high CO conversion activity. On the other hand, as compared with the conventional two-stage process, the direct production of high-quality liquid fuels from syngas without refining stage would be more energy- and cost-efficient, and could increase the competitiveness of FT technology for the production of liquid fuels (Vliet et al., 2009). This

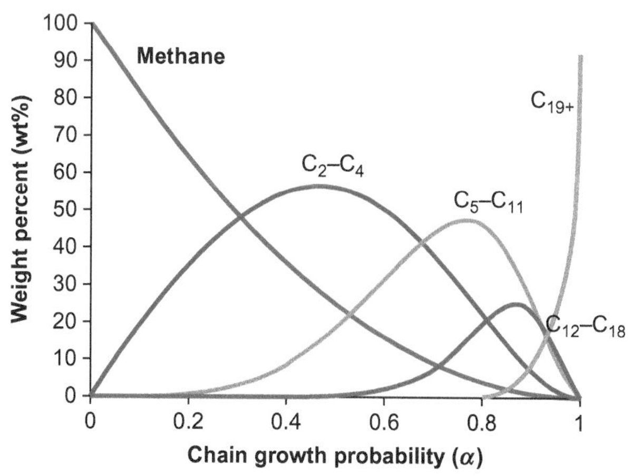

Figure 5 ASF product distribution of FTS (Zhang et al., 2013).

requires the design of novel FT catalysts with high selectivities to a desired range of hydrocarbons (Bao et al., 2008; Cheng et al., 2012; Kang et al., 2009, 2011; Tavakoli et al., 2008; Xiao et al., 2011; Yu et al., 2009; Zhang et al., 2012, 2013; Zhou et al., 2011).

With the development of FTS, more and more researchers found that FTS product distribution do not obey ASF model. Many explanations toward the modified ASF distribution have been proposed (Liu et al., 2011), among which there are some main ones: (1) limits of experimental conditions and experimental skills; (2) the cause of the gas–liquid equilibrium; (3) some intrinsic interpretation of olefin readsorption reactions, "dual active center" theory, and hydrocracking reactions; (4) diffusion inside the catalyst particles; (5) the integral effect of the reactor bed. For the first one, the experimental conditions and experimental skills effect on products analysis, such as control of reaction conditions and measurement of the dynamics parameters for FTS reaction (Donnelly et al., 1988; Farias et al., 2006). For the second, the gas–liquid equilibrium would influence the scale of bubbles and the reactor (Font et al., 2006; Shi et al., 2013), leading to the influence of mass transfer or heat transfer during the FTS reaction; For the third and fourth problems, the diffusion (including internal and external diffusion) combined the properties of catalyst active sites could significantly influence the reaction performance and product distribution, as stated in Section 1.4.2; for the fifth problem, the size of catalyst pellet and reactor also affect the results of FTS reaction (Anderson et al., 1964; Xu et al., 2005). Except experiment method, as mentioned above, the combination of the molecular scale, the particle scale of active phase, and the catalyst bed or reactor scale results in very complex reaction performance and modified ASF product distribution of FTS. Although there have been some divergence of these explanations, it can be clearly seen that without the coupling of different scales, the product distribution cannot be explained properly. It is considered that only solving the mesoscale problems can improve the development of industrial FTS.

However, the molecular reaction did not consider the temperature variation, as well as the internal and external diffusion during the reaction. Therefore, due to the integral accumulation effect along the catalyst bed, the modified ASF distribution has appeared, even though the reactor is strict isothermal and the internal and external diffusion of catalyst can be ignored. Therefore, if we want to control the product distribution of FTS, we must clarify the detailed dynamic information on the particle-reactor level, which needs to couple the catalyst particle model and the reactor flow model.

2.3 Kinetic Model of FTS

Kinetic model is a key point to describe complex FTS reaction. Although much research has been done to deal with FTS, the kinetic model research still stay at lumped kinetic model (Anderson, 1956; Derosset et al., 1976; Feimer et al., 1981; Frohning and Cornils, 1974; Karn et al., 1960; Thomas and Eckert, 1984), because the reaction mechanism of FTS is very complex containing series of surface reactions. The lumped kinetics can only explain syngas conversion rate, the distribution of products usually be described by the semiempirical model (like ASF model) (Wang et al., 1999). However, if the lumping kinetics use in the reactor simulate, only total conversion and distribution of temperature can be obtained. Table 2 (Wang et al., 1999) shows the lumped kinetics for fix-bed FTS on iron-based catalyst.

The researches of FTS kinetics can be divided into two parts: synthetic gas consumption dynamic model and modified kinetics model. Because the main catalysts in FTS are Co- or Fe-based catalysts, the research of FTS kinetics mainly aims at these two kinds of catalysts.

The major problem in describing the FT reaction kinetics is the complexity of its reaction mechanism and the large number of species involved. As discussed above, the mechanistic proposals for the FTS used a variety of surface species and different elementary reaction steps, resulting in empirical power law expressions for the kinetics. However, the rate equations of Langmuir–Hinshelwood–Hougen–Watson (LHHW) have been applied based on a reaction mechanism for the hydrocarbon-forming reactions. In most cases, the rate-determining step was assumed to be the formation of the monomer.

2.3.1 Synthetic Gas Consumption Kinetics Model

In general, for iron catalysts, the FT reaction rate increases with partial pressure of H_2 and decreases with partial pressure of water.

Anderson (Anderson and Karn, 1960) proposed a rate equation which included water inhibition as Eq. (2), where P is the partial pressure of CO, H_2, and H_2O, k is the correlation coefficient:

$$R_{FT} = \frac{kP_{CO}P_{H_2}}{P_{CO} + aP_{H_2O}} \tag{2}$$

Dry (1976) and Huff and Satterfield (1984) derived the same equation from the combined enol/carbide mechanism, assuming strong adsorption of CO and water relative to H_2 and CO_2 as shown in Eq. (2).

Table 2 The Lumped Kinetics Models for Fixed-Bed FTS over Iron-Based Catalyst (Wang et al., 1999)

Catal.	Temperature (°C)	Pressure (MPa)	Syngas H_2/CO	Rate Equation	Activation Energy
Iron	–	–	–	$kP_{H_2}^2 P_{CO}$	88
Fused iron (Fe–K)	250–320	2.2–2.4	2.0	kP	79
Fused iron				kP_{H_2} (syngas conversion $<60\%$) $kP_{H_2}P_{CO}/(P_{CO}+bP_{H_2O})$ (syngas conversion $\geq 60\%$)	84
Precipitated Fe/Cu/K	220–280	10	$CO/H_2=3/1$	$kP_{H_2}^{1.5}P_{CO}^{0.2}+bP_{H_2}^{0.2}P_{CO}^{0.5}$	105
Fused iron (Fe–K)	225–255	2.2	0.25–2.0	$kP_{H_2}^{0.66}P_{CO}^{0.34}$	71–100
Iron				$kP_{H_2}P_{CO}/(P_{CO}+bP_{H_2})$	63
Fused iron	250–315	2.0	2.0	$kP_{H_2}P_{CO}/(P_{CO}+bP_{H_2})$	84
Iron	250–300	0.77–3.1	1.5–3.9	$kP_{H_2}P_{CO}/(P_{CO}+bP_{H_2})$	37
Precipitated Fe/Cu/K	220–270	1.0–2.0	1.0–6.0	$kP_{H_2}P_{CO}^{-0.26}$	79–92
Fused iron	225–245	2.2	0.25–2.0	$kP_{H_2}^{0.6}P_{CO}^{0.4}-bP_{H_2}^{0.5}$	84
Precipitated Fe/Cu/K	503–538	1.5–3.0	1.0–2.5	$kP_{H_2}P_{CO}/(aP_{CO}+bP_{H_2})$	55.8
Precipitated Fe/Cu/K	493–573	1.0–3.0	1.07–2.79	$kP_{H_2}P_{CO}/(aP_{CO}+bP_{H_2})$	66.52

At CO conversion of lower than 60% and in the case of a high shift activity of the catalyst, Eq. (2) can be simplified to a first-order dependency in H_2, due to low partial pressures of water in Eq. (3):

$$R_{FT} = kP_{H_2} \tag{3}$$

Huff and Satterfield (1984) observed a linear decrease in the adsorption parameter a in Eq. (2) with hydrogen pressure on a fused iron catalyst and incorporated this by modifying Eq. (2) to Eq. (4):

$$R_{FT} = \frac{kP_{CO}P_{H_2}^2}{P_{CO}P_{H_2} + a' P_{H_2O}} \tag{4}$$

Deckwer et al. (1986) used Eq. (5) to describe the kinetic results for H_2/CO feed ratios between 0.8 and 2.0 on a potassium-promoted iron catalyst in the slurry phase. However, at low H_2 to CO feed ratios, this equation was not able to describe the results due to high water–gas shift activity. The water–gas shift reaction can increase or decrease the FTS reaction rate by altering the concentrations of the reactants and products. Generally, CO_2 inhibition is not as strong as water inhibition due to the large difference in adsorption coefficients. However, iron catalysts with a high activity of the water–gas shift reaction convert a significant amount of water into CO_2. Ledakowicz et al. (1985), Nettelhoff et al. (1985), and Deckwer et al. (1986) reported the following equation including CO_2 (P_{CO_2}) inhibition (Eq. 5).

$$R_{FT} = \frac{kP_{CO}P_{H_2}}{P_{CO} + aP_{CO_2}} \tag{5}$$

However, cofeeding of CO_2 with syngas showed that the CO_2 addition does not alter the reaction rate of the syngas significantly. Yates and Satterfield (1989) suggested that the inhibition attributed to CO_2 on catalysts with a high shift activity is caused by H_2O.

2.3.2 Modified Kinetics Model

The selected parameters of the syngas consumption kinetics model aim at calculating syngas conversion. Due to the lumped kinetics model excessively concerned about the syngas, this kinetics model may give a good forecast for designing a reactor. However, it cannot precisely simulate the reaction rate of each product. Therefore, the modified kinetics, which couples the syngas consumption kinetics model with the diffusion model of FTS, is very important and impendency. Lox and Froment (1993) developed a reaction network for the formation of linear hydrocarbons on a commercial precipitated iron catalyst (as shown in Eqs. 6–8, in which k is the correlation coefficient of each hydrocarbon, P is the pressure of CO and H_2, and α is the chain-growth possibility.). For each reaction path, one or more elementary reaction steps were assumed to be slower than the other steps. Based on the carbide mechanism, they developed several kinetic models; those give product distribution along with ASF model.

$$R_{C_nH_{2n+2}} = \frac{k_{HC5}P_{H_2}\left(\frac{k_{HC1}P_{CO}}{k_{HC1}P_{CO}+k_{HC5}p_{H_2}}\right)\alpha^{n-1}}{1+\left(\frac{k_{HC1}P_{CO}}{k_{HC1}P_{CO}+k_{HC5}P_{H_2}}\right)\frac{1}{1-\alpha}} \tag{6}$$

$$R_{C_nH_{2n}} = \frac{k_{HC6}\left(\frac{k_{HC1}P_{CO}}{k_{HC1}P_{CO}+k_{HC5}P_{H_2}}\right)\alpha^{n-1}}{1+\left(\frac{k_{HC1}P_{CO}}{k_{HC1}P_{CO}+k_{HC5}P_{H_2}}\right)\frac{1}{1-\alpha}} \tag{7}$$

$$\alpha = \frac{k_{HC1}P_{CO}}{k_{HC1}P_{CO}+k_{HC5}P_{H_2}+k_{HC6}} \tag{8}$$

2.4 Reactor Types and Its Flow Model of FTS

Great efforts have been devoted to development of reactor to satisfy the requirements of different GTL processes. Several reactor types are currently used for FTS. For example, reactors for FTS include the multitubular fixed-bed, gas–solid fluidized-bed, and slurry bubble column reactors (Hussain et al., 2015). The differences between these reactors are largely related to different approaches to temperature control and the choice of catalyst.

2.4.1 Fix-Bed Reactor

Fix-bed reactor is a typical reactor in FTS process, which is firstly applied in industrial FTS process. As early as 1979, Atwood and Bennett (Albal et al., 1984) have built a 0.25 billion c.f./d syngas management FT tubular reactor one-dimensional pseudo homogeneous piston flow model, based on dynamic model of molten iron catalysts as shown in Eqs. (9) and (10), in which dx is the occupancy of CO, d_t is the diameter of the tank, r is the reaction rate, F is the inlet amount of the gas, C_p is the specific heat capacity of the system, h_w is the heat transfer coefficient, and z is the axial coordination of the tank.

$$\frac{\mathrm{d}x_{\mathrm{CO}}}{\mathrm{d}z}=\pi\frac{d_t^2}{4}\frac{r_{\mathrm{CO}}}{3600F_{\mathrm{CO}}} \tag{9}$$

$$C_P G\frac{\mathrm{d}T}{\mathrm{d}z}=\frac{r_{\mathrm{CO}}(-\Delta H)}{3600}\frac{4h_w(T-T_w)}{\mathrm{d}t} \tag{10}$$

Bub and Baerns (Bub et al., 1980) proposed a two-dimensional homogeneous model to describe FTS fix-bed reactor as shown in Eqs. (11) and (12), where the u and c represents the viscosity and molarity, respectively, and α is the cancellation factor of each component, ε is the porosity of the catalysts, D is the diffusion coefficient, and λ is the efficiency factor.

$$\frac{\partial\left(u_g c_i\right)}{\partial z}=\rho_b\sum_{j=1}^{\mathrm{NR}}\alpha_{ij}r_j+\varepsilon D_{\mathrm{er}}\frac{1}{r}\frac{\partial}{\partial r}\left(r\frac{\partial c_i}{\partial r}\right) \tag{11}$$

$$\frac{\partial\left(u_g\rho_f C_p T\right)}{\partial z}=\rho_b\sum_{j=1}^{N}r_j\left(-\Delta H_f\right)+\lambda_{\mathrm{er}}\frac{1}{r}\frac{\partial}{\partial r}\left(r\frac{\partial T}{\partial r}\right) \tag{12}$$

The boundary conditions are:

$$\begin{aligned}
&z=0;\quad c_i=c_{i,0};\quad T=T_0\\
&r=0;\quad \frac{\partial c_i}{\partial r}=\frac{\partial T}{\partial r}=0\\
&r=R_t;\quad \frac{\partial c_i}{\partial r}=0;\quad -\frac{\lambda_{\mathrm{er}}T}{r}=h_w(T-T_w)
\end{aligned}$$

Recently, Liu and coworkers (Liu et al., 1999) analyzed steady state and dynamic behavior of FTS fixed-bed reactor based on lumped kinetic model, where a is the heating area and η is the diffusion coefficient.

gaseous phase:

conservation of mass:

$$\varepsilon\frac{\partial c}{\partial t}=\frac{1}{r}\frac{\partial}{\partial r}\left(rD_{\mathrm{rf}}\frac{\partial c}{\partial r}\right)+\frac{\partial}{\partial z}\left(rD_{\mathrm{zf}}\frac{\partial c}{\partial z}\right)-\frac{\partial(uc)}{\partial z}+a_v k_c\left(c_p^s-c\right) \tag{13}$$

heat balance:

$$\varepsilon\rho_f C_f\frac{\partial T}{\partial t}=\frac{1}{r}\frac{\partial}{\partial r}\left(rk_{\mathrm{rf}}\frac{\partial T}{\partial r}\right)+\frac{\partial}{\partial z}\left(rk_{\mathrm{zf}}\frac{\partial T}{\partial z}\right)-\frac{\partial\left(u\rho_f C_f T\right)}{\partial z}+a_v h\left(T_p^s-T\right) \tag{14}$$

solid phase:
conservation of mass:

$$(1-\varepsilon)\frac{\partial c_p}{\partial t} = a_v k_c \left(c_p^s - c\right) + (1-\varepsilon)\eta a r_v \tag{15}$$

heat balance:

$$\begin{aligned}(1-\varepsilon)\rho_p C_p \frac{\partial T_p}{\partial t} = \frac{1}{r}\frac{\partial}{\partial r}\left(r k_{\mathrm{rp}} \frac{\partial T_p}{\partial r}\right) + \frac{\partial}{\partial z}\left(r k_{\mathrm{zp}} \frac{\partial T_p}{\partial z}\right) \\ + (1-\varepsilon)\frac{1}{x^2}\frac{\partial}{\partial x}\left(x^2 k_p \frac{\partial T_p}{\partial x}\right)\end{aligned} \tag{16}$$

2.4.2 Slurry Bed Reactor

Although different reactor types had used for GTL processes, the major attention during the past 30 years was paid to the slurry reactor from both academic and industrial interests. The slurry reactor presents the advanced reactor technology for GTL processes. Its advantages include simple construction, excellent heat transfer performance, online catalyst addition and withdrawal, and reasonable interphase mass transfer rates with low energy input. However, its multiphase flow behaviors are very complex and have some remarkable scale-up effects, due to the lack of detailed information on hydrodynamics and mass transfer over a wide range of industrial operating conditions. For design and scale-up, predictions of the gas holdup, liquid flow pattern, gas–liquid mass transfer rate, and heat transfer rate are desired. Knowledge of the bubble size distribution, coalescence, and breakup behaviors is also important, since the bubble behaviors are closely related to the hydrodynamics and heat and mass transfer (Dudukovic et al., 2002; Sie and Krishna 1999). Further, the following issues must be addressed for development and design of a slurry reactor for FTS processes: (1) to get an increased volumetric productivity, the reactor should operate at a high superficial gas velocity and a high solid concentration in the typical heterogeneous regime. However, the hydrodynamics and mass and heat transfer behaviors in the heterogeneous regime are much more complex than in the homogeneous regime, and studies on this flow regime are still limited. (2) The influences of the temperature and pressure must be reliably predicted since all slurry reactors for GTL processes operate under high temperature and pressure. (3) The gas–liquid mass transfer limitation may cause a decrease in the reaction conversion, especially at high solid concentrations and superficial gas velocities. (4) The gas distributor may affect the stable operation of the reactor and requires a proper design for robust and efficient performance (DOE/NETL-2004/1199). (5) The separation of fine particles from wax is difficult and requires special treatments (Wang et al., 2007a, 2007b).

Based on mentioned above, the slurry bubble reactor is very complex, such as the size of catalyst and bubble, the behavior of catalyst and bubble, as well as the contact of catalyst and bubble in the reactor, indicating that the mesoscale problem is also very important for designing the slurry bubble reactor.

2.4.2.1 Single- and Double-Bubble Model

Slurry reactor can be divided into three main types: homogeneous bed reactor (uniform particles distribution), heterogeneous bed reactor (the distribution of particles is thin under strong structure) and three-phase slurry reactor (accumulation of particles in below) (Tsutsumi et al., 1999). The FTS is mainly operated in the nonuniform bubble turbulent zone, and the visual observation and image method illustrate that the bubbles with different size exist in the turbulent zone (heterogeneous) (Grund et al., 1992; Hikita et al., 1981). The single-bubble model deals the bubble as a certain size, resulting in narrow range of application. To optimize the model, some researchers have developed double-bubble model. It has a more reasonable physical background, and the idea of the model is that the bubble slurry follows axial diffusion model, the particle behavior follows sedimentation-dispersion model, gas–liquid mass transfer resists in the liquid film, the reaction takes place in the liquid phase. The basic framework of steady state model is shown in Eqs. (17) and (18):

$$k_L \alpha_i \left(c_i^* - c_{i,L}\right) + (1 - \varepsilon_G) r_i = 0 \tag{17}$$

$$\frac{\partial}{\partial z}\left[(1-\varepsilon_G) D_S \frac{\partial \rho_S}{\partial z}\right] - \frac{\partial}{\partial z}\{[(1-\varepsilon_G)u_s - u_{SL}]\rho_S\} = 0 \tag{18}$$

The mass conservation equation of the double-bubble model was as shown in Eqs. (19)–(22):

$$\frac{\partial}{\partial z}\left(\varepsilon_{G,l} D_{G,l} \frac{\partial c_{i,G,l}}{\partial z}\right) - \frac{\partial\left(u_{G,l} c_{i,G,l}\right)}{\partial z} - k_L a_{i,l}\left(c_{i,l}^* - c_{i,L}\right) = 0 \tag{19}$$

$$\frac{\partial}{\partial z}\left(\varepsilon_{G,s} D_{G,s} \frac{\partial c_{i,G,s}}{\partial z}\right) - \frac{\partial\left(u_{G,s} c_{i,G,s}\right)}{\partial z} - k_L a_{i,s}\left(c_{i,s}^* - c_{i,L}\right) = 0 \quad (20)$$

$$k_L a_i\left(c_i^* - c_{i,L}\right) + (1 - \varepsilon_G) r_i = 0 \quad (21)$$

$$\frac{\partial}{\partial z}\left[(1 - \varepsilon_G) D_S \frac{\partial \rho_S}{\partial z}\right] - \frac{\partial}{\partial z}\{[(1 - \varepsilon_G) u_S - u_{SL}]\rho_S\} = 0 \quad (22)$$

Heat conservation equation:

$$\frac{\partial}{\partial z}\left[\rho_{SL} c_{pSL}(1 - \varepsilon_G) D_L \frac{\partial T}{\partial z}\right] - \frac{\partial\left(\rho_{SL}\, \rho_{pSL}\, u_{SL}\, T\right)}{\partial z}$$
$$-H_h(T - T_c) + (1 - \varepsilon_G)(-\Delta H_{FT} r_{FT}) = 0 \quad (23)$$

2.4.2.2 Multistage Tandem Stirred Tank Model

This model divides the slurry bed reactor into several CSTR. The model can be regarded as a quasi axial diffusion model, and the idea of the model was that the interfacial resistance between the gas and liquid can be omitted; fluid particle was considered well-distributed; the exit of pre-CSTR is the entrance of the latter stage CSTR. The model is shown in Eqs. (24) and (25):

$$\left(u_{GO} - u_{i,G}\right)\frac{c_i}{c_{GO}} - k_L a_{i,G}\left(c_{i,G} - c_{i,L}\right) = 0 \quad (24)$$

$$u_{SLO} c_{SLO} - u_{SL} c_{i,SL} + k_L a_{i,G}\left(c_{i,G} - c_{i,L}\right) + (1 - \varepsilon_G) r_i = 0 \quad (25)$$

This model supposed that the contact between gas and liquid was sufficient, and the increased number of reactor could improve accuracy. Although it can investigate species and the change of the heat, it is unable to be used in the gas–liquid phase and forecast the spatial distribution of the particle velocity. Furthermore, without considering the influence of air bubble, the application of this model is limited.

2.4.2.3 The CFD Model

CFD simulation considered momentum equation of all phases, which can provide the velocity of every phase and the flow field. However, big reactor requires a lot of grid, leading to significant increase of the computation time. Therefore, it is necessary to establish a mesoscale model which combines multiscale drag force with viscosity model. The CFD model consists of Euler–Euler, Euler–Lagrange, and Lagrange–Lagrange methods.

Euler–Euler method considers all phases are continuous medium with mutual penetration between different phases, known as the fluid mode, which is the most appropriate model for the industrial reactor simulations.

Euler–Lagrange model considered the fluid as continuous medium, and treated bubble and particle with a discrete process, which is suitable for study of small scale, known as particles simulation. Lagrange–Lagrange is suitable for microscale simulation.

From what we introduced above, for the industrial reactor, the method should consist of gas phase and the slurry phase (Baten and Krishna, 2003; Krishna et al., 1999, 2000, 2001; Troshko and Zdravistch, 2009), gas–liquid phase and particle phase (Eidus, 1967; Friedel and Anderson, 1950), meanwhile, the gas phase was treated as single bubble, double bubble, and multibubble phase (equilibrium state). The model considers two phase of gas and slurry as shown in Eqs. (26) and (27):

$$\frac{\partial(\varepsilon_k \rho_k)}{\partial t} + \nabla \cdot (\varepsilon_k \rho_k u_k) = \Gamma_k + m_k \tag{26}$$

$$\frac{\partial(\varepsilon_k \rho_k)}{\partial t} + \nabla \cdot (\varepsilon_k \rho_k u_k u_k) = \mu_{\text{eff}} \varepsilon_k \left[\nabla u_k + (\nabla u_k)^T\right] - \varepsilon_k \nabla p + M_{k,l} + \rho_k \varepsilon_k g + \Gamma_k u_k + m_k u_k \tag{27}$$

The model closes interaction phase $M_{k,l}$ with effective viscosity μ_{eff}, and it is considered that the slurry phase of FTS is in the turbulent zone. The effective viscosity can be considered as turbulence viscosity, which can usually be solved by standard model, where turbulence induced by bubbles is also taken into consideration. Interaction force between the phases consists of drag, lift, and virtual mass force. The main force of interaction between phases often is overlooked. Drag force, caused by gas, drives the serious movements and the type of which is written as Eq. (28):

$$M_{L,G} = -M_{G,L} = \frac{3}{4}\left(\frac{C_D}{d_b}\right)\varepsilon_G \rho_L (u_G - u_L)|u_G - u_L| \tag{28}$$

In Eqs. (17)–(28), the symbols index is as follow:

α	The unit volume of slurry gas–liquid contact area, m^{-1}
c	The reactant concentration, mol/m^3
C_D	The bubble populations drag coefficient
C_{D0}	The single-bubble drag coefficient

Continued

d_b	The bubble diameter, m
D	Diffusion coefficient, m^2/s
G	Acceleration of gravity, m^2/s
H_h	Heat transfer quantity, W/K
ΔH	The reaction heat, J/mol
K	The reaction rate constant, mol/(kg s Pa^2)
$K_{L,\alpha}$	Mass transfer coefficient, s^{-1}
$M_{k,I}$	Two-phase momentum exchange coefficient
m	The mass transfer phase
p	Pressure, Pa
r	Reaction rate, mol/(m^3 s)
T	Temperature, K
u	Superficial gas velocity, m/s
u_t	Terminal gas velocity, m/s
ρ	Density, kg/m^3
ρ_s	The solid mass concentration, kg/m^3
μ	Viscosity, kg/(m s)
μ_{eff}	The fluid effective viscosity, kg/(m s)
ε	The phase fraction
Γ	Source phase

Superscript and subscript index:

*	Gas saturation solubility conditions
c	Condenser pipe
i	Species
k	Phase number
L	Liquid
l	Large bubble
pSL	Slurry particles

S	The solid particles suspended
SL	Slurry
s	Small bubbles
0	The initial state

The double-bubble model is more useful than single-bubble model, and the key in the double-bubble model is to combine the diameter of each bubble. However, lots of empirical correlation and major parameters obtained by small test make it only be adapted to a narrow range. The accuracy of multistage tandem stirred tank model depends on the number of the used stirred tanks in experiment. Because the model mainly depends on empirical correlation, it is necessary to associate the structure parameter from the physical consideration. CFD model is able to study concentration distribution and the influence of the internals of the reactor. However, how to associate the important parameters limits its development, for example, the drag force of the turbulent viscosity is limited by calculating capacity.

2.4.3 FTS Product Distribution of Fixed-Bed Reactor

In Moutsoglou and coworker's (Moutsoglou and Sunkara, 2011) research, the effects of reactor length, inlet pressure, syngas ratio on paraffin and olefin selectivities, and mass flow rates in FTS have been documented. Figures 6–8 have shown variation of selectivity of olefins and paraffins with carbon number and other factors. The conversion rate is found to increase with inlet pressure, reactor length, and syngas molar ratio, while the usage ratio is shown to increase with inlet pressure and molar feed ratio and to decrease with the increased reactor length. The total selectivity of paraffins is shown to increase with the increased syngas ratio and reactor length and to decrease with the increased inlet pressure. On the other hand, the total selectivity of olefins is found to decrease with the increased syngas ratio and reactor length, while increasing with the increased inlet pressure. At low and moderate carbon atom numbers, the mass flow rate of paraffins increase with the increased syngas ratio, the decreased inlet pressure, and increased reactor length. These effects subside with the increased carbon number. The mass flow rate of methane is by far the greatest among paraffins. Mass flow rates of olefins exceed corresponding paraffin rates except for $n=2$, where the mass flow rate of ethylene is the lowest in olefins. The mass flow rates of olefins decrease with the increased syngas ratio at low carbon atom numbers, but increase with the feed ratio at large carbon atom numbers. Olefin mass flow

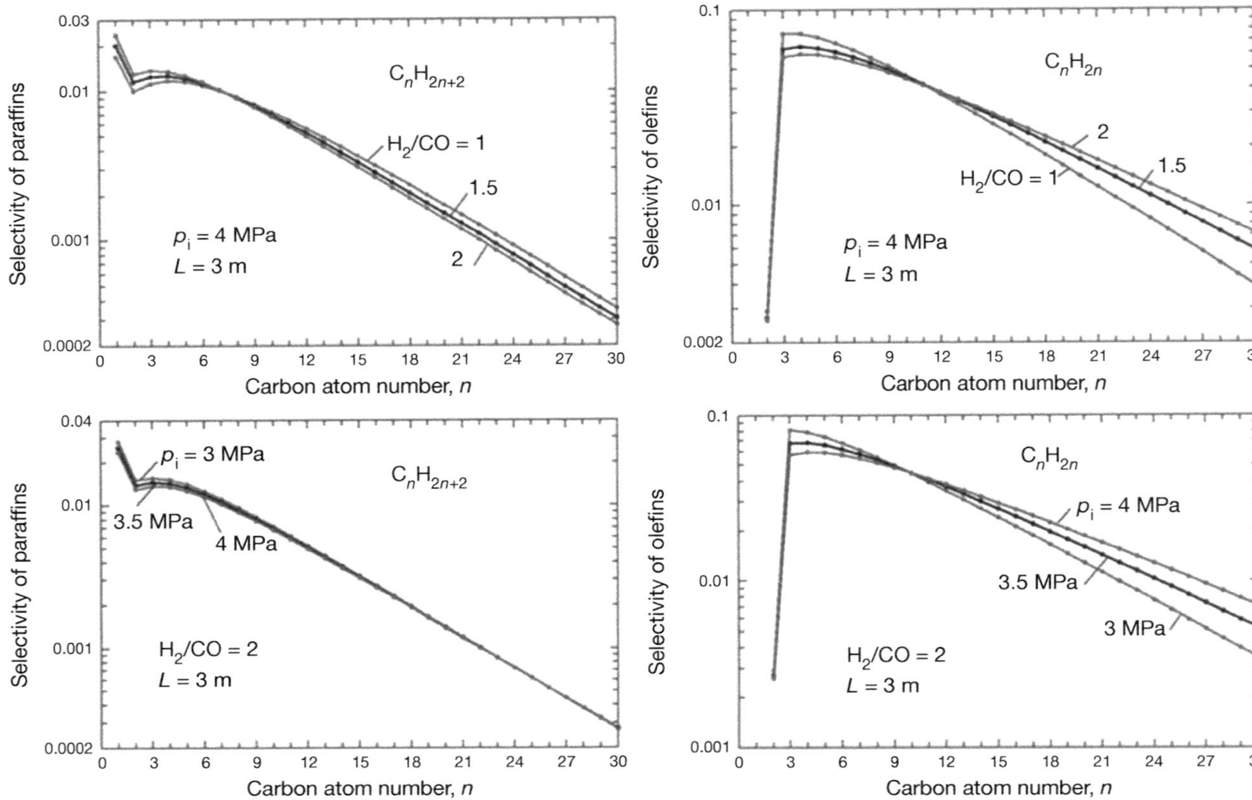

Figure 6 Variation of selectivity of olefins and paraffins with carbon number and other factors (feed ratio and inlet pressure) (Moutsoglou and Sunkara, 2011).

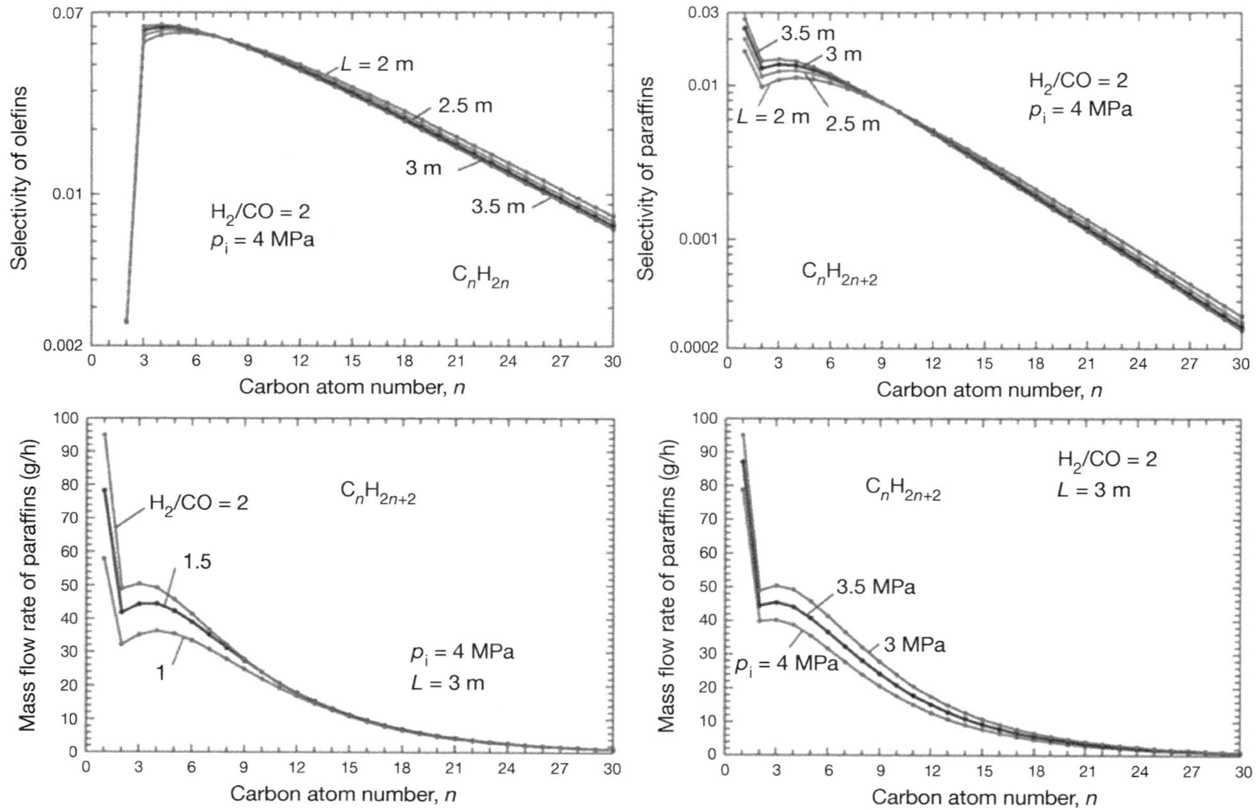

Figure 7 Variation of selectivity of olefins and paraffins with carbon number and other factors (feed ratio, inlet pressure, and reactor length) (Moutsoglou and Sunkara, 2011).

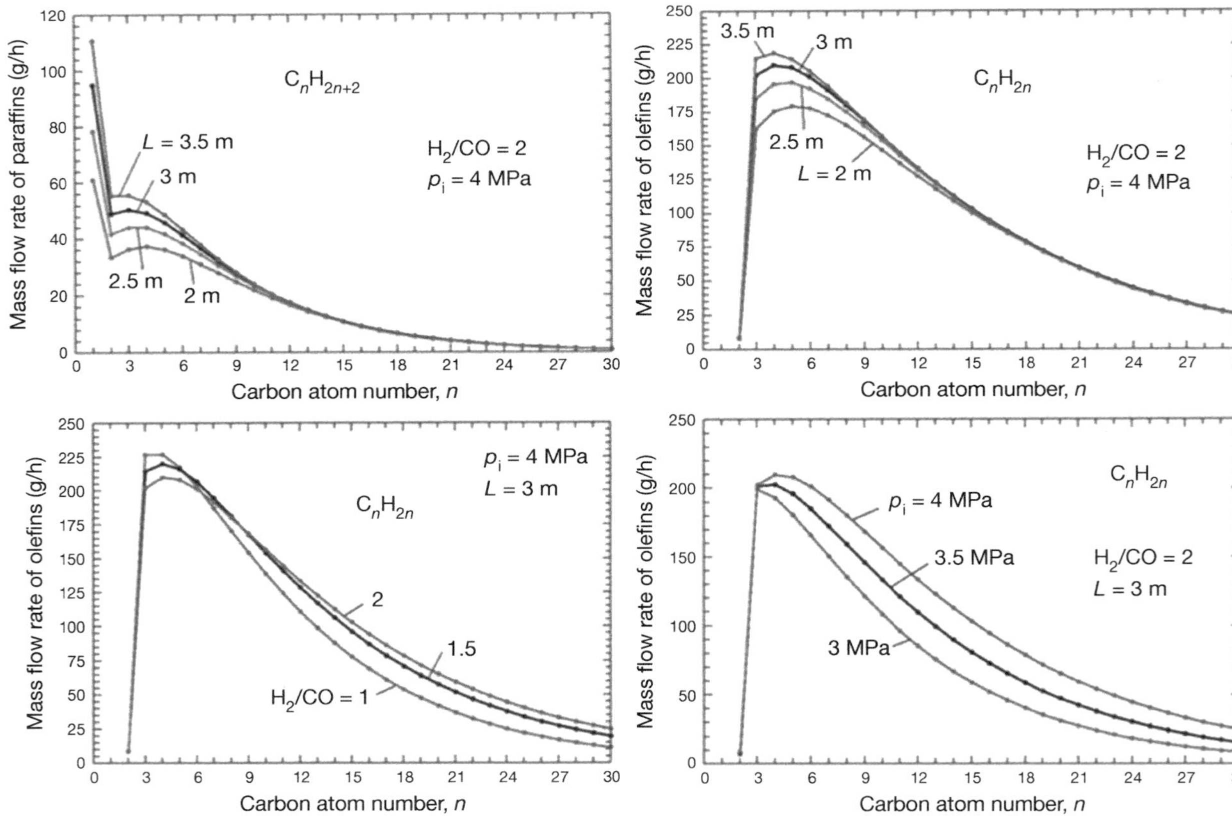

Figure 8 Variation of selectivity of olefins and paraffins with carbon number and other factors (feed ratio, inlet pressure, and reactor length) (Moutsoglou and Sunkara, 2011).

rates are shown to increase with the increased inlet pressure and reactor length. Finally, three distinct stages of the polymerization reactions are identified. At the initial stage, all of the adsorbed syngas is converted to propagating alkyl and alkenyl species with increasing carbon atom numbers, as chain termination reactions do not occur at this stage. The second stage is characterized with desorption of olefins and paraffins, as chain termination reactions become significant and decrease the formation rates of lower carbon number products. Eventually, the decreasing propagation reactions at lower carbon numbers in turn limit the termination reactions to lower carbon numbers themselves. The second stage ends with the ceasing of desorption of liquid olefin and paraffins. At the last stage, the absence of desorbed liquids results in constant gas flow rates, as any decrease in syngas results in the formation of low carbon number gaseous olefins and paraffins.

In this chapter, the multiscale analysis of FTS process is illustrated, and the effects of every scale are discussed on the product distribution of FTS reaction. However, constraint will be exploded in mesoscales viewpoint because there are many combined factors to influence FTS reaction, such as complex reaction mechanism, mass transfer of reactants and products, and flow type of reactor. Therefore, the ASF distribution is always improper in the FTS system. On the other hand, the flow model of the reactor has been investigated, but how the reactor influence the reaction performance of catalyst is still not clear. Therefore, if we want to know the product distribution of different reactors, a lot of experiment should be taken place. Based on these points, the mesoscale phenomena and effects for products of FTS reaction should be investigated in detail.

3. MESOSCALE PROBLEMS AND ITS EFFECTS IN FTS PROCESS

Because any individual boundary scale is in the domains of two disciplines, one as the element scale, and the other as the system, the experimental results, such as product distribution, would be influenced by both sides. As well known, the FTS process is very complex and includes several scales or levels. Except experiment method, as mentioned above, the combination of the molecular scale, the particle scale of active phase, and the catalyst bed or reactor scale results very complex reaction performance and modified ASF product distribution of FTS. Although there have been some divergence of these explanations, it can be clearly seen that without the coupling of different scales, the product distribution of FTS cannot be explained properly. It is

In Ali Nakhaei Pour and coworker's (Ali et al., 2013) work, the chain-growth probabilities of the FTS reaction are developed by LHHW modeling in gradientless reactor over Fe/Cu/La/Si catalyst, which can simultaneously provide a chain-growth probabilities α_1 and α_2. To calculate α_1 and α_2, they give a method as shown in Eqs. (32)–(38), where R and x are the reaction rate and molar fraction of each species, A and B are two important parameters which are related to the system.

$$x_i = A\alpha_1^{i-1} + B\alpha_2^{i-1} \tag{32}$$

$$A\alpha_1^{i-1} = B\alpha_2^{i-1}, \quad i = \xi \tag{33}$$

$$\frac{R_{C_nH_{2n+2}} + R_{C_nH_{2n}}}{R_{C_{n-1}H_{2n}} + R_{C_{n-1}H_{2n-2}}} = \alpha_1 \quad (n \leq \xi) \tag{34}$$

$$\frac{R_{C_nH_{2n+2}} + R_{C_nH_{2n}}}{R_{C_{n-1}H_{2n}} + R_{C_{n-1}H_{2n-2}}} = \alpha_2 \quad (n > \xi) \tag{35}$$

$$\sum_{i=1}^{\infty} x_i = \sum_{i=1}^{\infty} \left[A\alpha_1^{(i-1)} + B\alpha_2^{(i-1)}\right] = 1 \tag{36}$$

$$\begin{aligned}\sum_{i=3}^{\infty} x_i &= \sum_{i=1}^{\infty} \left[A\alpha_1^{(i-1)} + B\alpha_2^{(i-1)}\right] - A(1+\alpha_1) - B(1+\alpha_2) \\ &= 1 - x_1 - x_2\end{aligned} \tag{37}$$

$$\begin{aligned} A &= (1 - x_1 - x_2)\left[\frac{\alpha_2^1}{1-\alpha_1} + \left(\frac{\alpha_1}{\alpha_2}\right)^{\xi-1} \times \left(\frac{\alpha_2^2}{1-\alpha_2}\right)\right]^{-1} \frac{x_i}{1 - x_1 - x_2} \\ &= \left[\alpha_1^{i-1} + \left(\frac{\alpha_1}{\alpha_2}\right)^{\xi-1} \times \alpha_2^{i-1}\right] \times \left[\frac{\alpha_2^1}{1-\alpha_1} + \left(\frac{\alpha_1}{\alpha_2}\right)^{\xi-1} \times \left(\frac{\alpha_2^2}{1-\alpha_2}\right)\right]^{-1} \end{aligned} \tag{38}$$

In that work, the carbon number distribution of FT products on iron catalyst was studied by use of a modified ASF distribution with two chain-growth probabilities. Based on enol mechanism and two ASF distributions the complete set of elementary reactions is given. Using mechanistic kinetic studies of FTS reaction, the chain-growth probabilities (α_1 and α_2) for two ASF distributions formulated. The calculated two ASF model are carefully fitted with experimental results at low carbon monoxide. Thus, the two ASF model is a useful model for prediction of products distribution on lanthanum-promoted iron catalyst in their experimental conditions at low carbon monoxide conversions. The results for higher carbon monoxide conversions deviate substantially because the FTS reaction highly depends on the hydrogen formed by the WGS as the carbon monoxide conversion increases.

Meanwhile, "positive" deviations characterized by concave curvatures in the ASF plot, or by a "break" in the C_{10}–C_{13} range, have been firmly established and extensively studied (Puskas and Hurlbut, 2003). Working with Fe catalysts in a fixed-bed reactor, Koenig and Gaube (1983) had found that without alkali promotion, Fe gave ideal ASF distribution with low value of growth probability (0.65). With K-promoted Fe, the product distribution gave curved ASF plot. These authors assumed that alkalization of the catalyst created two different catalytic sites with differing chain-growth characteristics. The unpromoted sites produced low α-value, but the alkalized sites gave high α-value. Based on this concept, they developed a mathematical model ("two-site model" or two superimposed ASF distributions) and calculated that 61% of the product was produced on unalkalized sites with α of 0.57, and 39% on alkalized sites with α of 0.87. As an alternative to the two-site model, a "distributed site" model was suggested by Satterfield (Satterfield et al., 1985) and shown to be equivalent. The assumptions of this latter model are that the K at the Fe sites is randomly distributed with concentrations between zero and a maximum value.

3.1.2 Olefin Readsorption Mechanism Model

In 1993, Lox and Froment (1993) proposed detailed dynamic model mechanism which based on FTS carbide mechanism, the model clarify the association between operating conditions change and FTS product distribution under classical ASF frame. This was the first attempt to develop detailed dynamic model. Recently, Ma and coworkers (Ma et al., 1999a, 1999b) proposed a FTS detailed dynamic-olefin readsorption mechanism model, which indicates eigenactivity of olefin readsorption, the formation of which is shown in Eqs. (39)–(45), where R is the reaction rate of different hydrocarbons, K_i is the reaction rate constant of different reactions, α_n is the chain-growth possibility corrected by β_n, α_A is the chain-growth possibility, and β_n is a correction factor caused by olefin readsorption.

$$R_{C_nH_{2n+2}} = \frac{K_5(P_{H_2})\prod_{j=1}^{n}\alpha_j}{1+\left(1+\frac{1}{K_2K_3K_4}\frac{P_{H_2O}}{P_{H_2}^2}+\frac{1}{K_3K_4}\frac{1}{P_{H_2}}+\frac{1}{K_4}\right)\sum_{i=1}^{n}\left(\prod_{j=1}^{n}\alpha_j\right)} \quad (n \geq 1) \tag{39}$$

$$R_{C_nH_{2n}} = \frac{K_5(1-\beta_n)\prod_{j=1}^{n}\alpha_j}{1+\left(1+\frac{1}{K_2K_3K_4}\frac{P_{H_2O}}{P_{H_2}^2}+\frac{1}{K_3K_4}\frac{1}{P_{H_2}}+\frac{1}{K_4}\right)\sum_{i=1}^{n}\left(\prod_{j=1}^{n}\alpha_j\right)} \quad (n \geq 2) \tag{40}$$

$$R_{WGS} = \frac{K_V\left(\frac{P_{CO}P_{H_2O}}{P_{H_2}^{0.5}} - \frac{P_{CO}P_{H_2}^{0.5}}{K_{e,WGS}}\right)}{1+\frac{K_V P_{CO}P_{H_2O}}{P_{H_2}^{0.5}}} \tag{41}$$

$$\alpha_n = \frac{K_1P_{CO}}{K_1P_{CO}+K_5P_{H_2}+K_6(1-\beta_n)} \quad (n \geq 2) \tag{42}$$

$$\alpha_A = \frac{K_1P_{CO}}{K_1P_{CO}+K_5P_{H_2}+K_6} \tag{43}$$

$$\alpha_1 = \frac{k_1P_{CO}}{k_1P_{CO}+k_5P_{H_2}} \tag{44}$$

$$\beta_n = \frac{\frac{K_{-6}}{K_6} P_{C_nH_{2n}}}{\alpha_A^{n-1}\frac{K_1P_{CO}}{K_1P_{CO}+K_5P_{H_2}} + \frac{K_{-6}}{K_1P_{CO}+K_5P_{H_2}+K_6}\sum_{i=2}^{n}\left(\alpha_A^{i-2}P_{C(n-i+2)H_2(n-i+2)}\right)} \quad (n \geq 2) \tag{45}$$

Equilibrium constant of WGSR $K_{e,WGS}$ can be calculated by:

$$\ln K_{e,WGS} = \frac{5078.0045}{T} - 5.8972089 + 13.958689 \times 10^{-4}T - 27.592844 \times 10^{-8}T^2 \tag{46}$$

For this model, the product distribution of FTS should be determined or influenced by not only the chain-growth factor α_n but also the secondary reaction of olefins (β_n), which can direct the design of FTS catalyst and engineering enlargement. However, the olefin readsorption mechanism model needs to be studied further to clarify the intrinsic product distribution of FTS.

3.2 Diffusion-Reaction Model of FTS Process

3.2.1 Diffusion-Reaction Model of FTS Process

The investigation of catalytic reaction is complicated by the fact that the process ordinarily studied frequently involves diffusion as well as molecular reaction. So, this problem can be divided into two scales: one is the molecular reaction on the catalyst active sites and the other is the diffusion in the catalyst pore. However if the problem is studied separately, the experiment results do not agree with the actual results. Such as the olefins readsorption can result in the actual product distribution with a deviation from ASF model. In other words, the two scales are not easy to be separated as the factors affecting each other. One or the other may dominate any particular heterogeneous reaction; the relative importance of these quite different rate processes determines the proper design and the performance of catalytic reactor. Therefore, we need to couple two different scales of FTS to find out the real problem and then figure it out.

The general theoretical approach is to develop the mathematical equations for simultaneous mass transfer and chemical reaction, as the reactants and products diffuse into and out of the porous catalyst. When reaction occurs simultaneously with mass transfer within a porous structure, a concentration gradient is established. Since interior surfaces are thus exposed to lower reactant concentrations than surfaces near the exterior, the overall reaction rate throughout the catalyst particle under isothermal conditions is less than it would be if there were no mass transfer limitations. As will be shown, the apparent activation energy, the catalyst selectivity, and other important observed characteristics of a reaction are also dependent upon the structure of the catalyst and the effective diffusivity of reactants and products (Charles and Thomas, 1963).

The effects of mass transfer within a porous structure on observed reaction characteristics were apparently first recognized and mathematically analyzed by Thiele (1939) in the United States. The work has been extended and developed by Wheeler (1995) and many other researchers. The important result of these analyses is the quantitative description of the factors which determine the effectiveness of a porous catalyst. The "effectiveness factor," here symbolized by η, is defined as the ratio of the actual reaction rate to that which would occur if all of the surface throughout the inside of the catalyst particle were exposed to reactant of the same concentration and temperature as the existing at the outside surface of the particle.

Charles and Thomas (1963) in their paper considered firstly a spherical catalyst pellet of radius R and focus attention on a spherical shell of thickness dr and radius, as shown in Fig. 9. He assumes isothermal condition and that the complicated diffusion phenomena within the porous structure can be represented by a single overall effective diffusion coefficient, D_{eff}. In the catalyst pellet, reactants are transported to the shell by diffusion and consumed

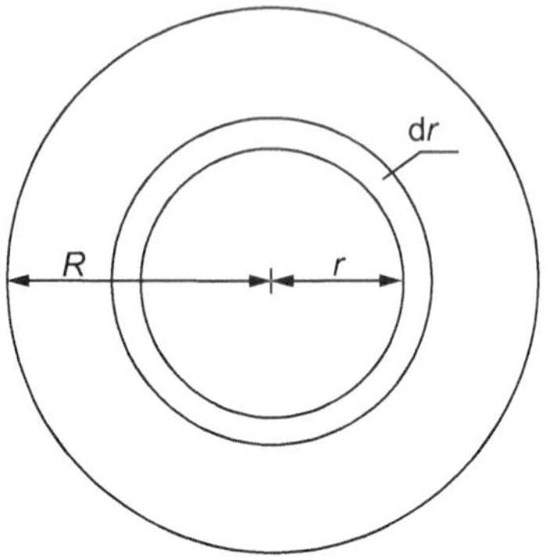

Figure 9 Spherical model for simultaneous diffusion and reaction (*R*: spherical catalyst pellet of radius, *dr*: spherical shell of thickness, *r*: spherical shell of radius) (Charles and Thomas, 1963).

within the differential shell by reaction. At steady state, a mass balance on the differential shell becomes: (rate of diffusion inward at $r = r + \mathrm{d}r$) − (rate of diffusion inward at $r = r$) = rate of reaction in shell.

According to the above mass balance, the Eq. (47) can be got as follows:

$$4\pi(r+\mathrm{d}r)^2 D_{i,\,\mathrm{eff}}\frac{d}{\mathrm{d}r}\left(C_i+\frac{\mathrm{d}c_i}{\mathrm{d}r}\right)-4\pi r^2 D_{i,\,\mathrm{eff}}\left(\frac{\mathrm{d}c_i}{\mathrm{d}r}\right)_r=\left(4\pi r^2\mathrm{d}r\right)\rho R_i \quad (47)$$

Where $4\pi r^2$ is the inner superficial area of the spherical shell, $D_{i,\mathrm{eff}}$ is the effective overall diffusion coefficient of i component through the porous structure, $\mathrm{d}C_i/\mathrm{d}r$ is the concentration gradient of i component at the spherical surface, ρ is the particle density of catalyst, and R_i is the rate of i component.

Eq. (47) becomes Eq. (48):

$$D_{i,\,\mathrm{eff}}\left(\frac{d^2c_i}{\mathrm{d}r^2}+\frac{2\,\mathrm{d}c_i}{r\,\mathrm{d}r}\right)=\rho R_i \quad (48)$$

This can be solved for the boundary condition of at $r = R$ (that is, the outside surface of the pellet) and at $r = 0$ (that is, no concentration gradient will exist at the center of the sphere).

Therefore, in order to figure out the whole process, three major areas should be understood: the structure of the catalyst, the effective diffusion coefficient of reactants and products, and the reaction or formation rate of reactants and products. Different catalysts have different particle density, particle size, and pore structure. Particle density corresponds to ρ in Eq. (49).

Particle size affects the diffusion distance of the reactants and products. The longer distance of the diffusion, the bigger concentration gradient of reactant and products is. The effective diffusion coefficient of reactants and products is a complex coefficient. It is subject to influence by many factors, such as pore structure of catalyst, porosity of catalyst, and kinds of reactants and products. The reaction or formation rate of reactants and products is determined by reaction kinetics. Therefore, the research results of catalyst particle size, catalyst pore size, and diffusivity of products should be discussed in detail.

3.2.1.1 Catalyst Particle Size Effects

Anderson et al. (1964) studied fused iron catalysts which had either been reduced or reduced and nitrided prior to use in fixed-bed reactors, determining reaction kinetics and the effects of the extent of reduction and particle size on catalyst activity. Particle sizes ranged from 42–60 mesh to 4–6 mesh. The catalyst activity increased with smaller particle size until the diameter reached about 0.3 mm for the most active catalysts tested. Catalyst particles were modeled as an active layer of catalyst surrounding an inert core, with the depth of the active layer governed by the reduction temperature. Their calculations allowed them to estimate the effective reactant diffusivity, and they were also able to quantify the depth of the active layer of catalyst. Variations in catalyst activity were attributed to the diffusion of reactant through a wax-filled pore and the depth of the active layer.

Atwood and Bennett (1979) estimated the effectiveness factor for a catalyst with wax-filled pores using nonlinear reaction kinetics, which included inhibition by product water, to estimate the Thiele modulus. The effectiveness factor was then calculated from an expression derived from first-order kinetics and flat-plate geometry. They showed that the catalyst effectiveness factor was quite low for 2–6 mm particles but stated that 0.3 mm particles would be free of heat and mass transfer effects. Dixit and Tavlarides (1982) derived an integral expression for the catalyst effectiveness factor and applied it by using nonlinear FTS kinetics including the concentrations of both H_2 and CO. Feimer et al. (1981) checked for intraparticle mass transfer limitations experimentally using 60/100 mesh precipitated iron catalyst in a fixed-bed reactor by varying the total system pressure at constant reactant partial pressure. They did not see evidence of intraparticle diffusion resistance; however, as the authors noted, this technique is not reliable for wax-filled pores. Liquid concentrations at the pore openings are dependent only on species partial pressures, and the total system pressure in this case will have little subsequent effect on the reaction rate.

Zimmerman and coworkers (Zimmerman et al., 1989) used catalysts with three different mesh size ranges in a fixed-bed reactor. The result showed that particle size has a dramatic effect on the apparent activity of a fused iron FTS catalyst. The criteria for intraparticle and interphase heat and mass transfer revealed that only the intraparticle concentration gradient in the catalyst particle was important under the conditions used in their study. For the smallest particle size (170/230 mesh), catalyst activity approached its intrinsic value and compared well with a previous fixed-bed test of the same catalyst (Satterfield et al., 1985). For larger particle sizes, the rate constants and activation energies were significantly diminished due to the diffusion resistance. The effectiveness factors were calculated assuming that the catalyst pores were filled with wax produced during the synthesis, using first-order kinetics and single-reaction stoichiometry. These estimates compared well with the experimental effectiveness factors. However, the calculations tended to underpredict the effect of intraparticle mass transfer. And the first order in H_2 rate expression used in the calculations may not be applicable when severe mass transfer limitations are present. The mass transfer resistance within the catalyst particle can lead to high local conversions where first-order kinetics is not obeyed, and inhibition by water can be magnified by an intraparticle concentration gradient. A more sophisticated treatment which accounts for CO and water mass transfer may be needed to reliably predict effectiveness factors for catalysts used for the FTS.

Post et al. (1989) prepared a series of iron and cobalt-base catalysts. The studies were performed in a fixed-bed micro reactor system at temperatures in the range 473–523 K. Variation of catalyst particle size in the range 0.2–2.6 mm showed that the conversion of synthesis gas decreases considerably when the average particle size was increased. Under reaction conditions, the major part of the hydrocarbon product would be the liquid. The liquid would fill the pores of the catalyst so that transport of hydrogen and carbon monoxide to the reactive sites occurred by diffusion of these reactants through this liquid medium inside the pores. The apparent effective diffusivity D, can be related to the molecular diffusivity, D_m (H_2) and solubility, H (H_2) of hydrogen in the paraffinic liquid by Eq. (49):

$$D = D_m(H_2)\frac{\varepsilon}{\tau}RT \cdot H(H_2) \tag{49}$$

where ε is the catalyst particle porosity and τ is the catalyst tortuosity. They used the correlation proposed by Wilke and Chang to predict the molecular diffusivity of hydrogen in the liquid phase. By the calculation, they thus arrived at a value for D in the range 0.9×10^{-9} to 1.8×10^{-9} m, which was in excellent agreement with the experimental value obtained from the Thiele relation. Wang et al. (2001) focused on the transfer and reaction phenomenon in a catalyst pellet for FTS. On the basis of the phenomena observed from experiments, a comprehensive pellet model was suggested for catalyst design, in which the detailed FTS kinetics was properly imbedded. The reaction and diffusion interact in a catalyst pellet and its effects on the product selectivity were further investigated. They simulated the concentration profiles of key components in wax-filled catalyst pores (see Fig. 10). Because hydrocarbon-forming reactions generally prevail in the FTS system, the concentration of the accompanying product, H_2O, exhibits a steadily increasing trend along the pellet dimension. The resulting relatively high liquid concentration of H_2O means that the WGS reaction will be driven gradually from a startup state to a fully developing state. Meanwhile, CO is consumed not only by the FTS but also by the WGS reaction, leading to the rapid consumption of CO. From an analysis of the

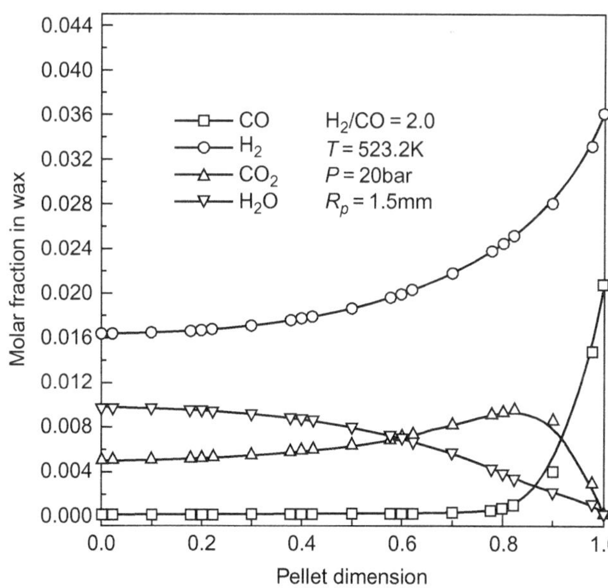

Figure 10 Concentration profiles of key components in catalyst pellet (Wang et al., 2001).

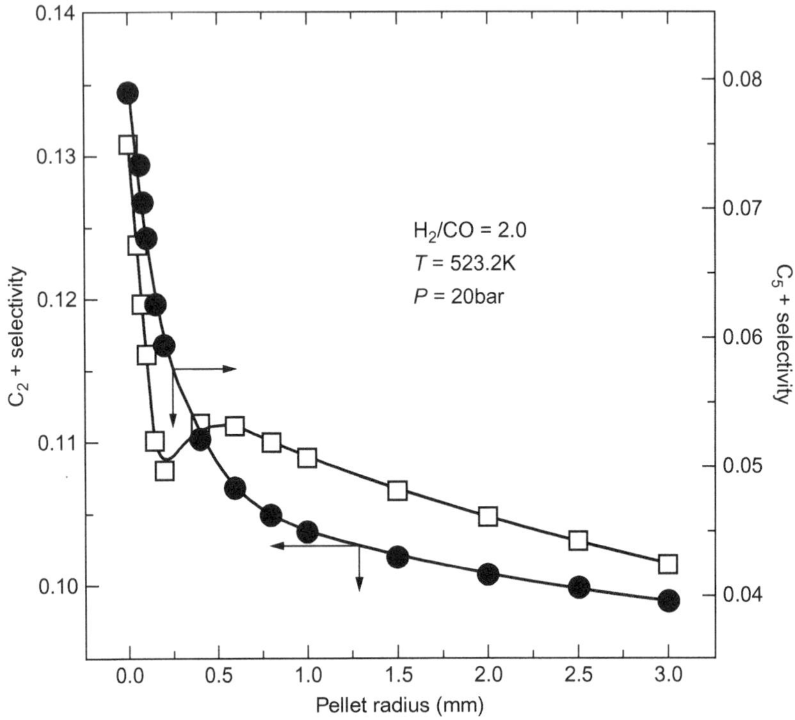

Figure 11 Selectivity variations of CO_2 and CH_4 with pellet radius (Wang et al., 2001).

concentration profiles, they concluded that a serious intraparticle diffusion limitation exists, especially for the reactant CO.

They proposed that with increasing pellet size, the effectiveness factor decreases continuously from a value of about 1 to a relatively low value. In particular, for industrial pellets with diameters of about 2–4 mm, the effectiveness factor is within a range of 0.14–0.28 in their case. They also analyzed that the selectivities of the undesired products increase with increasing pellet size, leading to general decreasing trends of the selectivities of C_2^+ and C_5^+ products (see Figs. 11 and 12; Wang et al., 2001).

3.2.1.2 Catalyst Pore Size Effects

Fan et al. (1995) studied how the pore size of the catalyst affected the transfer of reactants and products in the supercritical-phase FTS reaction. Alumina-supported ruthenium catalysts were prepared with different pore size. They conclude that the chain-growth probability of the Ru catalysts increased while the Ru dispersion decreased with the increase of the pore diameter.

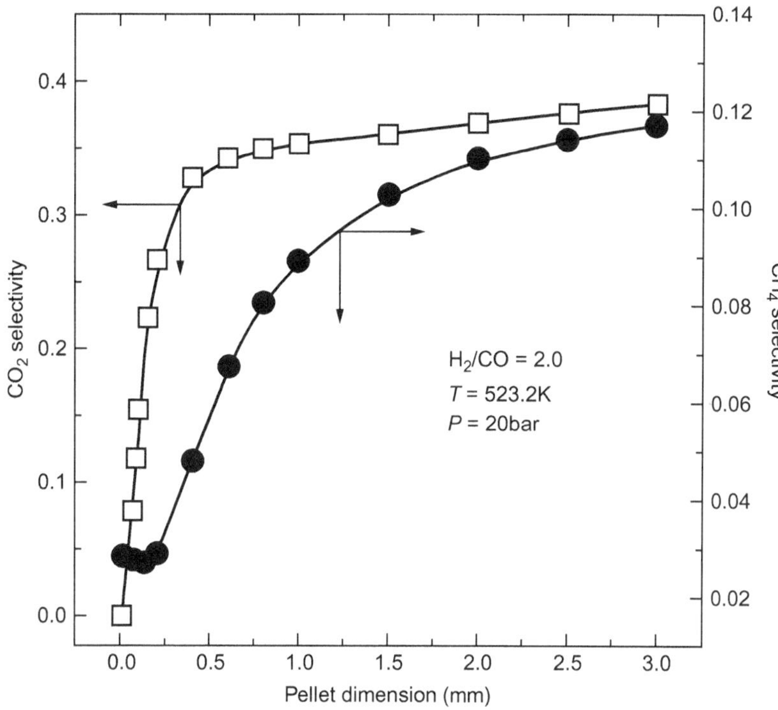

Figure 12 Selectivity variations of C_2+ and C_5+ products with pellet radius (Wang et al., 2001).

The increased chain-growth probability occurring on the catalyst of larger pore size might be attributed to the larger ruthenium particle size as well as the larger and uniform pore size. They also analyzed how pore size affected the product distribution. Due to the highly effective diffusion of the primary product (mainly 1-olefin) from the inside of the catalyst, especially in the case of the catalyst with the largest pores, the probability of the secondary reaction of 1-olefin was lower, thus the degree of the secondary hydrogenation and the hydrocracking was lower. As a result, a lower methane selectivity and a higher olefin selectivity can be achieved for the large pore size catalyst.

Many works were conducted on the liquid-phase and supercritical-phase FTS reaction, because of the advantages in temperature control, wax extraction, and catalyst lifetime extension (Bochniak and Subramaniam, 1988; Fan et al., 1995). However, the main drawback of the liquid-phase FTS reaction was the slow diffusion rate of the syngas and products inside the pores which filled with formed hydrocarbons and solvent (Fan et al., 1995; Post et al.,

1989). A good catalyst for liquid-phase FTS reaction must be with large pores to reduce the pore diffusion resistance. At the same time, the large surface area of catalyst support is favorable to increase CO conversion. On the other hand, the large surface area of catalyst support is favorable to increase metal dispersion, leading to high CO conversion of the supported FT metal catalysts. Unfortunately, a higher surface area means a smaller pore size and the stronger pore resistance. In order to meet these two requirements, Xu (Xu et al., 2005) prepared the bimodal catalyst as shown in Fig. 13, where both large pore and small pore coexist, can guarantee high-transport efficiency, because the large pore lead to a high-diffusion rate of reactants and products, while the small pores lead to a large surface area, and finally a high-metal surface area in the supported metal catalyst. Compared with the catalysts whose pore diameters are 3.7 and 54 nm, bimodal catalyst shows the highest CO conversion, and, meanwhile, selectivities of CH_4 and CO_2 were as low as those which prepared from support with 54 nm pore diameter (see Table 3; Xu et al., 2005).

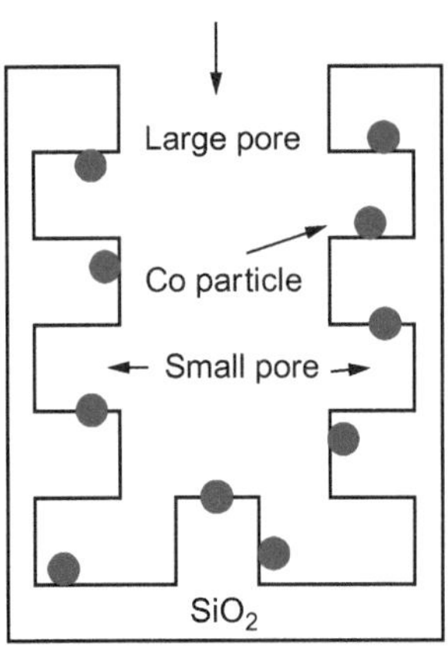

Figure 13 Diagram of bimodal catalysts.

Table 3 The Reaction Performance of Cobalt Catalysts Prepared from Different Supports (Xu et al., 2005)

Catalyst	Q-50	Bimodal	Q-3
CO conv. (%)	11.8	32	16.9
Methane sel. (%)	7.1	8.2	35.8
CO_2	3.2	2.8	33
Chain-growth probability	0.86	0.86	0.84

Also, a mathematical model was developed to elucidate the kinetics coupled with mass transfer effect inside the pores of 3.7 nm, 54 nm, and bimodal pore catalyst. The reactants concentration profiles inside the pores were given. The concentration profiles of reactants inside the pores in large pore and bimodal pore catalyst were slightly changed. However, in small catalyst, it decreased quickly. The CO concentration would decrease to 0 not far from surface of catalyst. The author thought that must be another reason which caused a higher CH_4 selectivity.

It is easy to find that both the dispersion of the supported metal and the pore size of the catalyst showed their close relationship with the propagation of the carbon chain and product distribution. A few studies were carried out on the influence of the iron particle size and the pore size of support on the FTS reaction activity and selectivity, controversy persists, because these observations were the results of complex interplay among many factors. In order to investigate the sole effect of pore size or iron particle size on formation of light olefins in FTS, Zhang and Liu (Liu et al., 2015) prepared various iron-supported catalysts with pore size of 5, 50, and 80 nm which were modified by ethylene glycol to realize the similar iron particle size. The similar properties of the iron active phase on different catalysts guaranteed solely comparing the effect of pore size on formation of light olefins. They concluded that the smaller iron particle was much beneficial to forming light hydrocarbons including methane. Furthermore, smaller iron particle was advantageous to form more light olefins. Meanwhile, the olefins to paraffin ratio (O/P) of C_2–C_4 hydrocarbons was more sensitive to pore size of catalysts due to suppressing the second reaction of formed olefins.

3.2.2 Diffusivity Effects

Slow removal of reactive products (for example, α-olefins) due to a decrease of diffusion coefficients with increasing chain length can influence the FTS reaction rate and selectivity. Erkey et al. (1990) measured the molecular diffusion coefficients of three paraffins (*n*-octane, *n*-dodecane, and *n*-hexadecane) in FT wax with an average chain length of C_{28} (see Fig. 13; Erkey et al., 1990; Madon and Iglesia, 1994). Using the correlation proposed by Wilke–Chang (Reid et al., 1987) to predict the molecular diffusivity of hydrocarbons in a heavy paraffinic FTS products (viscosity wax with $M=300$ g/mol, at $T=504$ K, $\mu=0.6\times10^{-3}$ Ns/m^2), it is easy to find a chain-length dependency of $D_n \propto n^{-0.5}$. The calculated values for the diffusivities of C_1–C_{17} are also plotted in Fig. 13.

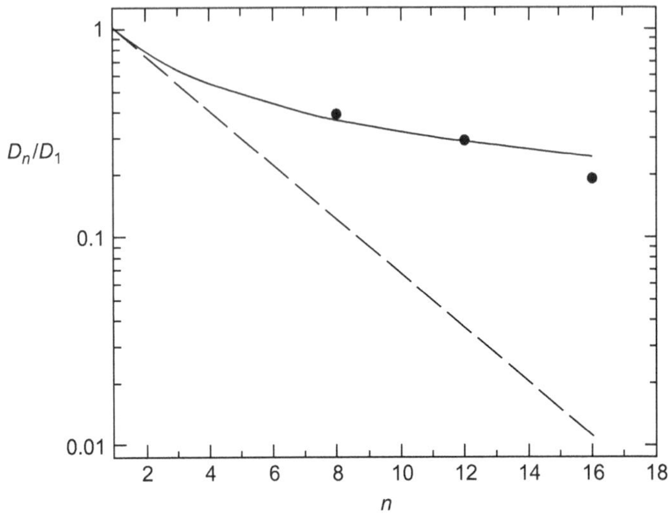

Figure 14 Diffusivities of *n*-paraffins in FT wax (•, experimental data at $T=504$ K; —, Wilke–Chang correlation at $T=540$ K; - - - correlation by Iglesia et al.; *n*, carbon number; D_n, diffusivities of *n*-paraffins) (Erkey et al., 1990).

Iglesia and coworkers studied the influence of chain-length-dependent diffusion coefficients on secondary reactions (Madon and Iglesia, 1993, 1994; Madon et al., 1991). They reported an empirical equation describing a strong influence of chain length on diffusivity for olefins and paraffins $D_n \propto e^{-0.3n}$, which was not verified by experimental data. This carbon number dependency is a factor of three higher than that determined by Erkey et al. (1990) and apparently is wrong (see Fig. 14).

Measurements of these coefficients at FTS operating conditions are scarce, so there are many uncertainties in the studies.

4. SUMMARY

As summaries, the influence of product distribution for FTS was discussed from two aspects of multiscales and mesoscale. It can be found that the multiscale problem widely appears on FTS, and one of the most prominent is the ASF distribution, which is derived based on the polymerization mechanism. However, as the research of mesoscale problems of FTS is just on a primary level, there are only several examples to describe mesoscale phenomena of FTS, as mentioned above, such as molecular reaction–diffusion, olefin readsorption. Therefore, the further study on mesoscale problem of FTS is quiet necessary.

However, in the complex FTS reaction system, clearly identifying the effects of different element (small scale) and system (large scale) is the key point to research the mesocale effects on FTS. It is considered that there are several systems in FTS reaction, those consist of a lot of elements which intersect or interact each other. For the basic system in FTS, the molecular reaction should consist of the active site, the catalyst structure which always influences the properties of active site and support, the concentration of reactants and products which are always influenced by the pore structure of catalyst. For the catalyst pellet system, there are the size and the morphology of catalyst which always influence the external and internal diffusion, and the flow type of reactor which determines the external diffusion and size or morphology of catalyst pellet. For the reactor system, it should consist of molecular reaction, catalyst pellet, type of reactor, and sometimes solvent. Inside these intersected or interacted elements, it is considered that clearly identifying the effects of single element on FTS would contribute to analysis of the mesocale effects on FTS reaction. For instance, using nanotechnology to preparing active phase with the same properties of active site would minimize the influence of pore structure on the properties of active site, contributing to sorely identifying the effects of internal diffusion by eliminating the influence of active site on reaction results. Development of the different pore sizes materials with ordered pore structure and homogeneous pore distribution as support would contribute to clearing the relationship between the external and internal diffusion. Therefore, analysis of mesoscale effects on complex FTS reaction needs a lot of researchers in different research fields to work in detail and cooperatively.

REFERENCES

Aasberg P: Fischer-Tropsch technology, *Stud Surf Sci Catal* 152:258–405, 2004.

Albal RS, Shah YT, Carr NL, Bell AT: Mass transfer coefficients and solubilities for hydrogen and carbon monoxide under Fischer-Tropsch conditions, *Chem Eng Sci* 39(5):905–907, 1984.

Ali NP, Hamideh K, Mohammad I, Mohammad RH: Mechanistic double ASF product distribution study of Fischer-Tropsch synthesis on precipitated iron catalyst, *J Nat Gas Sci Eng* 15(6):53–58, 2013.

Anderson RB: Catalysis. In: Emmett H, editor: *IV: hydrocarbon synthesis, hydrogenation and cyclization*, vol. 4, New York, 1956, Reinhold.

Anderson RB: Schulz-Flory equation, *J Catal* 55(1):114–115, 1978.

Anderson RB, Karn FS: A rate equation for the Fischer-Tropsch synthesis on iron catalysts, *J Phys Chem* 64(6):805–808, 1960.

Anderson RB, Karn FS, Schultz JF: Kinetics of the Fischer-Tropsch synthesis on iron catalysts. In *Bulletin 614*Washington DC, 1964, US Bureau of Mines.
Atwood HE, Bennett CO: Kinetics of the Fischer-Tropsch reaction over iron, *Ind Eng Chem Process Des Dev* 18:163–170, 1979.
Bao J, He J, Zhang Y, Yoneyama Y, Tsubaki N: A core/shell catalyst produces a spatially confined effect and shape selectivity in a consecutive reaction, *Angew Chem Int Ed* 120(2):359–362, 2008.
Baten JMV, Krishna R: Eulerian simulation strategy for scaling up a bubble column slurry reactor for Fischer-Tropsch synthesis, *Ind Eng Chem Res* 43(16):4483–4493, 2003.
Bochniak DJ, Subramaniam B: Fischer-Tropsch synthesis in near-critical n-hexane: pressure-tuning effects, *AIChE J* 44:1889–1896, 1988.
Bongiorno M, Thiringer TA: A generic DFIG model for voltage dip ride-through analysis, *IEEE Trans Energy Convers* 28(1):76–85, 2013.
Bub G, Baerns M, Büssemeier B, Frohning C: Prediction of the performance of catalytic fixed bed reactors for Fischer-Tropsch synthesis, *Chem Eng Sci* 35:348–355, 1980.
Cao G, Chen X: Buckling of single-walled carbon nanotubes upon bending: molecular dynamics simulations and finite element method, *Phys Rev B* 73(15):2952–2961, 2006.
Charles KS, Thomas NS: *The role of diffusion in catalysis,* Chicago, Illinois, 1963, Addison-Wesley Publishing Company, Inc.
Cheng K, Kang J, Huang S, You Z, Zhang Q, Ding J: Mesoporous beta zeolite-supported ruthenium nanoparticles for selective conversion of synthesis gas to C_5-C_{11} isoparaffins, *ACS Catal* 2(3):441–449, 2012.
Colley SE, Copperthwaite RG, Hutchings GJ, Terblanche SP, Thackeray MM: Identification of body-centred cubic cobalt and its importance in cohydrogenation, *Nature* 339(6220):129–130, 1989.
de Smit E, Weckhuysen BM: The renaissance of iron-based Fischer-Tropsch synthesis: on the multifaceted catalyst deactivation behavior, *Chem Soc Rev* 37:2758–2781, 2008.
Deckwer WD, Kokuun R, Sanders E, Ledakowicz S: Kinetic studies of Fischer-Tropsch synthesis on suspended Fe/K catalyst. Rate inhibition by CO_2 and H_2O, *Ind Eng Chem Process Des Dev* 25:643–649, 1986.
Derosset AJ, Neuzil RW, Korous DJ: Liquid column chromatography as a predictive tool for continuous countercurrent adsorptive separations, *Ind Eng Chem Process Des Dev* 15(2):261–266, 1976.
Dixit RS, Tavlarides LL: Integral method of analysis of Fischer-Tropsch synthesis reactions in a catalyst pellet, *Chem Eng Sci* 37:539–544, 1982.
DOE/NETL-2004/1199: *Commercial-scale demonstration of the liquid phase methanol (LPMEOHTM) process, a DOE assessment,* Washington, DC, 2003, Department of Energy.
Doering EL, Cremer GA: Advances in the shell coal gasification process, *Div Fuel Chem* 40(2):312–317, 1995. Preprints of Papers.
Donnelly TJ, Yates IC, Satterfield CN: Analysis and prediction of product distributions of the Fischer-Tropsch synthesis, *Energy Fuel* 2(6):734–739, 1988.
Dry ME: Advances in Fischer-Tropsch chemistry, *Ind Eng Chem Prod Res Dev* 15:282–286, 1976.
Dry ME: The Fischer-Tropsch process—commercial aspects, *Catal Today* 6(3):183–206, 1990.
Dudukovic MP, Larachi F, Mills PL: Multiphase catalytic reactors: a perspective on current knowledge and future trends, *Catal Rev* 44:123–246, 2002.
Dyk JC, Keyser MJ, Coertzen M: Syngas production from South African coal sources using Sasol-Lurgi gasifiers, *Int J Coal Geol* 65(3–4):243–253, 2006.

Eidus YT: The mechanism of the Fischer-Tropsch reaction and the initiated hydro-polymerisation of alkenes, from radiochemical and kinetic data, *Russ Chem Rev* 36:338–351, 1967.

Erkey C, Rodden JB, Akgerman A: Diffusivities of synthesis gas and n-alkanes in Fischer-Tropsch wax, *Energy Fuel* 4:275–276, 1990.

Fan L, Yokota K, Fujimoto K: Characterization of mass-transfer in supercritical-phase Fischer-Tropsch synthesis reaction, *Top Catal* 2:267–283, 1995.

Fan L, Yokota K, Fujimoto K: Supercritical phase Fischer-Tropsch synthesis: catalyst pore-size effect, *AICHE J* 38(10):1639–1648, 2004.

Farias FEM, Sales FG, Fernandes FAN: Effect of operating conditions and potassium content on Fischer-Tropsch liquid products produced by potassium-promoted iron catalysts, *J Nat Gas Chem* 17(2):175–178, 2006.

Feimer JL, Silveston PL, Hudgins RR: Steady-state study of the Fischer-Tropsch reaction, *Ind Eng Chem Prod Res Dev* 20:609–615, 1981.

Fischer F, Tropsch H: Synthesis of petroleum at atmospheric pressure from gasification products of coal, *Brennst-Chem* 7:97–104, 1926a.

Fischer F, Tropsch H: Uber die direkte synthese von erdol-kohlenwasserstoffen bei gewohnlichem druck. (erste mitteilung), *Ber Dtsch Chem Ges* 59(4):830–831, 1926b.

Font F, Josephus J, Helena MH, John R, Newton D: Fischer-Tropsch synthesis process carried out on a floatable structure, EPUS7037947 B2, 2006.

Friedel RA, Anderson RB: Composition of synthetic liquid fuels. I. Product distribution and analysis of C5–C8 paraffin isomers from cobalt catalyst, *J Am Chem Soc* 72(3):1212–1215, 1950.

Frohning CD, Cornils B: Chemical feedstocks from coal, *Hydrocarb Process* 53(11):143–146, 1974.

Ge W, Liu X, Ren Y, Xu J, Li J: From multi-scale to meso-scale: new challenges for simulation of complex processes in chemical engineering, *CIESC J* 61(7):1613–1620, 2010.

Grund G, Schumpe A, Deckwer WD: Gas–liquid mass transfer in a bubble column with organic liquids, *Chem Eng Sci* 47(92):3509–3516, 1992.

Heidemann RA: Wax precipitation modeled with many mixed solid phases, *AICHE J* 51(1):298–308, 2004.

Henrici-Olivé G, Olivé S: Reactions of carbon monoxide with transition metal-carbon bonds, *Transition Met Chem* 1(2):77–93, 1976.

Hikita H, Asai S, Kikukawa H, Zaike T, Ohue M: Heat transfer coefficient in bubble columns, *Ind Eng Chem Process Des Dev* 20(3):540–545, 1981.

Hilmen AM, Bergene E, Lindvog OA, Schanke D, Eri S, Holmen A: Fischer-Tropsch synthesis on monolithic catalysts of different materials, *Catal Today* 69:227–232, 2001.

Hoover W, Holian B, Moran B, Straub G: Shock-wave structure via nonequilibrium molecular dynamics and Navier–Stokes continuum mechanics, *Phys Rev A* 22(6):2798–2808, 1980.

Huff GA, Satterfield CN: Intrinsic kinetics of the Fischer-Tropsch synthesis on a reduced fused-magnetite catalyst, *Ind Eng Chem Process Des Dev* 23(4):696–705, 1984.

Hussain R, Blank JH, Elbashir NO: Modeling the fixed-bed Fischer-Tropsch reactor in different reaction media, *Comput Aided Chem Eng* 37:143–148, 2015.

Iglesia E, Reyes SC, Madon RJ, Soled SL: Selectivity control and catalyst design in the Fischer-Tropsch synthesis: sites, pellets, and reactors, *Adv Catal* 39(1):221–302, 1993.

Jager B: Developments in Fischer-Tropsch technology, *Stud Surf Sci Catal* 107:219–224, 1997.

Jager B, Espinoza R: Advances in low temperature Fischer-Tropsch synthesis, *Catal Today* 23(94):17–28, 1995.

Kang J, Zhang S, Zhang Q, Wang Y: Ruthenium nanoparticles supported on carbon nanotubes as efficient catalysts for selective conversion of synthesis gas to diesel fuel, *Angew Chem Int Ed* 48(14):2603–2606, 2009.
Kang J, Cheng K, Zhang L, Zhang Q, Ding J, Hua W: Mesoporous zeolite-supported ruthenium nanoparticles as highly selective Fischer-Tropsch catalysts for the production of C_5-C_{11} isoparaffins, *Angew Chem Int Ed* 50(22):5200–5203, 2011.
Karn FS, Shultz JF, Anderson RB: Kinetics of the Fischer-Tropsch synthesis on iron catalysts. Pressure dependence and selectivity of nitrided catalysts, *J Phys Chem* 64(4):446–451, 1960.
Khodakov AY, Wei C, Fongarland P: Advances in the development of novel cobalt Fischer-Tropsch catalysts for synthesis of long-chain hydrocarbons and clean fuels, *Chem Rev* 38(33):1692–1744, 2007.
Kibby CL, Pannell RB, Kobylinski TP: Hydrogenation of olefins in the presence of carbon monoxide on supported cobalt catalysts, *Div Pet Chem* 188(8):44, 1984.
Kim SS, Noh JW, Churn KS: Morphological stability of cylindrical pores in an external diffusion field, *J Mater Sci Lett* 8(11):1320–1322, 1989.
Klerk AD: *Fischer-Tropsch refining,* Weinheim, 2011, Wiley-VCH, pp. 1249–1279.
Koenig L, Gaube J: Fischer-Tropsch synthese—neuere untersuchungen und entwicklungen, *Chem Ing Tech* 55:14–22, 1983.
Krishna R, Urseanu MI, Baten JMV, Ellenberger J: Influence of scale on the hydrodynamics of bubble columns operating in the churn-turbulent regime: experiments vs. Eulerian simulations, *Chem Eng Sci* 54(21):4903–4911, 1999.
Krishna R, Baten JMV, Urseanu MI: Three-phase Eulerian simulations of bubble column reactors operating in the churn-turbulent regime: a scale up strategy, *Chem Eng Sci* 55(16):3275–3286, 2000.
Krishna R, Baten JMV, Urseanu MI, Ellenberger J: A scale up strategy for bubble column slurry reactors, *Catal Today* 66(2):199–207, 2001.
Kuipers EW, Scheper C, Wilson JH, Vinkenburg IH, Oosterbeek H: Non-ASF product distributions due to secondary reactions during Fischer–Tropsch synthesis, *J Catal* 158(1):288–300, 1996.
Kwauk M: Exploring complex systems in chemical engineering—the multi-scale methodology, *Chem Eng Sci* 58(2):521–535, 2003.
Ledakowicz S, Nettelhoff H, Kokuun R, Deckwer WD: Kinetics of the Fischer-Tropsch synthesis in the slurry phase on a potassium-promoted iron catalyst, *Top Catal* 24:1043–1049, 1985.
Lee SB, Kim SM, Ryu DDY: Effects of external diffusion and design geometry on the performance of immobilized glucose isomerase reactor system, *Biotechnol Bioeng* 21(11):2023–2043, 2004.
Li J, Huang W: *Towards mesoscience: the principle of compromise in competition,* London, 2014, Springer express.
Li J, Tung Y, Kwauk M: Method of energy minimization in multi-scale modeling of particle-fluid two-phase flow. In Circulating fluidized bed technology proceedings of the second international conference, 1988, pp 89–103.
Li J, Ge W, Wang W, et al: *From multiscale modeling to meso-science: a chemical engineering perspective; principles, modeling, simulation, and applications,* London, 2013, Springer.
Li J, Hu Y, Yuan Q: Mesoscience: exploring old problems from a new angle, *Sci Sin Chim* 44(3):277–281, 2014.
Liu W: Ministructured catalyst bed for gas–liquid–solid multiphase catalytic reaction, *AIChE J* 48(7):1519–1532, 2002.
Liu Q, Zhang Z, Zhou L: The steady-state and dynamic behavior of the fixed-bed catalytic reactor for Fischer-Tropsch synthesis, *J Petrochem Univ* 12(1):14–29, 1999.

Liu Z, Li X, Asami K, Fujimoto K: Iso-paraffins synthesis from modified Fischer-Tropsch reaction-insights into pd/beta and pt/beta catalysts, *Catal Today* 104(1): 41–47, 2005.

Liu W, Hu J, Wang Y: Fischer-Tropsch synthesis on ceramic monolith-structured catalysts, *Catal Today* 140:142–148, 2009.

Liu G, Larson ED, Williams RH, Kreutz TG, Guo X: Making Fischer-Tropsch fuels and electricity from coal and biomass: performance and cost analysis, *Energy Fuel* 25(1):415–437, 2010.

Liu X, Hamasaki A, Honma T, Tokunaga M: Anti-ASF distribution in Fischer-Tropsch synthesis over unsupported cobalt catalysts in a batch slurry phase reactor, *Catal Today* 175(1):494–503, 2011.

Liu J, Chen J, Zhang Y: Cobalt-imbedded zeolite catalyst for direct syntheses of gasoline via Fischer–Tropsch synthesis, *Catal Sci Technol* 3(10):2559–2564, 2013.

Liu Y, Chen J, Zhang Y: The effect of pore size or iron particle size on the formation of light olefins in Fischer–Tropsch synthesis, *RSC Adv* 5:29002–29007, 2015.

Lox ES, Froment GF: Kinetics of the Fischer-Tropsch reaction on a precipitated promoted iron catalyst. 2. Kinetic modeling, *Ind Eng Chem Res* 32:71–82, 1993.

Ma W, Li Y, Zhao Y, Xu Y, Zhou J: Kinetics of Fischer-Tropsch synthesis over Fe-Cu-K catalyst-kinetic model on the basis of mechanism (1), *J Chem Ind Eng (China)* 50(2):159–166, 1999a.

Ma WP, Li YW, Zhao YL, Xu YY, Zhou JL: Kinetics of Fischer-Tropsch synthesis over Fe-Cu-K catalyst-kinetic model on the basis of mechanism (2), *J Chem Ind Eng (China)* 50(2):167, 1999b.

Madon RJ: On the growth of hydrocarbon chains in the Fischer-Tropsch synthesis, *J Catal* 57(1):183–186, 1979.

Madon RJ, Iglesia E: The importance of olefin readsorption and H_2/CO reactant ratio for hydrocarbon chain growth on ruthenium catalysts, *J Catal* 139:576–590, 1993.

Madon RJ, Iglesia E: Hydrogen and CO intrapellet diffusion effects in ruthenium-catalyzed hydrocarbon synthesis, *J Catal* 149:428–437, 1994.

Madon RJ, Reyes SC, Iglesia E: Primary and secondary reaction pathways in ruthenium-catalyzed hydrocarbon synthesis, *J Phys Chem* 95:7795–7804, 1991.

Mikkola J, Warna J, Virtanen P, Salmi T: Effect of internal diffusion in supported ionic liquid catalysts: interaction with kinetics, *Ind Eng Chem Res* 46(12):3932–3940, 2007.

Moutsoglou A, Sunkara PP: Fischer–Tropsch synthesis in a fixed bed reactor, *Energy Fuel* 25(5):2242–2257, 2011.

National Natural Science Foundation of China: *Project guide of NSFC key research program on meso-scale mechanisms and manipulation in multiphase reaction systems [EB/OL], 2013:* Beijing, 2013, NSFC. http://www.nsfc.gov.cn/publish/portal0/zdyjjh/006/info195.htm.

Nettelhoff H, Kokuun R, Ledakowicz S, Deckwer WD: Studies on the kinetics of Fischer-Tropsch synthesis in slurry phase, *Ger Chem Eng* 8:177–185, 1985.

Ogata S, Lidorikis E, Shimojo F, Nakano A, Vashishta P, Kalia RK: Hybrid finite element molecular dynamics electronic density functional approach to materials simulations on parallel computers, *Comput Phys Commun* 138(2):143–154, 2001.

Pichler H, Schluz H: Neuere Erkenntnisse auf dem Gebiet der Synthese von Kohlenwasserstoffen aus CO und H_2 (New insights in the area of the synthesis of hydrocarbons from CO and H_2), *Chem Ing Tech* 42(18):1162–1174, 1970.

Post MFM, van Hoog AC, Minderhoud JK, Sie ST: Diffusion limitations in Fischer-Tropsch catalysts, *AIChE J* 35:1107–1114, 1989.

Puskas I: Unusual reactions on a cobalt-based Fischer-Tropsch catalyst, *Catal Lett* 22(4):283–288, 1993.

Puskas I, Hurlbut RS: Comments about the causes of deviations from the Anderson-Schulz-Flory distribution of the Fischer-Tropsch reaction products, *Catal Today* 84:99–109, 2003.

Reid RC, Prausnitz JM, Poling BE: *The properties of gases and liquids,* ed 4, New York, 1987, McGraw-Hill.

Rofer-Depoorter CK: A comprehensive mechanism for the Fischer-Tropsch synthesis, *Chem Rev* 5:447–474, 2002.

Rudd RE, Broughton JQ: Coarse-grained molecular dynamics and the atomic limit of finite elements, *Phys Rev B* 58(10):R5893, 1998.

Sarup B, Wojciechowski BW: Studies of the Fischer-Tropsch synthesis on a cobalt catalyst II: kinetics of carbon monoxide conversion to methane and to higher hydrocarbons, *Can J Chem Eng* 67:62–74, 1989.

Satterfield CN, Huff GA, Stenger HG, Carter JL, Madon RJ: A comparison of Fischer-Tropsch synthesis in a fixed-bed reactor and in a slurry reactor, *Ind Eng Chem Fundam* 24(4):450–454, 1985.

Schneider P, Mitschka P: Effect of internal diffusion on catalytic reactions, *Chem Eng Sci* 21:455–463, 1966.

Shi Z, Weng L, Men Z, Pu Y, Liu K: Progress in numerical simulation of Fischer-Tropsch synthesis slurry reactor, *Chem Eng (Chin)* 41(11):48–52, 2013.

Sie ST, Krishna R: Fundamentals and selection of advanced Fischer-Tropsch reactors, *Appl Catal A Gen* 186(1):55–70, 1999.

Sirignano WA, Reviewer ACFE: Fluid dynamics and transport of droplets and sprays, *J Fluids Eng* 122(1):190, 2000.

Smirnova JA, Zhigilei LV, Garrison BJ: A combined molecular dynamics and finite element method technique applied to laser induced pressure wave propagation, *Comput Phys Commun* 118(1):11–16, 1999.

Steynberg AP, Espinoza RL, Jager B, Vosloo AC: High temperature Fischer–Tropsch synthesis in commercial practice, *Appl Catal A Gen* 186(1):41–54, 1999.

Storch HH, Goulombic N, Anderson RB: *The Fischer-Tropsch and related syntheses,* New York, 1951, Wiley.

Tam W, Wong WK, Gladysz JA: Neutral metal formyl complexes: generation, reactivity, and models for Fischer-Tropsch catalyst intermediates, *J Am Chem Soc* 101(6):1589–1591, 1979.

Tavakoli A, Kargari A, Sohrabi M: Application of Anderson-Schulz-Flory (ASF) equation in the product distribution of slurry phase FT synthesis with nanosized iron catalysts, *Chem Eng J* 136(2–3):358–363, 2008.

Thiele EW: Relation between catalytic activity and size of particle, *Ind Eng Chem Res* 31(7):916–920, 1939.

Thomas ER, Eckert CA: Prediction of limiting activity coefficients by a modified separation of cohesive energy density model and UNIFAC, *Ind Eng Chem Process Des Dev* 23(2):194–209, 1984.

Travis KP, Evans TDJ: Departure from Navier–Stokes hydrodynamics in confined liquids, *Phys Rev B* 55(4):4288–4295, 1997.

Troshko AA, Zdravistch F: CFD modeling of slurry bubble column reactors for Fischer-Tropsch synthesis, *Chem Eng Sci* 64(5):892–903, 2009.

Tsubaki N, Zhang Y, Sun S, Mori H, Yoneyama Y, Li X: A new method of bimodal support preparation and its application in Fischer-Tropsch synthesis, *Catal Commun* 2:311–315, 2001.

Tsutsumi A, CHEN W, Kim YH: Classification and characterization of hydrodynamic and transport behaviors of three-phase reactor, *Korean J Chem Eng* 16(6):709–720, 1999.

Vannice MA: The catalytic synthesis of hydrocarbons from ja:math mixtures over the group VIII metals: I. The specific activities and product distributions of supported metals, *J Catal* 37:449–461, 1975.
Vliet OPRV, Faaij APC, Turkenburg WC: Fischer-Tropsch diesel production in a well-to-wheel perspective: a carbon, energy flow and cost analysis, *Energy Convers Manag* 50(4):855–876, 2009.
Wang Y, Li Y, Xu Y, Zhao Y, Zhang B: On development of engineering models for fixed bed Fischer-Tropsch synthesis: engineering problem and modelization, *J Fuel Chem Technol* 27(Suppl.):111–122, 1999.
Wang Y, Xu Y, Xiang H, Li Y, Zhang B: Modeling of catalyst pellets for Fischer-Tropsch synthesis, *Ind Eng Chem Res* 40:4324–4335, 2001.
Wang T, Wang J, Yong J: Slurry reactors for gas-to-liquid processes: a review, *Ind Eng Chem Res* 46(18):5824–5847, 2007a.
Wang X, Hu L, Shen J, Yu Z, Wang F: Multi-scale and multi-fractal analysis of pressure fluctuation in slurry bubble column bed reactor, *J Cent South Univ Technol* 14(5):696–700, 2007b.
Wheeler A: *Catalysis,* New York, 1995, Reinhold pressed.
Xiao S, Chun G, Zhao Y, Chen HJ: Effect of temperature fluctuation on hydrate-based CO_2 separation from fuel gas, *J Nat Gas Chem* 20(6):647–653, 2011.
Xu B, Zhang Y, Fan L, Tsubaki N: Pore diffusion simulation model of bimodal catalyst for Fischer-Tropsch synthesis, *AIChE J* 51(7):2068–2076, 2005.
Yates IC, Satterfield CN: Effect of carbon dioxide on the kinetics of the Fischer-Tropsch synthesis on iron catalysts, *Ind Eng Chem Res* 28:9–12, 1989.
Yu G, Sun B, Pei Y, Xie S, Yan S, Qiao M: Fe@C spheres as an excellent catalyst for Fischer-Tropsch synthesis, *J Am Chem Soc* 132(3):935–937, 2009.
Zhang Y, Yoneyama Y, Tsubaki N: Simultaneous introduction of chemical and spatial effects via a new bimodal catalyst support preparation method, *Chem Commun* 11:1216–1217, 2002.
Zhang Q, Kang J, Wang Y: Development of novel catalysts for Fischer-Tropsch synthesis: tuning the product selectivity, *ChemCatChem* 2(9):1030–1058, 2011.
Zhang L, Zhang Y, Deng J, Dai H: Surfactant-aided hydrothermal preparation of La2-xSrxCuO4 single crystallites and their catalytic performance on methane combustion, *J Nat Gas Chem* 21(1):69–75, 2012.
Zhang Q, Deng W, Wang Y: Recent advances in understanding the key catalyst factors for Fischer-Tropsch synthesis, *J Energy Chem* 22(1):27–38, 2013.
Zhou X, Chen Q, Tao Y, Weng H: Influence of ultrasound impregnation on the performance of $Co/Zr/SiO_2$ catalyst during Fischer-Tropsch synthesis, *Chin J Catal* 32(7):1156–1165, 2011.
Zimmerman WH, Rossin JA, Bukur DB: Effect of particle size on the activity of fused iron Fischer-Tropsch catalyst, *Ind Eng Chem Res* 28:406–413, 1989.

INDEX

Note: Page numbers followed by "*f*" indicate figures, and "*t*" indicate tables.

I

K

L

M

N

O

P

Q

R

U

V

W

Z

CONTENTS OF VOLUMES IN THIS SERIAL

Volume 1 (1956)
J. W. Westwater, *Boiling of Liquids*
A. B. Metzner, *Non-Newtonian Technology: Fluid Mechanics, Mixing, and Heat Transfer*
R. Byron Bird, *Theory of Diffusion*
J. B. Opfell and B. H. Sage, *Turbulence in Thermal and Material Transport*
Robert E. Treybal, *Mechanically Aided Liquid Extraction*
Robert W. Schrage, *The Automatic Computer in the Control and Planning of Manufacturing Operations*
Ernest J. Henley and Nathaniel F. Barr, *Ionizing Radiation Applied to Chemical Processes and to Food and Drug Processing*

Volume 2 (1958)
J. W. Westwater, *Boiling of Liquids*
Ernest F. Johnson, *Automatic Process Control*
Bernard Manowitz, *Treatment and Disposal of Wastes in Nuclear Chemical Technology*
George A. Sofer and Harold C. Weingartner, *High Vacuum Technology*
Theodore Vermeulen, *Separation by Adsorption Methods*
Sherman S. Weidenbaum, *Mixing of Solids*

Volume 3 (1962)
C. S. Grove, Jr., Robert V. Jelinek, and Herbert M. Schoen, *Crystallization from Solution*
F. Alan Ferguson and Russell C. Phillips, *High Temperature Technology*
Daniel Hyman, *Mixing and Agitation*
John Beck, *Design of Packed Catalytic Reactors*
Douglass J. Wilde, *Optimization Methods*

Volume 4 (1964)
J. T. Davies, *Mass-Transfer and Inierfacial Phenomena*
R. C. Kintner, *Drop Phenomena Affecting Liquid Extraction*
Octave Levenspiel and Kenneth B. Bischoff, *Patterns of Flow in Chemical Process Vessels*
Donald S. Scott, *Properties of Concurrent Gas–Liquid Flow*
D. N. Hanson and G. F. Somerville, *A General Program for Computing Multistage Vapor–Liquid Processes*

Volume 5 (1964)
J. F. Wehner, *Flame Processes—Theoretical and Experimental*
J. H. Sinfelt, *Bifunctional Catalysts*
S. G. Bankoff, *Heat Conduction or Diffusion with Change of Phase*
George D. Fulford, *The Flow of Lktuids in Thin Films*
K. Rietema, *Segregation in Liquid–Liquid Dispersions and its Effects on Chemical Reactions*

Volume 6 (1966)
S. G. Bankoff, *Diffusion-Controlled Bubble Growth*
John C. Berg, Andreas Acrivos, and Michel Boudart, *Evaporation Convection*
H. M. Tsuchiya, A. G. Fredrickson, and R. Aris, *Dynamics of Microbial Cell Populations*
Samuel Sideman, *Direct Contact Heat Transfer between Immiscible Liquids*
Howard Brenner, *Hydrodynamic Resistance of Particles at Small Reynolds Numbers*

FSC
www.fsc.org
MIX
Papier | Fördert gute Waldnutzung
FSC® C083411

Kolibri 360 GmbH
Ferdinand-Jühlke-Straße 7
99095 Erfurt, Deutschland
produktsicherheit@kolibri360.de